Optical Measurement of Surface Topography

Richard Leach (Ed.)

Optical Measurement of Surface Topography

Springer

Editor

Prof. Richard Leach
National Physical Laboratory
(NPL)
Industry & Innovation Div.
Hampton Road
TW11 0LW Teddington, Middlx.
United Kingdom
E-mail: richard.leach@npl.co.uk

ISBN 978-3-642-12011-4 e-ISBN 978-3-642-12012-1

DOI 10.1007/978-3-642-12012-1

Library of Congress Control Number: 2011924476

Typesetting: Data supplied by the authors

Cover Design: Scientific Publishing Services Pvt. Ltd., Chennai, India

Printed on acid-free paper

9 8 7 6 5 4 3 2 1

springer.com

Preface

The measurement and characterisation of areal surface topography is becoming crucial to many modern manufacturing methods. The control of areal surface structure allows a manufacturer to radically alter the functionality of a part. Examples include structuring to effect fluidics, optics, tribology, aerodynamics and biology. To control such manufacturing methods requires appropriate measurement strategies. There is also soon to be the introduction of a series of ISO specification standards in this area and this book will become a companion guide to these standards. These new standards are many and complex, as are the new measurement techniques, so industry will hopefully benefit from such a book.

There is now a wealth of new optical techniques on the market, or being developed in academia, that can measure areal surface topography. Each method has its strong points and limitations. This book will start with introductory chapters on optical instruments, their common language, generic features and limitations, and their calibration. Each type of modern optical instrument will then be described (in a common format) by experts in the field.

Acknowledgements

First and foremost I would like to thank all the chapter authors for their hard work and dedication to this book and David Flack (NPL) for reviewing some of the chapters. I also have to thank my beautiful wife to be for allowing me to spend hours writing and days travelling in order to become an expert in such an international field – thanks Sharmin. Last, but not least, my parents, sisters, son and step son also need to be thanked for their unwavering support.

This book is dedicated to the first engineer I ever met and the one that I want to please the most – thanks dad!

Contents

1 Introduction to Surface Texture Measurement

Richard Leach

Engineering Measurement Division, National Physical Laboratory
Hampton Road, Teddington
Middlesex TW11 0LW

Abstract. This chapter introduces some of the concepts used in the measurement and characterisation of surface texture that will be used throughout the rest of the book. A short history of optical measurement techniques will be given, followed by descriptions of surface profile and areal surfaces. How surface texture sits within the ISO Geometrical Product Specification is discussed along with the current position with surface texture specification standards. Non-optical and optical surface texture measurement instrument types are summarised and some general advice is given on instrument choice.

1.1 Surface Texture Measurement

Surface texture plays a vital role in the functionality of a component. It is estimated that surface effects cause 10 % of manufactured parts to fail and can contribute significantly to an advanced nation's GDP. In the last century, surface texture was primarily measured by a method that involved tracing a contacting stylus across the surface and measuring the vertical motion of the stylus as it traversed the surface features. In most cases only a single line, or surface profile, was measured and this gave rise to enough information to control production, but was limited to identifying process change. Optical instruments closely followed the development of stylus instruments and had the benefit of being non-contact and potentially faster than stylus instruments (see Leach 2009 for a short history of surface texture measurement and Hume 1980 for a more thorough history of engineering metrology).

In the early and middle of the twentieth century several different optical instrument designs were developed for measurement of surface texture and form. These included conventional Michelson and Twyman-Green interferometers (Leach 2009), Schmalz light sectioning microscopes (Schmalz 1929), Linnik microinterferometers (Miroshnikov 2010), Tolansky multiple beam interferometers (1960), and fringes of equal chromatic order (FECO) interferometers (Bennett 1976). The interferometers produced very good fringe images that accurately tracked the peaks and valleys of the surface texture with high vertical resolution. However, it was not until the development of high speed computing that the analysis of fringes could be automated to the point where digitised topography profiles and images could be produced in an automated way that could enable optical instruments to rival the usefulness of stylus instruments. The first of these was the phase shifting interfereometric (PSI) microscope in the early 1980s (Bhushan et al. 1985, Greivenkamp and Bruning 1992, see Chap. 8), which was primarily useful for measurement of smooth optical surfaces. This was followed in the early 1990s by the coherence scanning interferometric microscope (Caber 1993, Deck and de Groot 1994, see Chap. 9), also called the vertical scanning interferometer or scanning white light interferometer, which was aimed at overcoming the roughness limitations of the PSI technique. The success of interferometric microscopes was matched by confocal microscopes (Schmidt and Compton 1992, see Chap. 11) in the 1990s. Nowadays, these two types of instruments dominate the market for optical surface topography instruments, but a number of other techniques have been developed as well. Many of these different types of optical instruments are discussed in the remaining chapters of the book.

1.2 Surface Profile and Areal Measurement

Surface profile measurement is the measurement of a line across the surface that can be represented mathematically as a height function with lateral displacement, $z(x)$. Areal surface texture measurement is the measurement of an area on the surface that can be represented mathematically as a height function with displacement across a plane, $z(x,y)$. Surface texture characterisation, both profile and areal, are not discussed here – this book concentrates on the methods used to capture areal surface texture data. The subsequent characterisation methods are presented in detail elsewhere (Whitehouse 2010, Leach 2009, Muralikrishnan and Raja 2008).

1.3 Areal Surface Texture Measurement

Over the past three decades there has been an increased need to relate surface texture to surface function and, whilst a profile measurement may give some functional information about a surface, to really determine functional information, a three dimensional (3D), or areal, measurement of the surface is necessary. Control of the areal nature of a surface allows a manufacturer to alter the way a surface interacts with its surroundings. By controlling the areal nature of a surface optical, tribological, biological, aerodynamic and many other properties can be altered (Evans and Bryan 1999, Lonardo et al. 2002, de Chiffre et al. 2003, Bruzzone et al. 2008).

The measurement of areal surface texture has a number of benefits over profile measurement (Blunt and Jiang 2003). Areal measurements give a more realistic representation of the whole surface and have more statistical significance. There is also less chance that significant features will be missed by an areal method and the manufacturer, therefore, gains a better visual record of the overall structure of the surface. The need for areal surface texture measurements resulted in stylus instruments that could measure over an area (a series of usually parallel profiles) and optical techniques. Optical instruments either scan a beam over the surface akin to stylus instruments, or take an areal measurement by making use of the finite field of view of a microscope objective. There are currently many commercial instruments that can measure areal surface texture, both stylus and optical (see Sect. 1.5).

1.4 Surface Texture Standards and GPS

Surface texture documentary standards are part of the scope of the International Organization for Standardisation (ISO) Technical Committee 213 (TC 213), dealing with Dimensional and Geometrical Product Specifications and Verification as well as many national committees. ISO TC 213 has developed a wide range of standards for surface texture measurement for both profiling and areal methods and has an ambitious agenda for future standards. Some of these standards are listed below.

1.4.1 Profile Standards

There are nine ISO specification standards relating to the measurement and characterisation of surface profile. These standards only cover the use of stylus instruments. The details of most of the standards are presented in Leach (2001) and summarised in Leach (2009), and their content is not discussed in detail in this book. It should be noted that the current ISO plan for surface texture is that the profile standards will become a sub-set of the areal standards. Whilst the basic standards and details will probably not change significantly, the reader should keep abreast of the latest developments in standards. The following is a list of the profile specification standards as they stand at the time of writing of this book:

- Nominal characteristics of contact (stylus) instruments (ISO 3274, 1996)
- Rules and procedures for the assessment of surface texture (ISO 4288, 1996)
- Metrological characteristics of phase correct filters (ISO 11562, 1996)
- Motif parameters (ISO 12085, 1996)
- Surfaces having stratified functional properties — filtering and general measurement conditions (ISO 13565 part 1, 1996)
- Surfaces having stratified functional properties — height characterization using material ratio curve (ISO 13565 part 2, 1998)
- Terms, definitions and surface texture parameters (ISO 4287, 2000)
- Measurement standards — material measures (ISO 5436 part 1, 2000)
- Software measurement standards (ISO 5436 part 2, 2000)
- Calibration of contact (stylus) instruments (ISO 12179, 2000)
- Surfaces having stratified functional properties — height characterization using material probability curve (ISO 13565 part 3, 2000)
- Indication of surface texture in technical product documentation (ISO 1302, 2002)

Note that there is only one specification standard (ISO 25178: 602 2011) that relates to the measurement of surface profile using optical instruments. However, in many cases where a profile can be mathematically extracted from an areal optical scan, the profile characterisation standards can be applied. It is important, however, to understand how the surface data is filtered, especially when trying to compare contact stylus and optical results (Leach and Haitjema 2010).

1.4.2 Areal Specification Standards

In 2002, ISO technical committee 213 formed a working group (WG) 16 to address standardisation of areal surface texture measurement methods. WG 16 is developing a number of draft standards encompassing definitions of terms and parameters, calibration methods, file formats and characteristics of instruments. Several of these standards have been published and a number are at various stages in the review and approval process. The plan is to have the profile standards as a sub-set of the areal standards (with appropriate re-numbering). Hence, the profile standards will be re-published after the areal standards (with some omissions, ambiguities and errors corrected) under a new numbering scheme that is consistent with that of the areal standards. All the areal standards are part of ISO 25178, which will consist of at least the following parts, under the general title Geometrical product specification (GPS) — surface texture: areal:

- Part 1: Areal surface texture drawing indications (2011)
- Part 2: Terms, definitions and surface texture parameters (2011)
- Part 3: Specification operators (2011)
- Part 4: Comparison rules
- Part 5: Verification operators
- Part 6: Classification of methods for measuring surface texture (2010)
- Part 70: Measurement standards for areal surface texture measurement instruments (2011)
- Part 71: Software measurement standards (2011)
- Part 72: Software measurement standards – XML file format (2011)
- Part 601: Nominal characteristics of contact (stylus) instruments (2010)
- Part 602: Nominal characteristics of non-contact (confocal chromatic probe) instruments (2010)
- Part 603: Nominal characteristics of non-contact (phase shifting interferometric microscopy) instruments (2011)
- Part 604: Nominal characteristics of non-contact (coherence scanning interferometry) instruments (2011)
- Part 605: Nominal characteristics of non-contact (point autofocus) instruments (2011)
- Part 606: Nominal characteristics of non-contact (variable focus) instruments (2011)

- Part 607: Nominal characteristics of non-contact (imaging confocal) instruments (2011)
- Part 700: Calibration of non-contact instruments (2011)
- Part 701: Calibration and measurement standards for contact (stylus) instruments (2010)

At the time of writing, a general standard on the calibration of all areal surface topography measuring instruments is being drafted, but is not yet a committee draft. The American National Standards Institute has also published a comprehensive documentary specification standard, ANSI/ASME B46.1 (2010) that includes some areal analyses (mainly fractal based).

1.5 Instrument Types in the ISO 25178 Series

ISO 25178 part 6 (2010) defines three classes of methods for surface texture measuring instruments (and see Fig. 1.1):

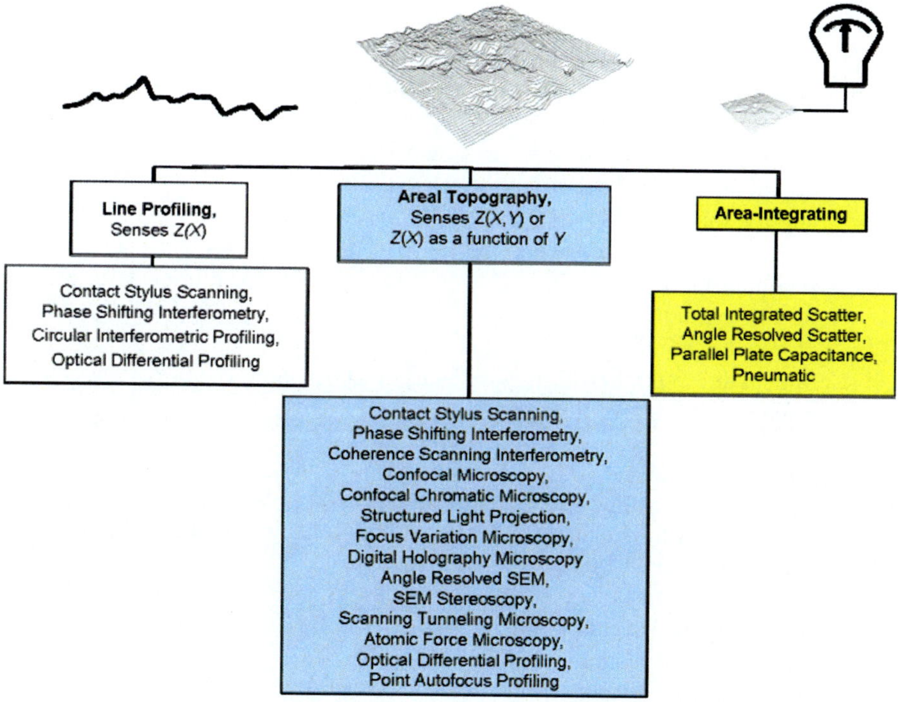

Fig. 1.1 A classification of surface texture measurement methods with examples, excerpted with permission (ISO 25178-6 2010)

Line profiling method
Surface topography method that produces a 2D graph or profile of the surface irregularities as measurement data that may be represented mathematically as a height function, $z(x)$. Examples given in ISO 25178 part 6 include:

- stylus instruments (see Sect. 1.5.1),
- phase shifting interferometry (in a line-scanning mode — see Chap. 8),
- circular interferometric profiling (this technique relies on circular scanning), and
- optical differential profiling (see Murphy 2001).

Areal topography method
Surface measurement method that produces a topographical image of the surface that may be represented mathematically as a height function, $z(x,y)$. Often, $z(x,y)$ is developed by juxtaposing a set of parallel profiles. Examples given in ISO 25178 part 6 include:

- stylus instruments (see Sect. 1.5.1),
- phase shifting interferometry (see Chap. 8),
- coherence scanning interferometry (see Chap. 9),
- confocal microscopy (see Chap. 11),
- confocal chromatic microscopy (see Chap. 5),
- structured light projection (see Chap. 3),
- focus variation microscopy (see Chap. 7),
- digital holography microscopy (see Chap. 10),
- angle resolved SEM (see Sect. 1.5.3),
- SEM stereoscopy (see Sect. 1.5.3),
- scanning tunnelling microscopy (see Leach 2009 for a brief description of this technique),
- atomic force microscopy (see Sect. 1.5.2),
- optical differential profiling (see Murphy 2001), and
- point autofocus profiling (see Chap. 6).

Area-integrating method
Surface measurement method that measures a representative area of a surface and produces numerical results that depend on area-integrated properties of the surface texture. Examples given in ISO 25178 part 6 include:

- total integrated scatter (see Chap. 12),
- angle resolved scatter (see Chap. 12),
- parallel plate capacitance (see Leach 2009 for a brief description of this technique), and
- pneumatic instruments (see Leach 2009 for a brief description of this technique).

All the optical techniques listed above are covered in this book with the exception of circular interferometric profiling and optical differential profiling. These techniques are not currently being covered in detail in ISO 25178 part 6 (circular interferometric profiling is similar to other interferometric techniques except that the scanning method is circular as opposed to linear).

1.5.1 The Stylus Instrument

The stylus instrument has been in use for around one hundred years and its principles of operation and limitations have been thoroughly discussed elsewhere (Thomas 1999, Leach 2009, Whitehouse 2010). In a typical stylus instrument a conispherical diamond tip is dragged across the surface being measured and the vertical motion of the tip in response to surface topography is measured. The vertical motion of the stylus tip and the displacement of the instrument parallel to the surface are used to determine a profile of the surface (see Sect. 1.2). Figure 1.2 is a schematic representation of a typical stylus instrument. Stylus instruments are the only instrument types that are fully covered by ISO specification standards at the time of writing of this book and there is good practice guidance available (Leach 2001, Giusca and Leach 2011).

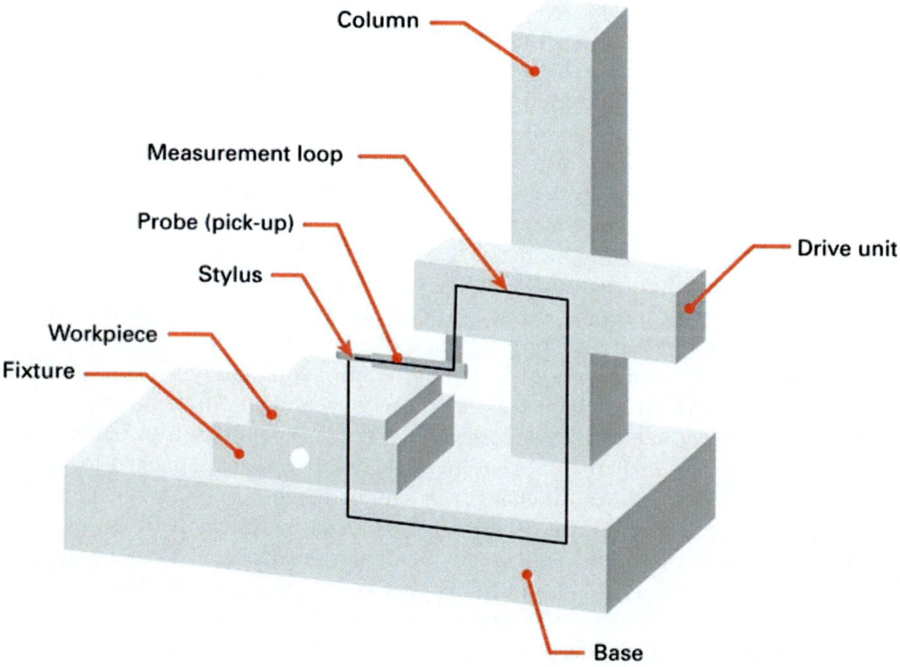

Fig. 1.2 Schematic representation of a typical stylus instrument (from Leach 2009)

Stylus instruments have a relatively simple operating principle and it is not difficult to calculate the trajectory of a ball-ended stylus over a surface. For this reason, it can be easier to predict the output of a stylus instrument compared to that of an optical instrument (Leach and Haitjema 2010). Hence, stylus instruments tend to be the instrument of choice when considering primary instruments that are designed with traceability in mind as opposed to usability. However, the stylus can damage the surface being measured, so the optical instruments have an advantage in this respect.

Stylus instruments can also be used to measure areal surface topography by moving the stylus across the surface in a raster fashion and building up an areal map. However, typical stylus speeds mean that it can take several tens of seconds or minutes to scan a few millimetres. When a square grid of data points is required for a high density areal scan with several thousand points in each profile, the total measurement time can be several hours. This is where the optical instruments have a clear advantage over stylus instruments, especially when measurement throughput is an issue, for example in-process or on-line applications.

1.5.2 Scanning Probe Microscopes

Scanning probe microscopes (SPMs) are a family of instruments that are used for measuring surface topography usually on a smaller scale than that of conventional stylus instruments and optical instruments. Along with the electron microscope (see Sect. 1.5.3), they are the instruments of choice when surface structure needs to be measured with spatial wavelengths smaller than the diffraction limit of an optical instrument (typically around 500 nm — see Sect. 2.6). The theory, principles, operation and limitations of SPMs are discussed thoroughly elsewhere (Weisendanger 1994, Meyer et al. 2003, Leach 2009). The SPM is a serial measurement device that uses a nanometre-scale probe to trace the surface of the sample based on local physical interactions (in a similar manner to a stylus instrument – see Sect. 1.5.1). While the probe scans the sample with a predefined pattern, the signal of the interaction is recorded and is usually used to control the distance between the probe and the sample surface. This feedback mechanism and the scanning of a nanometre-scale probe form the basis of all SPMs.

There are many different types of SPM but the most common type is the atomic force microscope (AFM) (Maganov 2008). In a conventional AFM the sample is scanned continuously in two axes underneath a force-sensing probe consisting of a tip that is attached to, or part of, a cantilever. A scanner is also attached to the z axis (height) and compensates for changes in sample height, or forces between the tip and the sample. The presence of attractive or repulsive forces between the tip and the sample will cause the cantilever to bend and this deflection can be monitored in a number of ways. The most common system to detect the bend of the cantilever is the optical beam deflection system, wherein a laser beam reflects off the back of the cantilever onto a photodiode detector. Such an optical beam deflection system is sensitive to sub-nanometre deflections of the cantilever. In this way an areal map of the surface is measured, usually over a few tens to hundreds of micrometres square, although longer range instruments are available (for example, Dai et al. 2009).

Some SPMs are simply listed as an instrument type in the ISO 25178 series of standards, but at the time of writing of this book, there is only limited good practice guidance available (VDI/VDE 2656-1 2008) and aspects of their calibration are covered by specification standards from ISO technical committee 201. It is rare to see comparisons of SPM with optical or stylus instruments because they tend to operate over a different spatial bandwidth. One example of a comprehensive comparison of results for SPM, stylus, optical profilers and scatterometers is that by Marx et al. (2002). SPMs tend to be difficult instruments to set up and use, and measurements over relatively large areas can be very time consuming.

A further form of SPM is the scanning near-field optical microscope (SNOM) that probes the surface using near-field optics (sometimes referred to as electromagnetic tunneling) (Courion 2003). Whilst the SNOM is an optical instrument, it is not covered in detail in this book as it is also sensitive to optical properties and material variations of the surface and is primarily used to study those qualities.

1.5.3 Scanning Electron Microscopes

The scanning electron microscope (SEM) uses a very fine beam of electrons, which is made to scan the specimen under test as a raster of parallel contiguous lines (see Egerton 2008, Goodhew et al. 2000 for thorough descriptions of electron microscopy). Upon hitting the specimen electrons will be reflected (backscattered electrons) or generated by interaction of the primary electrons with the sample (secondary electrons). The specimen is usually a solid object and the number of secondary electrons emitted by the surface will depend upon its topography or nature. These are collected, amplified and analysed before modulating the beam of a cathode ray tube scanned synchronously with the scanning beam. The image resembles that seen through an optical lens but at a much higher resolution.

SEMs typically measure surface topography on a much smaller spatial wavelength scale to stylus and optical instrument, but can be used over relatively large ranges. The main drawback with SEM is that it is essentially a two-dimensional technique, although 3D information can be obtained from some surfaces by tilting the sample and using a stereo imaging or angle-resolved scanning techniques (Goldstein et al. 2003). SEM stereoscopy and angle-resolved SEM are listed as techniques in ISO 25178 part 6 (2010) but at the time of writing of this book there are no plans for standards on metrological characteristics and calibration.

1.5.4 Optical Instrument Types

There are many different types of optical instruments that can measure surface texture. The techniques can be broken down into two major classes — those that measure the actual surface topography by either scanning a beam or using the field of view (profile or areal methods), and those that measure a statistical parameter of the surface, usually by analysing the distribution of scattered light (area-integrating methods). Whilst both these methods operate in the optical far field, there is a third area of instruments that operate in the near field — these are not discussed in this book.

The instruments that are discussed in Chap. 5 through Chap. 12 are the most common instruments that are available commercially. There are many more types of optical instruments, or variations on the instruments presented here, most of which are listed in ISO 25178 part 6 (2010) with appropriate references. At the time of writing, only the methods described in Chap. 5 through Chap. 11 are being actively standardised in the appropriate ISO committee (ISO TC 213 WG 16).

1.6 Considerations When Choosing a Method

There are many factors that need to be considered when choosing a method for measuring surface texture. Some of the questions that need to be addressed are as follows:

- What type of surfaces do you need to measure? This question may not have a straightforward answer; you may need to measure a single type of surface for a production process, or you may need a generic instrument to measure a whole range of surfaces. Further questions in this category include:
    - What surface geometries need measuring? This includes the spatial frequency spectrum and the amplitude distribution. For spatial wavelengths less than around 500 nm only SPM or SEM methods can be used, but such methods will only have limited height ranges.
    - What surface materials do you need to measure? This includes the material hardness, optical characteristics, electrical characteristics and chemistry. For example, if the surface is soft, then this may prohibit the use of a stylus instrument or if the surface has limited reflectivity, then an optical method may not be an option.
- What overall size of object do you need to measure the surface of? Some instruments will have a limited size of object that can be placed on the measurement table, for example some optical instruments will have limited object height due to the finite stand-off of the objective lens. Modern instruments tend to have the ability to accommodate relatively large object sizes, but in the case of some older SPMs, the object needs to be less then 1 mm in height with a base area of a few square millimetres.
- How fast do you need the measurement to be? This is an important consideration because the measurement times for the various instruments vary considerably. Optical methods tend to have shorter measurement times than stylus instruments, especially where areal measurements are required. The time to prepare the sample may also be important, for example, when using an SEM, it is necessary to apply a conducting coating to a dielectric sample and allow for time to pump the instrument down to the required level of vacuum.

- What sort of measurement uncertainty do you require? The answer to this question will depend on such things as what sort of manufacturing tolerances you need to meet and what type of quality system you need to comply with. Uncertainty analysis is very complex for surface texture measurement (see Chap. 4) but good practice should always be applied.
- What characterisation options do you require? As with the first question, this question is often not simple to answer. It will depend on what types of surfaces need measuring, what types of process you are trying to control and what sort of quality system you need to comply with. In some cases just a single profile parameter may be required with default filter settings, but in another case, complex areal analysis may be required with multiple filtering options.
- What is your financial budget? Surface texture measuring instruments can range from a few thousand euros for a simple hand-held stylus instrument to several hundred thousand euros for a high-end SPM or optical instrument with full product support and characterisation software.

Even when the questions above have indicated that an optical instrument is required, there are still many decisions to be made. It is one of the central aims of this book to help metrologists with these complicated choices. There is rarely an obvious frontrunner when considering the different optical techniques, so it will be useful to have a single source that can allow informed comparisons to be made.

Acknowledgements

The author would like to thank Dr Ted Vorburger (National Institute of Standards and Technology) for help with the historical section of this chapter.

References

ANSI/ASME B46.1, Surface texture, surface roughness, waviness and lay. American National Standards Institute (2010)

Bennett, J.M.: Measurement of the RMS roughness, autocovariance function and other statistical properties of optical surfaces using a FECO scanning interferometer. Appl. Opt. 15, 2705–2721 (1976)

Bhushan, B., Wyant, J.C., Koliopoulis, C.L.: Measurement of surface topography of magnetic tapes by Mirau interferometry. Appl. Opt. 24, 1489–1497 (1985)

Blunt, L.A., Jiang, X.: Advanced techniques for assessment surface topography. Kogan Page Science (2003)

Bruzzone, A.A.G., Costa, H.L., Lonardo, P.M., Lucca, D.A.: Advances in engineered surfaces for functional performance. Ann. CIRP 57, 750–769 (2008)

Caber, P.J.: Interferometric profiler for rough surfaces. Appl. Opt. 32, 3438–3441 (1993)

de Chiffre, L., Kunzmann, H., Peggs, G.N., Lucca, D.A.: Surfaces in precision engineering, micro-engineering and nanotechnology. Ann. CIRP 52, 561–577 (2003)

Courion, D.: Near-field microscopy and near-field optics. Imperial College Press, London (2003)

Dai, G., Wolff, H., Pohlenz, F., Danzebrink, H.-U.: A metrological large range atomic force microscope with improved performance. Rev. Sci. Instrum. 80, 043702 (2009)

Deck, L., de Groot, P.: High-speed noncontact profiler based on scanning white-light interferometer. Appl. Opt. 33, 7334–7388 (1994)

Egerton, R.F.: Physical principles of electron microscopy: an introduction to TEM, SEM and AEM, 2nd edn. Springer, Heidelberg (2008)

Evans, C.J., Bryan, J.B.: "Structured", "textured", "engineered" surfaces. Ann. CIRP 48, 541–556 (1999)

Giusca, C.L., Leach, R.K.: Calibration of stylus instruments for areal surface texture measurement. NPL Good practice guide, National Physical Laboratory, UK (2011)

Goldstein, J., Newbury, D.E., Joy, D.C., Lyman, C.E., Echlin, P., Lifshin, E., Sawyer, L.C., Michael, J.R.: Scanning electron microscopy and x-ray microanalysis, 3rd edn. Springer, Heidelberg (2003)

Goodhew, P.J., Humpheys, F.J., Beanland, R.: Electron microscopy and analysis. Taylor & Francis, Abington (2000)

Greivenkamp, J.E., Bruning, J.H.: Phase shifiting interferometry. In: Malacara, D. (ed.) Optical shop testing, 2nd edn., Wiley, New York (1992)

Hume, K.J.: A history of engineering metrology. Mechanical Engineering Publications Ltd (1980)

ISO 3274, Geometrical product specification (GPS) – Surface texture: Profile method – Nominal characteristics of contact (stylus) instruments. International Organization of Standardization (1996)

ISO 4288, Geometrical product specification (GPS) – Surface texture: Profile method – Rules and procedures for the assessment of surface texture. International Organization of Standardization (1996)

ISO 11562, Geometrical product specification (GPS) – Surface texture: Profile method – Metrological characteristics of phase correct filters. International Organization of Standardization (1996)

ISO 12085, Geometrical product specifications (GPS) – Surface texture: Profile method – Motif parameters. International Organization for Standardization (1996)

ISO 13565 part 1, Geometrical product specification (GPS) – Surface texture: Profile method – Surfaces having stratified functional properties – Filtering and general measurement conditions. International Organization for Standardization (1996)

ISO 13565 part 2, Geometrical product specification (GPS) – Surface texture: Profile method – Surfaces having stratified functional properties – Height characterization using material ratio curve. International Organization for Standardization (1998)

ISO 4287, Geometrical product specification (GPS) – Surface texture: Profile method – Terms, definitions and surface texture parameters. International Organization of Standardization (2000)

ISO 5436 part 1, Geometrical product specification (GPS) – Surface texture: Profile method – Measurement standards – Material measures. International Organization of Standardization (2000)

ISO 5436 part 2, Geometrical product specification (GPS) – Surface texture: Profile method – Software measurement standards. International Organization of Standardization (2000)

ISO 12179, Geometrical product specification (GPS) – Surface texture: profile method – Calibration of contact (stylus) instruments. International Organization for Standardization (2000)

ISO 13565 part 3, Geometrical product specification (GPS) – Surface texture: Profile method – Surfaces having stratified functional properties – Height characterization using material probability curve. International Organization for Standardization (2000)
ISO 1302, Geometrical product specification (GPS) - Indication of surface texture in technical product documentation. International Organization of Standardization (2002)
ISO/CD 25178 part 1, Geometrical product specification (GPS) – Surface texture: Areal – Part 1: Indication of surface texture. International Organization for Standardization (2011)
ISO/FDIS 25178 part 2, Geometrical product specification (GPS) – Surface texture: Areal – Part 2: Terms, definitions and surface texture parameters. International Organization for Standardization (2011)
ISO/DIS 25178 part 3, Geometrical product specification (GPS) – Surface texture: Areal – Part 3: Specification operators. International Organization for Standardization (2011)
ISO 25178 part 6, Geometrical product specification (GPS) – Surface texture: Areal – Part 6: Classification of methods for measuring surface texture. International Organization for Standardization (2010)
ISO/CD 25178 part 70, Geometrical product specification (GPS) – Surface texture: Areal – Part 70: Material measures. International Organization for Standardization (2011)
ISO/DIS 25178 part 71, Geometrical product specification (GPS) – Surface texture: Areal – Part 71: Software measurement standards. International Organization for Standardization (2011)
ISO/WD 25178 part 72, Geometrical product specification (GPS) – Surface texture: Areal – Part 72: XML softgauge file format. International Organization for Standardization (2011)
ISO 25178 part 601, Geometrical product specification (GPS) – Surface texture: Areal – Part 601: Nominal characteristics of contact (stylus) instruments. International Organization for Standardization (2010)
ISO 25178 part 602, Geometrical product specification (GPS) – Surface texture: Areal – Part 602: Nominal characteristics of non-contact (confocal chromatic probe) instruments. International Organization for Standardization (2010)
ISO/DIS 25178 part 603, Geometrical product specification (GPS) – Surface texture: Areal – Part 603: Nominal characteristics of non-contact (phase shifting interferometric microscopy) instruments. International Organization for Standardization (2011)
ISO/CD 25178 part 604, Geometrical product specification (GPS) – Surface texture: Areal – Part 604: Nominal characteristics of non-contact (coherence scanning interferometry) instruments. International Organization for Standardization (2011)
ISO/CD 25178 part 605, Geometrical product specification (GPS) – Surface texture: Areal – Part 605: Nominal characteristics of non-contact (point autofocusing) instruments. International Organization for Standardization (2011)
ISO/WD 25178 part 606, Geometrical product specification (GPS) – Surface texture: Areal – Part 605: Nominal characteristics of non-contact (point autofocusing) instruments. International Organization for Standardization (2011)
ISO/WD 25178 part 607, Geometrical product specification (GPS) – Surface texture: Areal – Part 607: Nominal characteristics of non-contact (focus variation) instruments. International Organization for Standardization (2011)
ISO/WD 25178 part 608, Geometrical product specification (GPS) – Surface texture: Areal – Part 608: Nominal characteristics of non-contact (imaging confocal microscope) instruments. International Organization for Standardization (2011)

ISO 25178 part 700, Geometrical product specification (GPS) – Surface texture: Areal – Part 700: Calibration of non-contact instruments. International Organization for Standardization (2011)

ISO 25178 part 701, Geometrical product specification (GPS) – Surface texture: Areal – Part 701: Calibration and measurement standards for contact (stylus) instruments. International Organization for Standardization (2010)

Leach, R.K.: The measurement of surface texture using stylus instruments. NPL Good practice guide No. 37. National Physical Laboratory, UK (2001)

Leach, R.K.: Fundamental principles of engineering nanometrology. Elsevier, Amsterdam (2009)

Leach, R.K., Haitjema, H.: Limitations and comparisons of surface texture measuring instruments. Meas. Sci. Technol. 21, 32001 (2010)

Lonardo, P.M., Lucca, D.A., de Chiffre, L.E.: Emerging trends in surface metrology. Ann. CIRP 51, 701–723 (2002)

Magonov, S.: Atomic force microscopy. John Wiley & Sons, Chichester (2008)

Marx, E., Malik, I.J., Strausser, Y.E., Bristow, T., Poduje, N., Stover, J.C.: Power spectral densities: a multiple technique study of different Si wafer surfaces. J. Vac. Sci. Technol. 20, 31–41 (2002)

Meyer, E., Hug, H.J., Bennewitz, R.: Scanning probe microscopy: the lab on a tip. Springer, Heidelberg (2003)

Miroshnikov, M.M.: Academician Vladimir Pavlovich Linnik—The founder of modern optical engineering (on the 120th anniversary of his birth). J. Opt. Technol. 77, 401–408 (2010)

Muralikrishnan, B., Raja, J.: Computational surface and roundness metrology. Springer, Heidelberg (2008)

Murphy, D.B.: Fundamentals of light microscopy and electronic imaging. Wiley-Liss Inc, Chichester (2001)

Schmaltz, G.: Über Glätte und Ebenheit als physikalisches und physiologisches Problem. Zeitschrift des Vereines deutcher Ingenieure 73, 1461 (1929)

Schmidt, M.A., Compton, R.D.: Friction, lubrication, and wear technology. In: ASM Handbook, vol. 18, ASM International (1992)

Thomas, T.R.: Rough surfaces, 2nd edn. Imperial College Press, London (1999)

Tolansky, S.: Multiple-beam interference microscopy of metals. Academic Press Inc., London (1970)

Tolansky, S.: Surface microtopography. Wiley Interscience, New York (1960)

VDI/VDE 2656 part 1, Determination of geometrical quantities by using scanning probe microscopes – calibration of measurement systems (2008)

Weisendanger, R.: Scanning probe microscopy and spectroscopy: methods and applications. Cambridge University Press, Cambridge (1994)

Whitehouse, D.J.: Handbook of surface and nanometrology, 2nd edn. CRC Press, Boca Raton (2010)

2 Some Common Terms and Definitions

Richard Leach

Engineering Measurement Division, National Physical Laboratory
Hampton Road, Teddington
Middlesex TW11 0LW, UK

Abstract. This chapter presents a number of terms and definitions that are common to the instruments presented in Chap. 5 to Chap. 12. The most common forms of optical aberration are described followed by a discussion of microscope objective lenses. The concepts of magnification, numerical aperture, lateral resolution, spot size and depth of field are discussed in limited detail to allow the reader to use them throughout the rest of the book.

2.1 Introduction

This chapter will present some terms and definitions that are common to many optical instruments that measure surface texture. It is assumed that the reader has a basic knowledge of geometrical and physical optics. There are several good texts on basic optics that can be consulted where necessary (see for example, Born and Wolf 1997, Hecht 2003). The chapters in this book that describe the instruments will present their associated basic theory, but there are some concepts that are common to most instruments. For example, many of the instruments use objective lenses that are common to the simple compound microscope (see Murphy 2001).

2.2 The Principal Aberrations

Aberrations are departures of the performance of an optical system from the predictions of paraxial optics (Welford 1986). Aberration leads to blurring of the image produced by an image-forming optical system. It occurs when light from one point of an object after transmission through the system does not converge into (or does not diverge from) a single point. Most optical instruments need to correct their optical components to compensate for aberration.

Most of the optical components in microscope-based instruments have spherical surfaces that can lead to a number of aberrations. These aberrations are corrected by the use of further optical components or systems, examples of which include compound lenses, elements with different refractive indexes, aspherical lenses and tube lenses. It is impossible to remove all aberrations, and many of the correcting components can also introduce further aberration, so optical design is always a compromise. This leads to a large array of different objective lenses.

The principal aberrations of lenses are breifly described below:

Chromatic aberration – This aberration, sometimes referred to as chromatism, is caused by a lens refracting light of different wavelengths by differing amounts (see Fig. 2.1). This causes blur in the image and, since each wavelength is focused at a different distance from the lens, there is also a difference in magnification for different colours. To compensate for chromatic aberration, compound lenses are made from glasses with different dispersion properties.

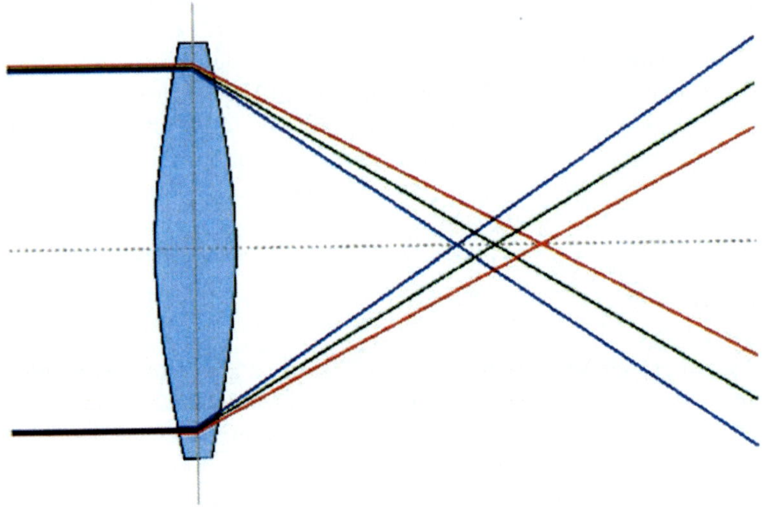

Fig. 2.1 Effect of chromatic aberration

Sperical aberration – This aberration occurs due to the increased refraction of light rays when they strike a lens (or a reflection of light rays from a mirror) near its edge, in comparison with those that strike nearer the centre. Hence there is not a well-defined focal spot for a point source and the sharpness of an image is affected. To compensate for spherical aberration, a compound lens can be used with concave and convex geometries of differing thicknesses.

Comatic aberration – This aberration, usually referred to as simply coma, is inherent to certain optical designs or due to imperfection in the lens, or other components which results in off-axis point sources appearing distorted. Coma causes a point object to appear as a comet with the tail extending towards the periphery of the field. Correction for coma is made to accommodate the diameter of the object field for a given lens.

Astigmatism – This aberration occurs when rays that propagate in two perpendicular planes have different foci. If an optical system with astigmatism is used to form an image of a cross, the vertical and horizontal lines will be in sharp focus at two different distances. As with coma, astigmatism is an off-axis aberration.

Curvature of field – This aberration occurs when the image plane is not flat but has the shape of a, usually concave, spherical surface as seen from the objective. Curvature of field is corrected by the design of the objective and by tube or relay lenses.

Distortion – This aberration causes the focus position of the optic image to shift laterally in the image plane with increasing displacement of the object from the optic axis. Distortion results in a nonlinear magnification in the image from the centre to the periphery of the field. Correction for distortion is as for curvature of field.

2.3 Objective Lenses

An objective lens is the part of a microscope that collects light reflected by the object surface and forms a magnified (inverted) real image at some convenient plane. In most of the instruments described in this book, the objective lens also focuses the light from the source onto the surface. Modern objective lenses are complicated arrangements of several optical elements that perform close to their theoretical limits and are, therefore, relatively expensive. The lens combination in an objective lens depends on the type of source used by the instrument and the type of image that is required. There are three main types of objective lens:

Achromats – These are corrected for chromatic aberration (usually in the red to blue end of the spectrum) and for spherical aberration (usually in the yellow to green end of the spectrum). The most common type of achromat is the achromatic doublet, which is composed of two individual lenses made from glasses with different amounts of dispersion. Usually one element is a concave lens made out of flint glass, which has relatively high dispersion, while the other, convex, element is made of crown glass, which has lower dispersion. The lens elements are mounted next to each other, typically cemented together, and shaped so that the chromatic aberration of one is counterbalanced by that of the other. Achromats perform well for white light or monochromatic radiation.

Semiapochromat – These have a high degree of correction for chromatic aberration and curvature of field. They are typically made from fluorite or fluorspar (or other synthetic materials) and have high contrast and transparency. Semiapochromats are used for biological, polarization and interference contrast microscopy.

Apochromats – These have a higher degree of correction for chromatic and spherical aberration than achromat lenses. Apochromats are typically used for colour photography and fluorescence microscopy.

2.4 Magnification and Numerical Aperture

The magnification of the lens system given in Fig. 2.2 is given by

$$M = {}^{b}\!/_{a} \tag{2.1}$$

where a is the object distance and b is the image distance defined in the figure. Also, the focal length, f, of the lens in Fig. 2.2 is given by

$$\frac{1}{f} = \frac{1}{a} + \frac{1}{b}. \tag{2.2}$$

The magnification of a compound microscope is given by the product of the magnification of the objective lens with that of the ocular lens (the lens that focuses the image onto a detector). A microscope image is located at a distance, a, between f and $2f$ and is real.

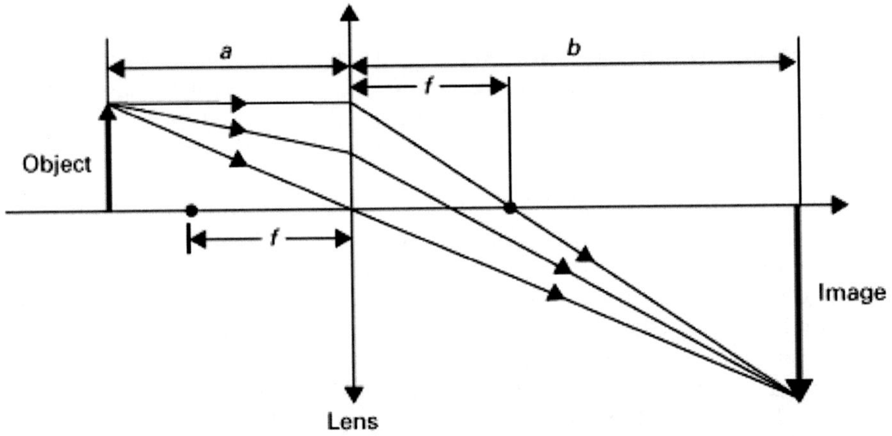

Fig. 2.2 Object-image relationship for a simple lens

The magnifications of typical instruments used to measure surface texture varies from 2.5× to 150× depending on the application and the type of surface being measured. The numerical (or angular) aperture, A_N, determines the largest slope angle on the surface that can be measured and affects the optical resolution. The numerical aperture of an objective is given by

$$A_N = n \sin \alpha \tag{2.3}$$

where n is the refractive index of the medium between the objective and the surface (usually air, so n can be approximated by unity) and α is the acceptance angle of the aperture (see Fig. 2.3 where the objective is approximated by a single lens). The acceptance angle will determine the slopes on the surface that can physically reflect light back into the objective lens and hence be measured (note that in some instruments the diffuse reflectance is used to increase the measurable slope angle).

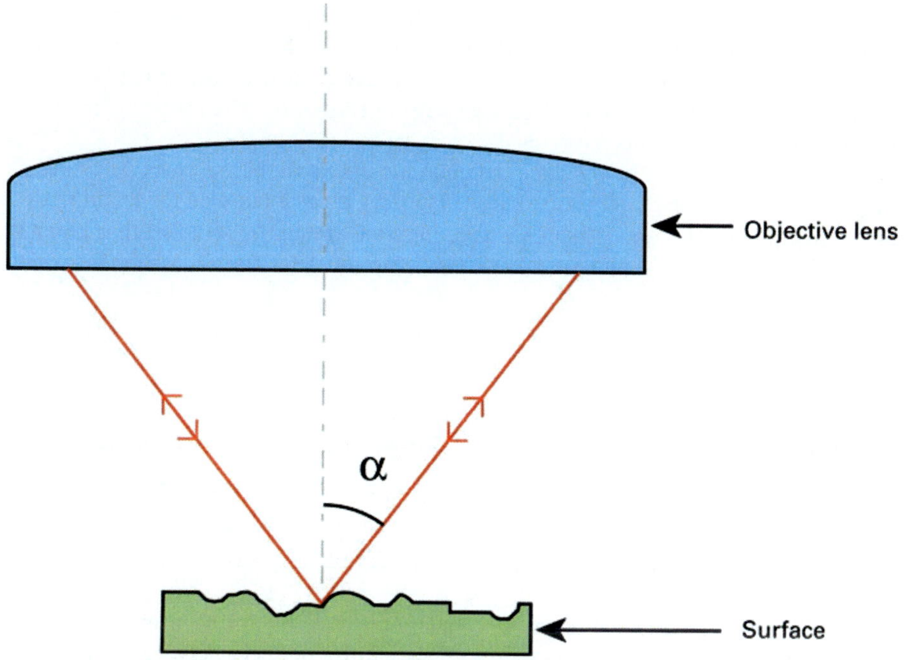

Fig. 2.3 Numerical aperture of a microscope objective lens

For instruments based on interference microscopy it may be necessary to apply a correction to the interference pattern due to the effect of the numerical aperture. Effectively the finite numerical aperture means that the fringe distance is not equal to half the wavelength of the source radiation (Schultz and Elssner 1991). This effect, often called the obliquity factor, also accounts for the aperture correction in gauge block interferometry (Leach 2009) and may cause a step height to be measured too short. This correction can usually be determined by measuring a step artefact with a calibrated height value or by using a grating (Greve and Krüger-Sehm 2004).

2.5 Spatial Resolution

The spatial resolution determines the minimum distance between two lateral features on a surface that can be distinguished. For a perfect optical system with a filled objective pupil, the resolution (or resolving power) is given by the Rayleigh criterion

$$r = 0.61\frac{\lambda}{A_N} \tag{2.4}$$

where λ is the wavelength of the incident radiation. Yet another measure of the optical resolution is the Sparrow criterion (the spatial wavelength where the instrument response drops to zero) where the factor of 0.61 in equation (2.4) is replaced by 0.82. The Rayleigh and Sparrow criterions are often used indiscriminately, so the user should always check which expression has been used in a given situation. Also, equation (2.4) sets a minimum value for the resolution. If the objective is not optically perfect (*i.e.* not aberration free), or if a part of the beam is blocked (for example, in a Mirau interference objective, or when a steep edge is measured) the resolution deteriorates.

For some instruments the distance between the pixels in the microscope camera array may determine the lateral resolution. Tab. 2.1 gives examples for a commercial microscope — for the 50× objective, it is the optical resolution that determines the minimum distance between features, but with the 10× objective, it is the pixel spacing.

Table 1 Minimum distance between features for different objectives

Magnification	A_N	**Resolution/µm**	**Pixel spacing/µm**
10×	0.3	1.00	1.75
20×	0.4	0.75	0.88
50×	0.5	0.60	0.35

When measuring surface texture, one must consider the ability to measure the spacing of points in an image together with the ability to accurately determine the heights of features. In this context, an optical equivalent of cut-off wavelength for stylus instruments is the lateral resolution or the wavelength at 50 % depth modulation. This is defined as one half the spatial period of a sinusoidal profile for which the instrument response (measured feature height compared to actual feature height) falls to 50 %. The instrument response can be found by direct measurement of the instrument transfer function (see de Groot and Colonna de Lega 2006). The value of the lateral (50 %) resolution will vary with the height of the features being measured. However, it is assumed to stabilise for very small heights, *i.e.* less than a quarter of a wavelength (see Sect. 4.5.4).

2.6 Optical Spot Size

Another important factor for optical instruments that magnify the surface being measured is the optical spot size. For scanning type instruments the spot size will determine the area of the surface measured as the instrument scans. To a first approximation, the spot size mimics the action of the tip radius on a stylus instrument, *i.e.* it acts as a low pass filter (Krüger-Sehm et al. 2006). The optical spot size is given by

$$d_0 = \frac{f\lambda}{w_0} \tag{2.5}$$

where f is the focal length of the objective lens and w_0 is the beam waist (the radius of the $1/e^2$ irradiance contour at the plane where the wavefront is flat).

2.7 Field of View

The field of view is the diameter of the circle of illumination on the object. The larger the field of view the smaller the magnification. In an instrument that measures areal surface texture without scanning, it is the field of view that determines the lateral area that is measured. In the example given in Tab. 2.1 the areas measured are 0.3 mm by 0.3 mm and 1.2 mm by 1.2 mm for the 50× and 10× objectives respectively.

2.8 Depth of Field and Depth of Focus

The wave nature of light and the phenomenon of diffraction cause the image of a point object to be a diffraction disk of finite diameter. The effect also occurs in the direction of propagation, i.e. the diffraction disk has a finite thickness. The depth of field in the object plane is the thickness of the optical section along the principle axis of the objective lens within which the object is in focus. The depth of field, Z, is given by

$$Z = \frac{n\lambda}{{A_N}^2} \tag{2.6}$$

where n is the refractive index of the medium between the lens and the object. Depth of field is effected by the optics used, lens aberrations and the magnification (see Pluta 1988 for a more thorough review). Note that as the depth of field increases, the resolution proportionally decreases. Depth of field can be improved by:

- Reducing the numerical aperture by making the aperture smaller, or using a lower numerical aperture objective lens.
- Lowering the magnification for a given numerical aperture.
- Reducing any zoom factor.
- Using longer illumination wavelengths.

Depth of field can be measured by imaging a periodic grating that has been tilted so that it is imaged obliquely.

The depth of focus is the thickness of the image plane, or the range of the image plane position at which the image can be viewed without appearing out of focus for a fixed position of object.

2.9 Interference Objectives

An interference objective is a microscope objective adapted to generate an interference signal in accordance with wave optics, resulting in an image or electronic interference signal that can be interpreted to determine areal topography. Typical interference objectives split the illumination into two paths, which after recombination, result in an interference effect that is sensitive to surface height or other sample characteristics. Chap. 9 and Chap. 10 provide further details of configurations and uses of interference objectives.

Acknowledgements

The author would like to thank Dr Peter de Groot (Zygo) for help with this chapter.

References

Born, M., Wolf, E.: Principles of optics: electromagnetic theory of propagation, interference and diffraction of light, 6th edn. Cambridge University Press, Cambridge (1997)

de Groot, P., Colonna de Lega, X.: Interpreting interferometric height measurements using the instrument transfer function. In: Proc FRINGE, pp. 30–37 (2006)

Greve, M., Krüger-Sehm, R.: Direct determination of the numerical aperture correction factor of interference microscopes. In: Proc. XI Int. Colloq. Surfaces, Chemnitz, Germany, pp. 156–163 (February 2004)

Hecht, E.: Optics (international edition), 4th edn. Pearson Education, London (2003)

Krüger-Sehm, R., Frühauf, J., Dziomba, T.: Determination of the short wavelength cutoff for interferential and confocal microscopes. Wear 264, 439–443 (2006)

Leach, R.K.: Fundamental principles of engineering nanometrology. Elsevier, Amsterdam (2009)

Murphy, D.B.: Fundamentals of light microscopy and electronic imaging. Wiley-Liss Inc., Chichester (2001)

Pluta, M.: Advanced light microscopy. In: Principles and basic optics, vol. 1. Elsevier, Amsterdam (1988)

Schultz, G., Elssner, K.–E.: Errors in phase measurement interferometry with high numerical aperture. Appl. Opt. 30, 4500–4505 (1991)

Welford, W.T.: Aberrations of optical systems. CRC Press, Boca Raton (1986)

3 Limitations of Optical 3D Sensors

Gerd Häusler and Svenja Ettl

Institute for Optics, Information and Photonics, University Erlangen-Nuremberg
Staudtstr. 7/B2
91058 Erlangen, Germany

Abstract. This chapter is about the physical limitations of optical 3D sensors. The ultimate limit of the measurement uncertainty will be discussed; in other words: "How much 3D information are we able to know?" The dominant sources of noise and how this noise affects the measurement of micro-scale topography will be discussed. Some thoughts on how to overcome these limits will be given. It appears that there are only four types of sensors to be distinguished by the dominant sources of noise and how the physical measurement uncertainty scales with the aperture or working distance. These four types are triangulation, coherence scanning interferometry at rough surfaces, classical interferometry and deflectometry. 3D sensors will be discussed as communication channels and considerations about information-efficient sensors will be addressed.

3.1 Introduction: What Is This Chapter About?

Is the visually accessible world 3D? Fortunately, we do not have an x-ray tomographic view of the world which would swamp our brain storage capacity. Only surfaces, embedded in 3D space, can be seen. The projection of these surfaces on to our more or less planar retina is visible. Under very restricted conditions (structured surfaces, close distance) stereo vision allows for some 3D perception. Quasi-3D information is given by shading with proper illumination. Stereo vision and shading information make it easier to play tennis and not to hit the garage wall with the car. But to acquire the 3D shape of a surface, or its 3D micro-scale topography, 3D sensors are needed.

This chapter is about the limitations of optical 3D sensors. Rather than discussing the technical implementations of 3D sensors, a specific viewpoint is taken and the physical limits of those sensors will be discussed. This makes it necessary to understand the underlying principles of signal formation, independent from the technical realisation of a given sensor. It appears that behind the hundreds of different available sensors there are only four fundamental physical principles of signal generation. Further, 3D sensors will be considered as communication systems and information-theoretical limits will be discussed. Of course, some technical limitations have to be discussed, but from the scientific point of view, a sensor is only perfect if it is not limited by technology, but rather by the fundamental limits of nature.

This viewpoint will allow insight in to the ultimate limits that yields many options, for example, knowledge about sensor limitations allows a judgment to be made as to whether a sensor can be improved by better technology or not. Expensive investments can be avoided that would be unnecessary if a sensor is already limited by nature.

Limitations are often given by an uncertainty product. For many sensors, the laterally resolvable distance δx is connected with the distance measurement uncertainty δz via an uncertainty product given by

$$\delta x \cdot \delta z = h\,, \tag{3.1}$$

where h is a system-dependent constant. Such a product gives useful insight: if some lateral resolution is sacrificed, the distance z can be acquired with less uncertainty. The most well-known uncertainty product is Heisenberg's uncertainty relation. It turns out that many of the observed limits directly stem from Heisenberg's ultimate limit, or at least by Fourier uncertainty products (for example, see Häusler and Leuchs 1997).

So the questions to be addressed in this chapter are straight-forward. What are the ultimate physical limitations of different sensor principles? How many sensor principles exist? How much information about the surface topography can be known? How much effort (i.e. channel capacity or technology development) has to be invested? How can information-efficient sensors be made? How can technical space-bandwidth limitations be overcome? Also, there might be questions still unanswered, for example, can the Abbe limit be overcome?

3.2 The Canonical Sensor

Fig. 3.1 schematically shows an optical 3D sensor in its canonical physical form. Considering each component's principle thoroughly, insight about the limitations can be obtained. It will become clear that the design and the proper choice of an optical sensor are difficult, since there are so many options for the illumination, the interaction of the light with the object and for the measured modality.

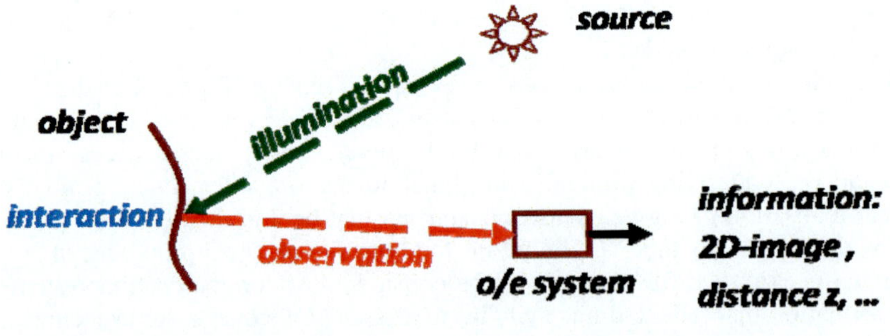

Fig. 3.1 The physical model of a canonical optical sensor

The illumination can be implemented in many different ways and can have a large effect on an instrument's performance. Illumination may be spatially coherent or incoherent, directed or diffuse, structured or homogeneous, monochromatic or coloured, polarised or un-polarised, and temporally continuous or pulsed.

The light hits the object surface under test. The interaction of light with the atoms of the object may be coherent or incoherent. Commonly, there is coherent (Rayleigh) scattering, which means that the scattered light is in phase with the exciting light field. Coherent scattering leads to a fundamental limitation: speckle noise (see for example, Häusler and Herrmann 1988, Häusler 1990, Baribeau and Rioux 1991, Dorsch et al. 1994). There are only a few practical options for incoherent scattering: fluorescence or the emission of thermal excitation. Several microscopic methods for medical imaging exploit the important advantage of fluorescence, i.e. avoidance of speckle noise. In laser material processing, the incoherence of thermal excitation can be exploited for accurate metrology (Häusler and Herrmann 1993, Häusler and Hermann 1995).

The interaction of light with the surface may cause diffuse reflection (as from ground glass) or specular reflection (as from a mirror), and the interaction may occur at the surface or within the object volume, for example volume scattering at human skin, teeth or plastics.

The information content contained in the light emanating from the surface is now considered. The information content is made up from the intensity and colour (as with photography), the complex amplitude (as with interferometry or holography), polarisation, the coherence properties and/or the time of flight. These features may be combined and exploited to get the 3D shape of a surface. In this chapter shape is conveniently defined as the local distance $z(x,y)$ at the location (x,y), which includes the micro-scale topography. Let $z_m(x,y)$ denote the measured shape, which may differ from the true shape $z(x,y)$. Now the question is posed: where are the limits of the measurement uncertainty δz? Here, δz represents the random error of the measured distance z_m. Systematic errors, such as those due to calibration, mechanical imperfections and thermal deformation of the sensor geometry, will not be discussed. As discussed in Chap. 4, an experiment to determine δz is to measure a planar object such as an optical flat (or a very fine ground glass plate) and calculate the standard deviation of the measured surface data $z_m(x,y)$.

As discussed above, all existing sensors can be categorised with only four types of physical principle (for a short overview see Knauer et al. 2006). More details are given in Sect. 3.7, where Tab. 3.1 explains what sensor type means and indicates the physical origin of the dominant sources of noise.

3.3 Optically Rough and Smooth Surfaces

Optical measurements strongly depend on the micro-scale topography of the surface — whether the surface is 'rough' or 'smooth' makes a big difference. With reference to Fig. 3.2, a surface is optically smooth if the surface height variation within the diffraction limited lateral resolution of the observation is smaller than $\lambda/4$ (Fig. 3.2, top). In this case, no wholly destructive interference can occur in the image of this surface. A surface is optically rough if the surface height variation is larger than $\lambda/4$. For height variations considerably greater than $\lambda/4$ (and for coherent illumination) full contrast speckles may occur in the image of the surface (Fig. 3.2, bottom).

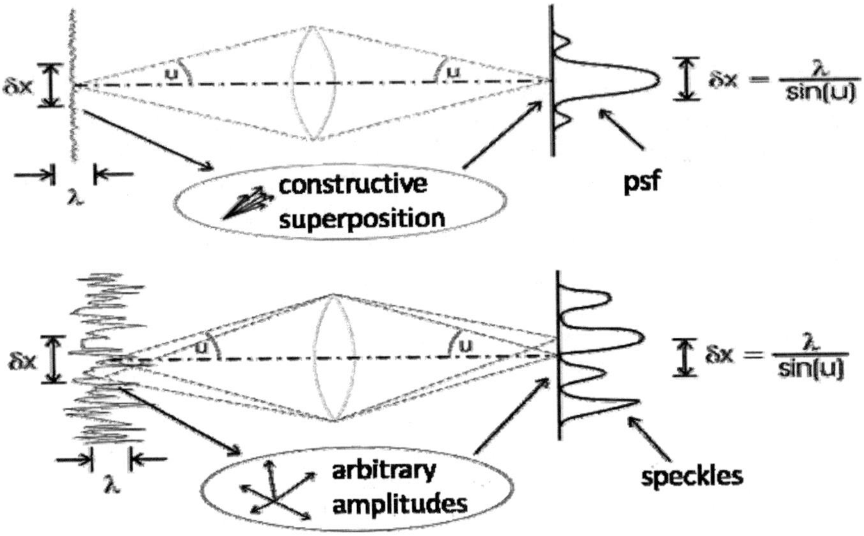

Fig. 3.2 Optically smooth and rough surfaces

Whether a surface is optically smooth or rough does not only depend on the surface itself but also on the observation aperture, sin u_{obs}. If ground glass is observed through a microscope with a very high numerical aperture, the ground glass surface will appear as if it is composed of a large number of tiny mirror surfaces. If the ground glass is observed by the naked eye or a low numerical aperture instrument, the ground glass will appear matt. The distinguishing property is the phase change introduced from the reflecting surface. If this phase variation within the laterally resolved distance given by equation (3.3)

$$\delta x = \bar{\lambda} \Big/ \sin u_{obs} \tag{3.3}$$

is smaller than ±90°, the reflected wavelets can never display fully destructive interference within the optical resolution, as they would result in fully developed speckle. The phase variation of ±90° corresponds to a surface variation of $\lambda/4$ within that resolution distance, considering that the light is travelling back and forth, by reflection. So, for high resolution, the surface height variation may be smaller than $\lambda/4$, while for low resolution (low aperture) the same surface may display a much larger height variation. In the image plane of the observing lens, this has a large effect.

In the smooth case, the wave that travels to the aperture and is focused at some position in the image plane is not affected by a large amount of random phase variation and generates a relatively bright spot. This is not the case for the rough surface, where the different areas of the aperture contribute to wavelets displaying large random phase differences. If these wavelets sum coherently, speckles will be seen.

From coherence theory a rule of thumb can be derived: speckles can be observed at rough surfaces if the illumination aperture is smaller than the observation aperture (Häusler 2004). In this case, the width of the spatial coherence function at the object is larger than the distance resolved by the detector.

Measuring in the smooth surface regime will allow for measurements with low optical noise — independent from the width of the spatial coherence function — i.e. speckle is not the dominant source of noise. In this case the most common dominant source of noise will be the detector, or with high-quality detectors under good conditions, it is photon noise.

The limitations of the four sensor types will now be examined in more detail.

3.4 Type I Sensors: Triangulation

Fig. 3.3 is a schematic representation of a triangulation sensor for measuring surface topography. A laser spot is focused along a projection axis onto the surface. The diffusely scattered spot is observed via some observation optics, along the observation axis. The projection axis and observation axis enclose the triangulation angle θ. From the location of the observed spot image and the known sensor geometry, the distance z can be calculated. Due to spatial coherence, the localisation accuracy of the spot image on the detector suffers from a physical uncertainty that cannot be overcome. The spot image displays speckle noise, as shown in Fig. 3.3, which introduces uncertainty about the knowledge of the spot image position.

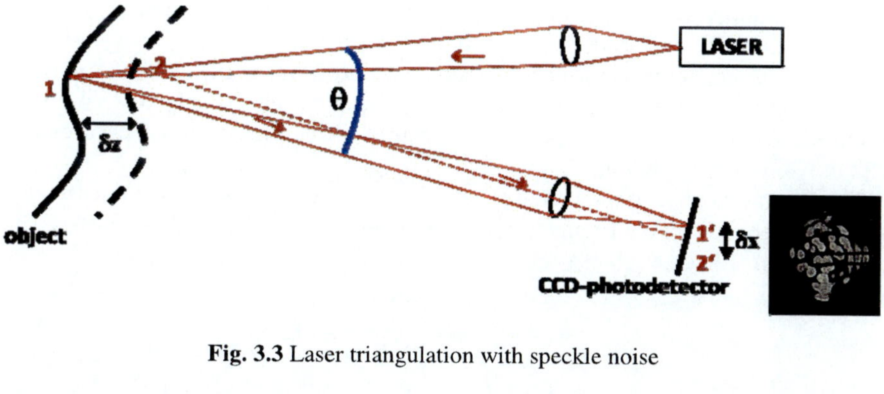

Fig. 3.3 Laser triangulation with speckle noise

In Fig. 3.4 a laser is focused on to a ground glass plate that is observed by a camera. It would be expected that the brightness maximum of the observed spot corresponds to the position of the illuminated spot on the plate. However, if the plate is moved laterally, the brightness maximum will start to wander because different lateral shifts of the plate cause light from different parts of the micro-scale topography to generate the image. The varying phase differences in the pupil of the observing camera cause the wandering spot image.

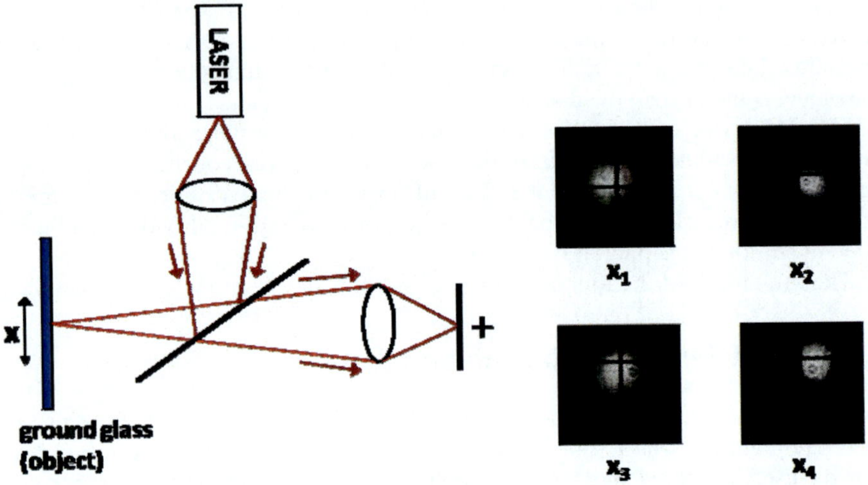

Fig. 3.4 The position of an observed spot image from a rough object is uncertain

The uncertainty of the spot image localisation is given by equation (3.4)

$$\delta x \sim \bar{\lambda} \sin u, \tag{3.4}$$

which is equal to the size of the diffraction spot. The relationship between the spot image uncertainty and the diffraction spot size can also be highlighted by considering that the individual photons emanating from the surface are indistinguishable, due to spatial incoherence. Therefore, there is no profit from any spatial or temporal averaging. The measurement will display the same uncertainty that could be achieved, according to Heisenberg's Uncertainty Principle, from a measurement with only one single photon.

This lateral localisation uncertainty leads to a distance measurement uncertainty δz, which depends on the triangulation angle θ and on the observation aperture sin u, as given in Dorsch et al. 1994,

$$\delta z = \frac{1}{2\pi} \frac{C\bar{\lambda}}{\sin\theta \sin u}, \tag{3.5}$$

where C is the speckle contrast ($C = 1$ for coherent illumination and $C < 1$ for partially incoherent illumination).

Equation (3.5) shows that for the case of laser triangulation ($C = 1$), δz decreases with larger observation apertures. Unfortunately, technical limitations and requirements for a large depth of field frequently prohibit the choice of a large aperture. There may be ways to overcome this difficulty in the presence of volume scattering, using white light illumination. In this case the temporal incoherence of the backscattered light can be exploited to reduce the speckle contrast. Several approaches to reduce speckle contrast are described in Dorsch et al. 1994 and in Ettl et al. 2009. Without any reduction of speckle contrast, and for sensors with a large stand-off distance, δz may have quite large values, for example, δz is approximately 100 µm, for $z = 1$ m, $\sin u = 0.01$ and $\theta = 5°$.

Fig. 3.5 (top) displays the result of a laser triangulation measurement (profile along a milled metal surface) with a standard deviation of 16.5 µm. Fig. 3.5 (bottom) displays a profile along the same object, measured with fluorescent light. The object was covered with a very thin fluorescent layer. The laser was used to excite fluorescence and the measurement was carried out with the fluorescent light. The measured standard deviation amounts to 1.1 µm showing that it is the spatially coherent interaction that introduces the ultimate source of noise.

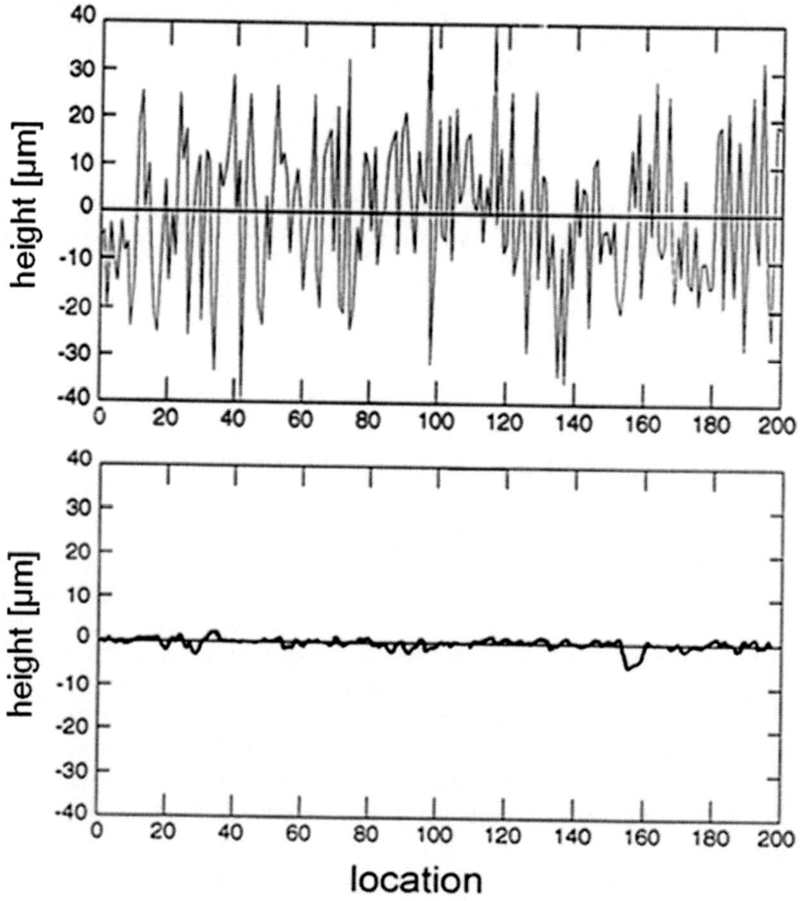

Fig. 3.5 Shape measured by laser triangulation (top) against triangulation with fluorescence (bottom)

In order to illustrate the severe influence of speckle noise, Fig. 3.6 shows a coin, illuminated both with a laser and with spatially incoherent light. The image of the laser illuminated coin clearly shows speckles and it is obviously problematic to acquire accurate data from this noisy image.

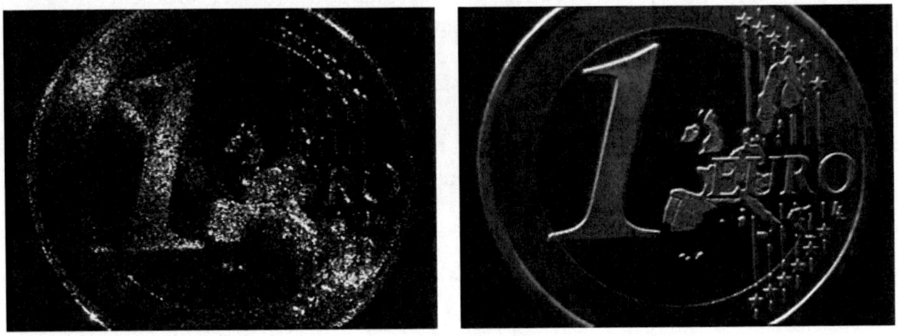

Fig. 3.6 Coin illuminated (left) by a laser and (right) by a spatially incoherent light source

In the rough surface regime, all triangulation systems (not only laser triangulation sensors) suffer from speckle. An observation aperture can be chosen that is much smaller than the illumination aperture, as explained above. One sensor principle that allows this is the so-called fringe projection or phase measuring triangulation (see for example, Halioua et al. 1984).

In fringe projection, a sinusoidal grating is projected onto the object surface, while a camera is used to observe the object with the grating image. The camera axis and the projector axis span the angle of triangulation. Again, the measurement uncertainty is given by equation (3.5), but now there is the potential to use an illumination aperture larger than the observation aperture, to reduce the speckle contrast. In addition, a source with low temporal coherence (white light illumination) further helps at volume scatterers. With fringe projection it is possible to achieve a dynamic range (range divided by noise) of one thousand or even up to ten thousand with very careful design. Unfortunately, increasing the illumination aperture often results in further limitations such as the available space, the cost of large lenses or the requirement for a large depth of field. Another example of low coherence triangulation is stereo photogrammetry, where the illumination is carried out using a ring shaped flash light positioned around the camera lens.

This section will finish with some practical hints when using triangulation. There is a large range of 3D sensors that are based on triangulation. However, the user or even the manufacturers may not be aware that their sensor is of type I. All focus sensors, such as confocal microscopy (see for example, Hamilton and Wilson 1982, Juškaitis et al. 1996, Semwogerere and Weeks 2005 and Chap. 11), structured-illumination microscopy (see for example, Engelhardt and Häusler 1988, Neil et al. 1997, Körner et al. 2001), chromatic confocal microscopy (see for example, Molesini et al 1984, Ruprecht et al. 2004, Fleischle et al 2010 and Chap. 5), fringe projection, and stereo photogrammetry (see for example, Chen et al. 2000), are all type I sensors. If rough surfaces are measured, equation (3.5) limits the measurement uncertainty of all of these instruments.

As mentioned above, instruments that use a focus search at rough surfaces belong to type I sensors. For focus detection with a microscope, there is no formal triangulation angle θ anymore; it is replaced by the aperture angle u. So the measurement uncertainty at a rough surface will be approximately given by equation (3.6), which, as δz scales with the square of the distance z, gives

$$\delta z \sim \frac{C\overline{\lambda}}{2\pi \sin^2 u}. \tag{3.6}$$

Depending on the chosen implementation for passive focus search or structured-illumination microscopy, there is an additional factor in equation (3.6) of the order of unity.

For a number of years, structured-illumination microscopy (SIM) has been used together with fluorescence for medical applications (see for example, Gustafsson et al. 2008, Kner et al. 2009, Fitzgibbon et al. 2010). The basic idea of SIM is to project a grating into the object volume and measure locally the grating contrast $C(z)$ in the observed image, while the object is scanned through the focal plane of the microscope. The maximum contrast occurs at the position $z_m(x,y)$, when the object location (x,y) is in focus. The basic SIM idea was described by Engelhardt and Häusler 1988, and Neil et al. (1997) augmented the concept by introducing phase shifting for the contrast measurements.

Recently, SIM has been implemented for engineering applications (see for example, Kranitzky et al. 2009, Kessel et al. 2010, Vogel et al. 2010, Häusler et al. 2010). For engineering applications, the advantage of fluorescence cannot be exploited, so the influence of speckle has to be considered, if rough surfaces are measured.

Fig. 3.7 displays the measurement uncertainty δz for SIM at rough surfaces against numerical aperture. The results clearly display the aperture dependency of δz, according to equation (3.5), typical for triangulation (type I) sensors. The true roughness of the measured surface is approximately 0.8 μm, but for small apertures the coherent noise causes artefacts, feigning a much larger roughness, similar to the effect shown in Fig. 3.5 (top).

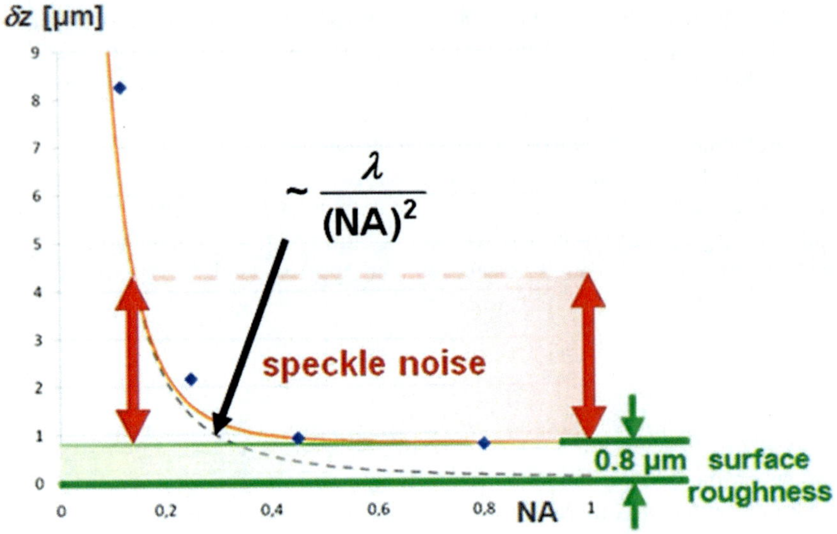

Fig. 3.7 Measurement uncertainty for SIM at rough surfaces against numerical aperture

SIM has some advantages over other methods. Smooth surfaces can be measured up to an inclination angle equal to the aperture angle u of the objective. Also, the depth of field is not limited by the Rayleigh depth, so images much like those from SEM can be acquired, with even nanometre 3D depth resolution. One example is shown in Fig. 3.8.

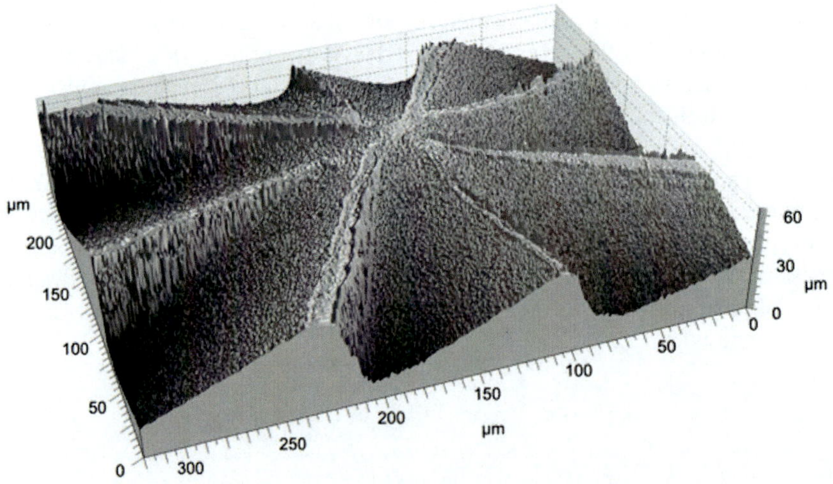

Fig. 3.8 Micro-milling tool measured with SIM

SIM can also be used to measure smooth objects. One example is shown in Fig. 3.9. The measurement uncertainty is better than 10 nm for smooth objects, since now the dominant source of noise is photon noise instead of speckle noise, which allows for a very low measurement uncertainty.

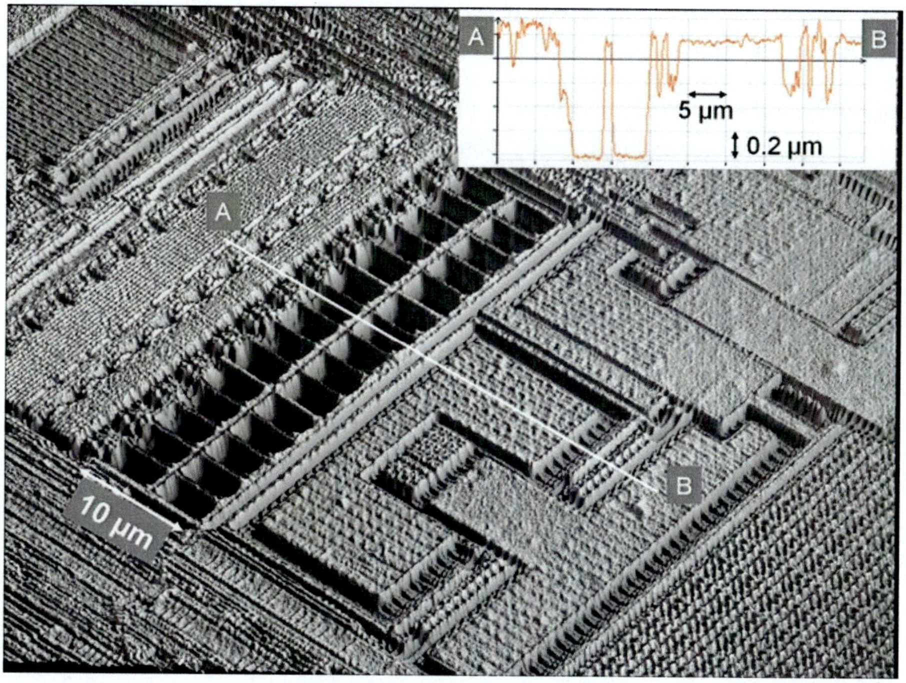

Fig. 3.9 SIM measurement of a wafer

3.5 Type II and Type III Sensors: Interferometry

Type II sensors include coherence scanning interferometry (CSI – see Chap. 9) on rough surfaces and time-of-flight sensors. Type III sensors include classical phase measuring interferometry (see Chap. 8). CSI at rough surfaces is essentially interferometry in individual speckles. The phase is not measured, as in classical interferometry, because this phase is random and uncorrelated in different speckles. Classical interferometry at rough surfaces does not display any useful fringes. In CSI broad-band illumination is used and the location $z_m(x,y)$ of the maximum of the envelope of the so-called correlogram (see below) has to be found. Commonly this is implemented by scanning the optical path length of one of the two interferometer arms. In practical embodiments, one of the interferometer mirrors is replaced by the object under test, which is positioned outside the mechanical body of the interferometer. The interferometer body is moved along the optical axis of the observation, while the object surface is observed and the interference signal $C(z)$ (which is commonly called the correlogram) is observed at each object location at (x,y), i.e. for each pixel separately.

A high-contrast correlogram needs high-contrast speckles. Therefore, three conditions have to be met. First, the illumination aperture must be smaller than the observation aperture. This implies spatially coherent illumination which is necessary to see speckles. Within each speckle, the phase is approximately constant. As discussed above, the maximum of the correlogram occurs at the position $z_m(x,y)$, where the optical path length is equal in the two arms of the interferometer, and $z_m(x,y)$ is the measured topography. Second, high-contrast speckles can be seen only if the pixel size of the camera is not significantly larger than the size of the subjective speckles at the video target. Third, the coherence length of the illumination must be larger than the roughness of the surface (more accurately, than the path length variation within the resolution cell, see Sect. 3.3). The third condition may be difficult to achieve for very steep surfaces and for volume scatterers. These three conditions are essential for CSI, described first in (Häusler 1991).

To understand the limitations of CSI, the physical source of noise for $z_m(x,y)$ must be identified. In classical interferometry, the ultimate source of noise is photon noise. This is different for CSI if speckles are involved. The maximum of the correlogram envelope does not exactly determine the real object topography $z(x,y)$ and the measured topography $z_m(x,y)$ contains statistical errors. The full derivation is given elsewhere (Häusler et al. 1999, Ettl 2001). The measured topography data $z_m(x,y)$ contain statistical noise, which depends on the standard deviation σ_o of the object surface

$$\sigma_c = \frac{1}{\sqrt{2}}\sqrt{\frac{\langle I \rangle}{I}}\sigma_0. \tag{3.7}$$

where I is the individual speckle intensity and $<I>$ is the average intensity. From equation (3.7) it can be seen that the standard deviation σ_c of the correlogram location depends on the standard deviation σ_o of the object micro-scale topography, and on I. To calculate the statistical error of the measured correlogram, the first order statistics of the speckle intensity I have to be introduced. The result is given by equation (3.8)

$$<| z_m(x,y) - \langle z_m \rangle |> = \sigma_o, \tag{3.8}$$

where $<z_m>$ is the local average of the measured height. From equation (3.8) it can be seen that, for rough objects, CSI does not measure the micro-scale topography itself, because the micro-scale topography is not resolved. What is measured in this case is a statistical representation of the micro-scale topography. The arithmetic mean of the (zero mean) magnitude of z_m displays the same value as the standard deviation σ_o of the surface micro-scale topography $z(x,y)$ itself. This result has been experimentally confirmed by optical measurements and by measurements with a stylus instrument (Ettl et al. 1998) as shown in Fig. 3.10. In this figure, the values calculated using equation (3.8), from optical measurements, are close to the mechanically measured roughness parameter Rq.

Note that equation (3.8) does not include any aperture dependence. Although the optically unresolved micro-scale topography cannot be accessed, nevertheless the surface roughness can be measured, independent of the magnitude of the optical resolution cell. In other words, the unresolved micro-scale topography introduces noise in

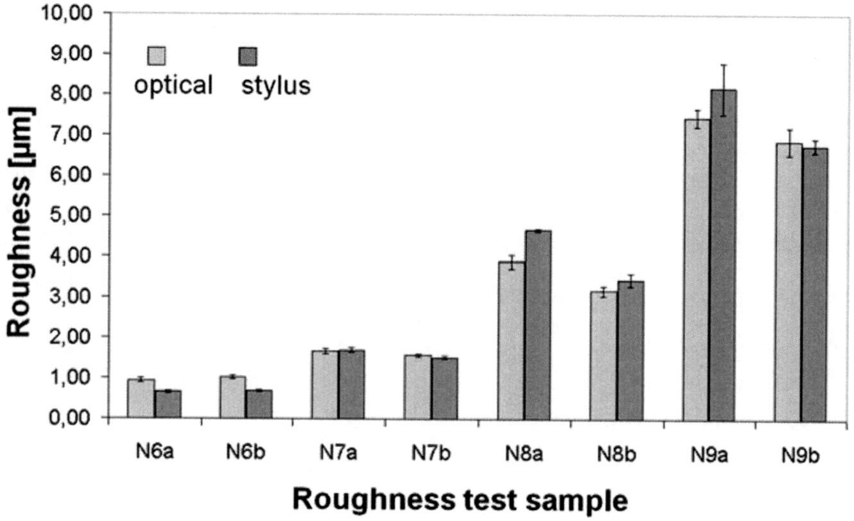

Fig. 3.10 Roughness measurement: stylus instrument against CSI

to the measured signal which is, after evaluation according to equation (3.8) equal to the surface roughness σ_0 – which corresponds to the roughness parameter Rq.

The measured roughness is independent of the observation aperture, as shown in Fig. 3.11. This result has been confirmed by simulations (Ettl 2001).

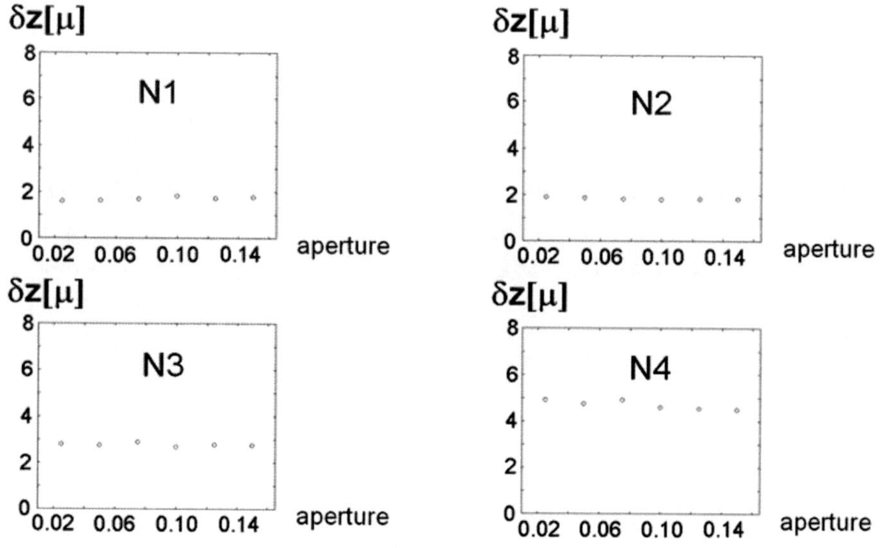

Fig. 3.11 Optically measured roughness against aperture for different PTB standard artefacts

As the results shown in Fig. 3.11 and simulations show, there is no lateral averaging of the distance over the optical resolution cell. This effect is different from measuring with a stylus instrument - the radius of the probe acts as a low-pass filter and averaging over a certain area of the surface is performed. Also, this effect is different from triangulation, where there is no surface smoothing, but where the noise scales with the aperture (see equations (3.5) and (3.6)).

How can these properties of the micro-scale topography beyond the resolution limit be "seen"? To understand this, it is necessary to refer back to Sect. 3.3. The field amplitude $u(x,y)$ that is reflected from the object surface $z(x,y)$ is given by

$$u(x, y) \sim \exp[2ikz(x, y)]. \tag{3.9}$$

Firstly, for smooth surfaces, where the variation of the surface height $z(x,y)$ is much smaller than $\lambda/4$, within the optical resolution cell, the reflected field amplitude u can be approximated by

$$u(x, y) \sim 1 + 2ikz(x, y). \tag{3.10}$$

Equation (3.10) infers that the field u and the height z are linearly related. For classical (smooth surface) interferometry, the low-pass filter introduced by the aperture causes an averaging over the field and at the same time a linear averaging over the height $z(x,y)$ given by

$$< u(x, y) > \sim < 1 + 2ikz(x, y) > = 1 + 2ik < z(x, y) > . \tag{3.11}$$

A classical interferometer averages over the local surface topography, as illustrated in Fig. 3.12. The small numerical aperture interferometry clearly displays local height averages while interferometry with large numerical aperture resolves the local height topography.

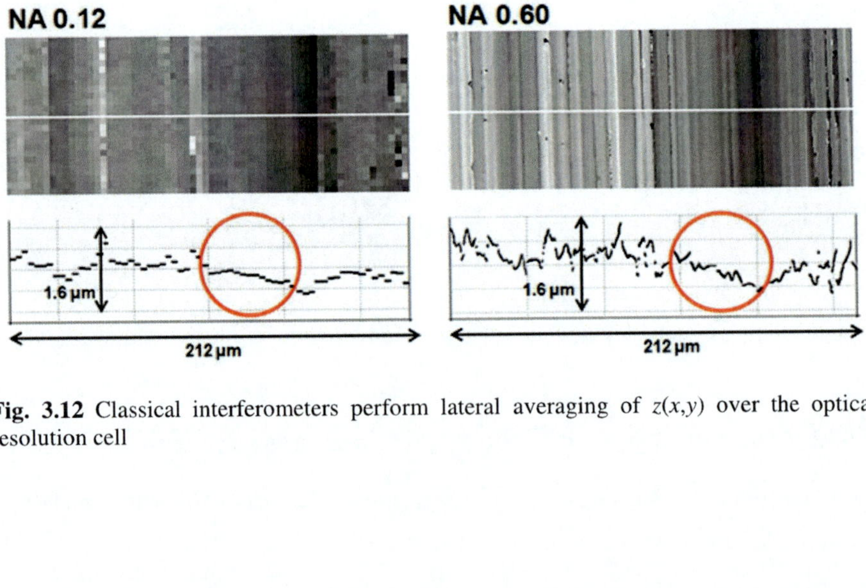

Fig. 3.12 Classical interferometers perform lateral averaging of $z(x,y)$ over the optical resolution cell

Secondly, in rough surface mode (non-resolved micro-scale topography), however, the field amplitude is a strongly non-monotonic nonlinear function of the height z, see equation (3.9). So the optical low-pass filter no longer averages over the height z, in a linear manner, but now over the exponential of the height. As a consequence, there is frequency mixing in the Fourier domain and some of these frequencies will be sub-harmonics. The frequencies that are down converted and transmitted through the aperture are "visible" to the CSI instrument. It is suggested this as an explanation why properties of the micro-scale topography are "seen" beyond the Abbe limit – but not the topography itself.

Can the intrinsic noise of rough surface interferometry be reduced? Equation (3.7) offers this option - use only bright speckles. Wiesner et al. (2006) have implemented this by illuminating the object from four different directions, thus generating four nearly independent speckle patterns. For each pixel only the brightest of the four speckles was chosen to evaluate the correlogram. This increased the quality of the measurements significantly.

A comparison of triangulation systems and CSI at rough surfaces is now presented. The measurements were carried out under very similar physical conditions such as working distance and aperture.

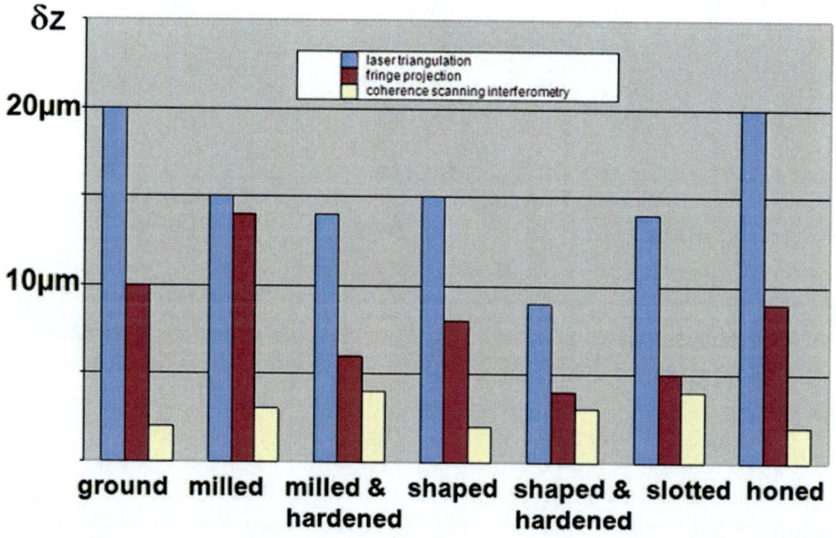

Fig. 3.13 Comparison of sensors and differently machined surfaces

Figure 3.13 shows the noise, or the measurement uncertainty, for different sensor principles. Laser triangulation displays the largest measurement uncertainty, fringe projection is marginally better, because a high illumination aperture can reduce speckle noise. The noise for the CSI is essentially due to, and equal to, the roughness of the object.

3.6 Type IV Sensors: Deflectometry

Type IV sensors use the so-called deflectometric principle to measure the local slope of smooth objects. Deflectometry is not new (see Ritter and Hahn 1983), but in the last ten years, highly accurate quantitative measurements have become possible (Häusler et al. 1999, Knauer et al. 2004, Kammel and Leon 2005, Bothe et al. 2004). The basic principle of deflectometry is depicted in Fig. 3.14.

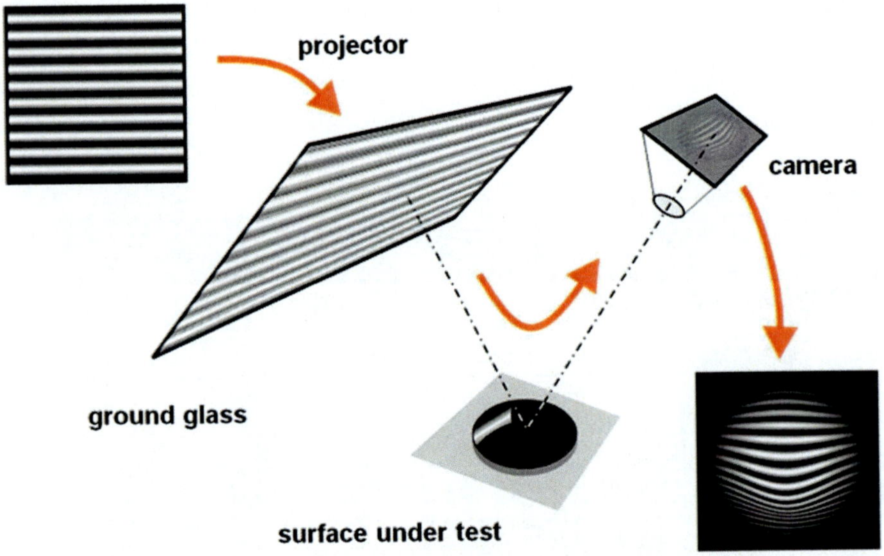

Fig. 3.14 Principle of deflectometry for macroscopic surfaces

A grating at a remote distance from the object is observed via the surface under test, which acts as a mirror. The observed mirror image of the grating is distorted. From this distortion, the local slope of the object surface can be calculated. Deflectometry is well established for the measurement of, for example, aspheric eye glasses (see Fig. 3.15), car windows and other aspheric optical elements.

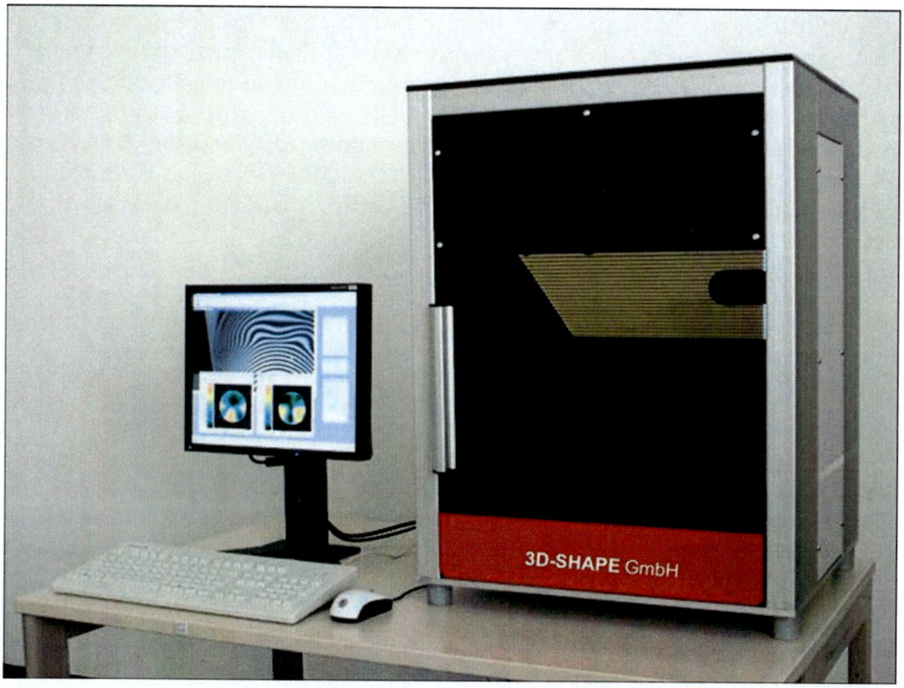

Fig. 3.15 Deflectometer to measure aspheric eye glasses

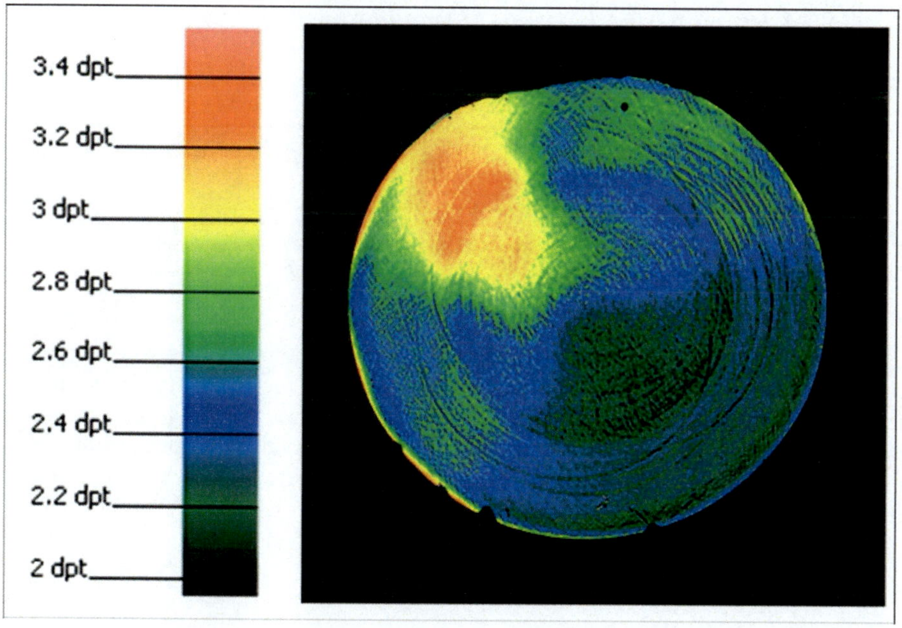

Fig. 3.16 Local refractive power of eye glass, with tool traces

Figure 3.16 shows a colour-coded local refractive power map of an eye glass surface. Since deflectometry is highly sensitive to local slope variations, even relatively small irregularities such as tool traces of only a few nanometres depth are visible.

Recently, a modification of deflectometry was introduced to measure the micro-scale topography of smooth surfaces (Häusler et al. 2008). One result is shown in Fig. 3.17, where the measurement of a micro-scale cylinder lens array is displayed.

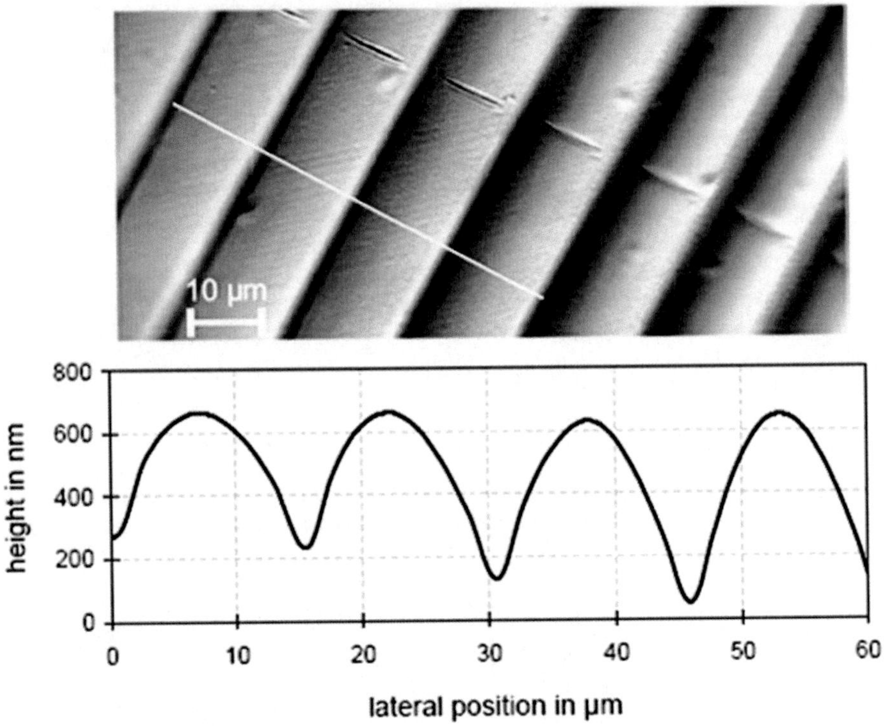

Fig. 3.17 Cylinder lens array measured with micro-scale deflectometry

Further details on deflectometry can be found in the references above. What are the physical limitations of the method? The camera is focused onto the surface under test. The mirror image of the grating, however, is not located at this surface. Therefore, this mirror image is defocused. By choosing an appropriate imaging aperture and an appropriate grating period, a trade off between the lateral resolution δx and the angular measurement uncertainty $\delta\alpha$ can be found. The optimum choice of the parameters leads again to an uncertainty product

$$\delta x \cdot \delta\alpha = \bar{\lambda} / Q. \tag{3.13}$$

In equation (3.13), Q is approximately the signal-to-noise ratio for the fringe phase and the major source of noise is camera noise. For high-quality cameras, the ultimate source of noise is photon noise. Q is in the range of 500 for measurements with standard video cameras. So the product, $\delta x \delta \alpha = \delta z$, which represents the height uncertainty within the lateral resolution cell, is only about one nanometre. After integration of the slope data local height variations can be detected with a sensitivity of only one nanometre, and with very simple incoherent technology (no interferometer). Also, the aperture is not included explicitly in equation (3.13). This implies that the measurement uncertainty in relation to the resolution cell is independent of the aperture or the distance. A trade off between lateral resolution and angular measurement uncertainty can be made (again an example for the usefulness of uncertainty products). A large δx allows a very high angular resolution. One last advantage of deflectometry is that since Q is limited only by photon noise it can be increased by measuring with many photons. The two latter properties are used by PTB to measure large planar surfaces with sub-nanometre accuracy (Geckeler and Just 2007).

A further illustrative measurement using deflectometry is shown in Fig. 3.18, where the height map of a cylinder tread surface is depicted.

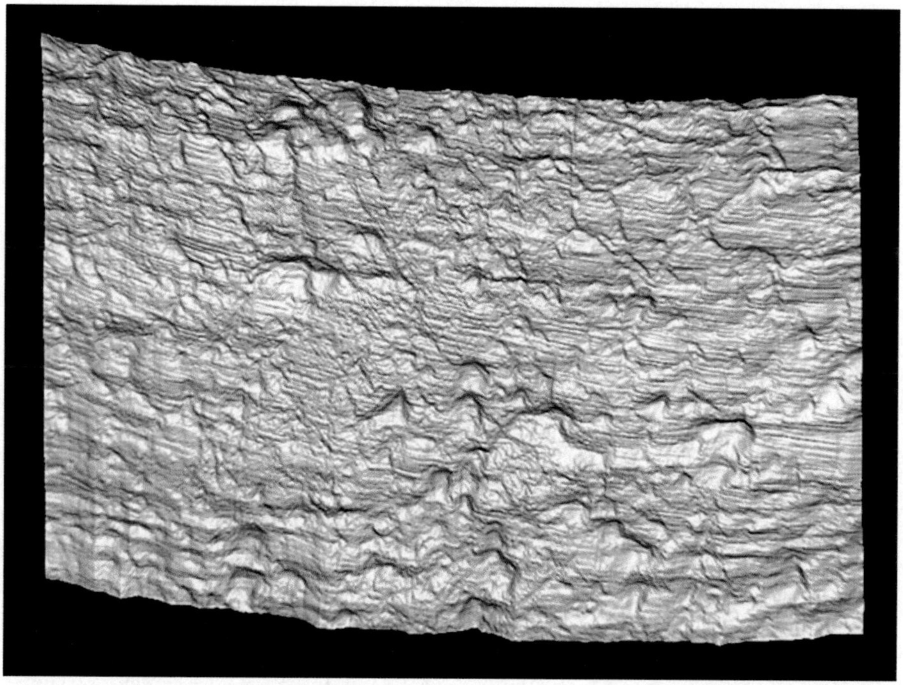

Fig. 3.18 Cylinder tread, measured with micro-scale deflectometry

3.7 Only Four Sensor Principles?

Optical 3D sensors can measure the distance of close stars (see for example, Baldwin and Haniff 2001) and the distance of atomic layers (for an overview on optical range sensors, see for example Beraldin et al. 2000 and 2003, Besl 1988, Blais 2004, Häusler 1999). Also, it appears that all available optical 3D sensors can be classified into one of the classes type I, II, III or IV (for short overviews see Häusler 1999, Häusler and Leuchs 1997, Knauer et al. 2006). Tab. 3.1 shows the different principles and indicates the physical origin of the dominant sources of noise. Tab. 3.1 also shows how the noise scales with the aperture and if there is lateral averaging of $z(x,y)$, across the optical resolution cell.

Table 3.1 Classification of optical 3D sensors

type	Principle	dominant noise	scaling behavior
Ia	triangulation at rough surfaces	speckle noise	$\delta z \sim \lambda/(\sin u \sin \theta)$ no z-averaging over the resolution cell
Ib	triangulation at smooth surfaces, triangulation with fluorescence	photon noise	$\delta z \sim \lambda/\sin^2 u$
II	CSI at rough surfaces	surface roughness	δz is independent from aperture, no z averaging
III	classical interferometry	photon noise	$\delta z \sim \sin u$, due to z averaging over the resolution cell
IV	Deflectometry	photon noise	$\delta x \cdot \delta\alpha \sim \lambda/Q$ $\delta x = \lambda/\sin u$

Type I includes all sensors based on triangulation: laser triangulation, fringe projection, stereo photogrammetry, focus search, confocal microscopy, chromatic confocal microscopy and SIM. Triangulation sensors, in general, measure the lateral perspective shift of local details. The measurement uncertainty is determined by the uncertainty of this shift measurement. It can easily be seen from geometrical considerations that the measurement uncertainty scales with the inverse square of the distance. The scaling factor depends on the dominant source of noise. For triangulation measurements at rough objects, the dominant source of noise is speckle noise. This kind of triangulation sensor is called type Ia. For smooth surface triangulation or for triangulation with fluorescence, the dominant source of noise is photon noise. This kind of triangulation sensor is called type Ib.

Type II sensors include CSI at rough surfaces, sometimes called coherence radar (Häusler 1991, Dresel et al. 1992), and time-of-flight methods (at rough surfaces). For the coherence radar, the location of the temporal coherence function is measured for each image pixel. For rough surfaces the major source of noise is the micro-scale topography of the surface itself. An interesting and useful result is that the measurement uncertainty does not scale with the distance or the aperture of the observation. Therefore, measurements within deep holes or at a large distance are possible, without increasing the measurement uncertainty. CSI on rough surfaces is now an established method and widely commercialised (see Petzing et al. 2010).

It should be noted that commonly hundreds of exposures have to be made to scan an object of several hundred micrometres depth. Therefore, CSI is not an efficient method and considerable effort has been invested to make CSI more efficient (Hýbl and Häusler 2010).

Type III sensors include the classical interferometers to measure smooth (specular) surfaces. The major source of noise is photon noise. Classical interferometry in an appropriate environment can have a measurement uncertainty down to sub-nanometre levels (with more sophistication, as in gravitational wave interferometry (Quetschke 2010) an uncertainty of 10^{-18} m is reported).

Tab. 3.1 shows that that there is a significant difference between classical interferometry and CSI at rough surfaces. Classical interferometers perform lateral averaging of the topography $z(x,y)$, while the coherence radar does not.

Type IV sensors include deflectometry and micro-scale deflectometry. Type IV is a special class of sensors that intrinsically measures the local slope; instead of the height (whether shearing interferometry belongs to type IV will be left open here). The spatial derivative is generated optically, i.e. no posterior (software) derivation is necessary. This is why the topography data $z(x,y)$ acquired by numerical integration (Ettl et al. 2008) can have low noise in the range of only a few nanometres. In the language of information theory, the optical differentiation corresponds to a source encoding, with strong reduction of redundancy. Therefore, deflectometry is highly information efficient. This means that considerably less channel capacity (expensive technology) has to be provided, than for a conventional sensor. Deflectometry displays a large dynamic range with simple and non-expensive hardware. It should also be noted that deflectometry can measure much steeper slopes than interferometry.

3.8 Conclusion and Open Questions

The length of the bars in Fig. 3.19 shows the dynamic range of different sensor principles and the left edge displays the ultimate physical limit of the measurement uncertainty. This limit is of course subject to different scaling factors as explained above. Fig. 3.19 suggests that deflectometry may play an important role in the future. Tab. 3.2 presents a range of sensor principles for a range of surfaces.

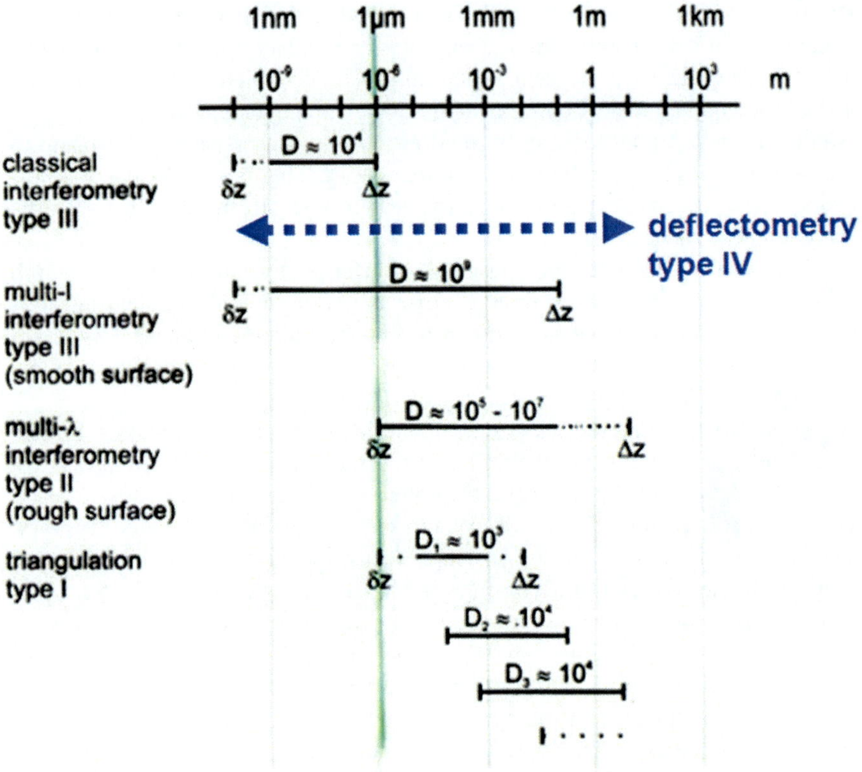
1nm
1µm
1mm
1m
1km
10^-9
10^-6
10^-3
1
10^3
m
classical
interferometry
type III
D ≈ 10^4
δz
Δz
deflectometry
type IV
multi-l
interferometry
type III
(smooth surface)
D ≈ 10^9
δz
Δz
multi-λ
interferometry
type II
(rough surface)
D ≈ 10^5 - 10^7
δz
Δz
triangulation
type I
D1 ≈ 10^3
δz
Δz
D2 ≈ .10^4
D3 ≈ 10^4

Fig. 3.19

Table 3.2 Range of sensor principles

surface/ sensor type	**laser triang. (Ia)**	**fringe project. (Ia)**	**SIM (Ib)**	**rough-surf. CSI (II)**	**classical interfer-ometry (III)**	**deflec-tometry (IV)**
specular, planar	– –	– –	+ +	+ +	+ +	+ +
specular, curved	– –	– –	+ +	0	0	+ +
matt, Lambertian	+	+ +	+	+ +	– –	– –
machined surface	0	0	+	+ +	– –	–
tilted, ma-chined	–	0	+	+	– –	–
deep boreholes	– –	– –	–	+	– –	– –
volume scatterer	–	0	–	+	– –	– –

This chapter finishes with two unanswered questions. The most important question, from the experience of the authors, is: how can the ever increasing demand for more space bandwidth (larger field and higher lateral resolution at the same time) be satisfied? Second, are there any more options to "see" details beyond the Abbe limit of lateral resolution? The last question is not a hopeless question considering new developments like SIM and other new microscope techniques exploiting fluorescence.

References

Baldwin, J.E., Haniff, C.A.: The application of interferometry to optical astronomical imaging. Phil. Trans. A360, 969–986 (2001)

Baribeau, R., Rioux, M.: Influence of speckle on laser range finders. Appl.Opt. 30, 2873–2878 (1991)

Beraldin, J.-A., Blais, F., Cournoyer, L., Godin, G., Rioux, M.: Active 3D sensing. In: Modelli e metodi per lo studio e la conservazione dell'architettura storica. Scuola Normale Superiore, pp. 22–46 (2000)

Beraldin, J.-A., Blais, F., Cournoyer, L., Godin, G., Rioux, M., Taylor, J.: Active 3D sensing for heritage applications. In: The e-Way into Four Dimensions of Cultural Heritage Congress, Vienna, Austria, pp. 340–343 (2003)

Besl, P.J.: Active, optical range imaging sensors. Mach. Vision. Appl. 1, 127–152 (1988)

Blais, F.: Review of 20 years of range sensor development. J. Electron Imaging 13, 231–240 (2004)

Bothe, T., Li, W., Kopylow, C., Jüptner, W.: High resolution 3D shape measurement on specular surfaces by fringe reflection. P. Soc. Photo-Opt. Ins. 5457, 411–422 (2004)
Chen, F., Brown, G.M., Song, M.: Overview of three-dimensional shape measurement using optical methods. Opt. Eng. 39, 10–22 (2000)
Dorsch, R.D., Häusler, G., Herrmann, J.M.: Laser triangulation: fundamental uncertainty in distance measurement. Appl. Opt. 33, 1306–1314 (1994)
Dresel, T., Häusler, G., Venzke, H.: 3D-sensing of rough surfaces by coherence radar. Appl. Opt. 31, 919–925 (1992)
Engelhardt, K., Häusler, G.: Acquisition of 3D-data by focus sensing. Appl. Opt. 27, 4684–4689 (1988)
Ettl, P.: Über die Signalentstehung in der Weisslichtinterferometrie. Dissertation University of Erlangen (2001)
Ettl, P., Schmidt, B., Schenk, M., Laszlo, I., Häusler, G.: Roughness parameters and surface deformation measured by 'Coherence Radar'. In: Proc. SPIE, vol. 3407, pp. 133–140 (1998)
Ettl, S., Kaminski, J., Knauer, M.C., Häusler, G.: Shape reconstruction from gradient data. Appl. Opt. 47, 2091–2097 (2008)
Ettl, S., Arold, O., Vogt, P., Hýbl, O., Yang, Z., Xie, W., Häusler, G.: "Flying Triangulation": A motion-robust optical 3D sensor principle. In: Proc. FRINGE 2009, pp. 768–771 (2009)
Fitzgibbon, J., Bell, K., King, E., Oparka, K.: Super-resolution imaging of plasmodesmata using three-dimensional structured illumination microscopy. Plant. Physiol. 153, 1453–1463 (2010)
Fleischle, D., Lyda, W., Mauch, F., Haist, T., Osten, W.: Untersuchung zum Zusammenhang von spektraler Abtastung und erreichbarer Messunsicherheit bei der chromatisch-konfokalen Mikroskopie an rauen Objekten. DGaO Proceedings 2010, A14 (2010)
Geckeler, R., Just, A.: Optimaler Einsatz und Kalibrierung von Autokollimatoren zur Formmessung mittels hochgenauer Deflektometrie. DGaO Proceedings 2007, A3 (2007)
Gustafsson, M.G., Shao, L., Carlton, P.M., Wang, C.J.R., Golubovskaya, I.N., Cande, W.Z., Agard, D.A., Sedat, J.W.: Three-dimensional resolution doubling in wide-field fluorescence microscopy by structured illumination. Biophys. J 94, 4957–4970 (2008)
Häusler, G.: About fundamental limits of three-dimensional sensing, or: nature makes no presents. In: Proc. SPIE, vol. 352 (1990)
Häusler, G.: Verfahren zur dreidimensionalen Vermessung eines diffus streuenden Objektes. German Patent, DE 41 08 944 (1991)
Häusler, G.: Three-dimensional sensors – potentials and limitations. In: Jähne, B., Haußecker, H., Geißler, P. (eds.) Handbook of computer vision and applications, vol. 1, pp. 85–506. Academic Press, Boston (1999)
Häusler, G.: Speckle and coherence. In: Encyclopedia of modern optics, vol. 1, pp. 4–123. Elsevier Ltd, Oxford (2004)
Häusler, G., Herrmann, J.M.: Range sensing by shearing interferometry: influence of speckle. Appl. Opt. 27, 4631–4637 (1988)
Häusler, G., Herrmann, J.M.: Optischer Sensor wacht über. Lasermaterialbearbeitung, Feinwerktechnik, Mikrotechnik und Messtechnik 103, 540–542 (1995)
Häusler, G., Leuchs, G.: Physikalische Grenzen der optischen Formerfassung mit Licht. Physikal Blätter 53, 417–421 (1997)
Häusler, G., Herrmann, J.M.: Procedure and device to measure distances. European Patent, EP 55 91 20 (1993)

Häusler, G., Ettl, P., Schenk, M., Bohn, G., Laszlo, I.: Limits of optical range sensors and how to exploit them. In: Asakura, T. (ed.) Trends in optics and phototonics, Ico IV. Springer Series in Optical Sciences, vol. 74, pp. 328–342 (1999)

Häusler, G., Richter, C., Leitz, K.-H., Knauer, M.C.: Microdeflectometry – a novel tool to acquire three-dimensional microtopography with nanometer height resolution. Opt. Lett. 33, 396–398 (2008)

Häusler, G., Vogel, M., Yang, Z., Kessel, A., Faber, C., Kranitzky, C.: Microdeflectometry and structural illumination microscopy – new tools for 3D-metrology at nanometer scale. In: Proc Precision Interferometric Metrology, ASPE 2010 Summer Topical Meeting, Asheville, North Carolina, USA, vol. 49, pp. 46–51 (2010)

Halioua, M., Liu, H., Srinivasan, V.: Automated phase-measuring profilometry of 3-D diffuse objects. Appl. Opt. 23, 3105–3108 (1984)

Hamilton, D.K., Wilson, T.: Surface profile measurement using the confocal microscope. J. Appl. Phys. 53, 5320–5322 (1982)

Hýbl, O., Häusler, G.: Information efficient white-light interferometry. In: Proc Precision Interferometric Metrology, ASPE 2010 Summer Topical Meeting, Asheville, North Carolina, USA, vol. 49, pp. 81–84 (2010)

Juškaitis, R., Wilson, T., Neil, M.A.A., Kozubek, M.: Efficient real-time confocal microscopy with white light sources. Nature 383, 804–806 (1996)

Kammel, S., León, F.P.: Deflectometric measurement of specular surfaces. In: IEEE Proc. Instrumentation and Measurement, pp. 108–117 (2005)

Kessel, A., Vogel, M., Yang, Z., Faber, C., Seraphim, M.C., Häusler, G.: Information efficient and accurate structured illumination microscopy (SIM). DGaO-Proceedings 2010, P9 (2010)

Knauer, M.C., Kaminski, J., Häusler, G.: Phase measuring deflectometry: a new approach to measure specular free-form surfaces. In: Proc. SPIE, vol. 5457, pp. 366–376 (2004)

Knauer, M.C., Richter, C., Häusler, G.: 3D sensor zoo – species and natural habitats. Laser Technik Journal 3, 33–37 (2006)

Kner, P., Chhun, B., Griffis, E.R., Winoto, L., Gustafsson, M.G.: Super-resolution video microscopy of live cells by structured illumination. Nature Methods 6, 339–342 (2009)

Körner, K., Windecker, R., Fleischer, M., Tiziani, H.J.: One-grating projection for absolute three-dimensional profiling. Opt. Eng. 40, 1653–1660 (2001)

Kranitzky, C., Richter, C., Faber, C., Knauer, M., Häusler, G.: 3D-microscopy with large depth of field. DGaO Proceedings 2009, A12 (2009)

Molesini, G., Pedrini, G., Poggi, P., Quercioli, F.: Focus-wavelength encoded optical profilometer. Opt. Commun. 49, 229–233 (1984)

Neil, M.A.A., Juškaitis, R., Wilson, T.: Method of obtaining optical sectioning by using structured light in a conventional microscope. Opt. Lett. 22, 1905–1907 (1997)

Petzing, J., Coupland, J.M., Leach, R.K.: The measurement of rough surface topography using coherence scanning interferometry. NPL Good practice guide, vol. 116. National Physical Laboratory (2010)

Quetschke, V.: LIGO – A look behind attometer (10^{-18} m) sensitivity and beyond. In: Proc ASPE, Summer Topical Meeting on Precision Interferometric Metrology, pp. 67–68 (2010)

Ritter, R., Hahn, R.: Contribution to analysis of the reflection grating method. Opt. Lasers Eng. 4, 13–24 (1983)

Ruprecht, A.K., Wiesendanger, T.F., Tiziani, H.J.: Chromatic confocal microscopy with a finite pinhole size. Opt. Lett. 29, 2130–2132 (2004)

Semwogerere, D., Weeks, E.R.: Confocal microscopy. In: Encyclopedia of biomaterials and biomedical engineering, London (2005)

Vogel, M., Kessel, A., Yang, Z., Faber, C., Seraphim, M.C., Häusler, G.: Tuning structured illumination microscopy (SIM) for the inspection of micro optical components. DGaO Proceedings 2010, A22 (2010)

Wiesner, B., Berger, A., Groß, R., Hýbl, O., Richter, C., Häusler, G.: Faster and better white light interferometry. In: Proc. ODIMAPV, pp. 216–221 (2006)

4 Calibration of Optical Surface Topography Measuring Instruments

Richard Leach and Claudiu Giusca

Engineering Measurement Division, National Physical Laboratory
Hampton Road, Teddington
Middlesex TW11 0LW

Abstract. The increased use of areal surface topography measuring instruments in the past ten years has led to a range of optical instruments being developed and becoming commercially available. Such instruments make use of sophisticated mathematical algorithms to process the raw height information and transform it into topography data that are used for visualisation and calculation of areal surface texture parameters. Optical areal surface topography measuring instruments are powerful and flexible tools of a complex design, which makes them difficult to calibrate. Because the calibration process is an essential part of quality control during production it is very difficult for an industrial user to exploit the benefits of these instruments. This chapter endeavours to provide some guidance on calibrating optical areal surface topography measuring instruments according to the ISO 25178 suite of specification standards on areal surface topography measurement. Mainly the calibration of the geometrical characteristics of the instruments is discussed with the addition of some brief comments on the effect of filtration and areal surface texture parameters calculation.

4.1 Introduction to Calibration and Traceability

The concept of traceability is one of the most fundamental in metrology and is the basis upon which all measurements can be claimed to be accurate. Traceability is defined as follows:

Traceability is the property of the result of a measurement whereby it can be related to stated references, usually national or international standards, through a documented unbroken chain of comparisons all having stated uncertainties (BIPM 2008a).

When measuring surface topography, all the measurements are related to length – profile measurements require two length measurements (x and z) and areal measurements require three (x, y and z). The traceability of length measurements can be illustrated best by the example of a simple length measurement in industry.

Consider the use of a micrometer to measure the dimension of a part. To ensure that the micrometer measurement is "correct", the instrument must be checked or calibrated against a more accurate displacement measuring system or compared to a calibrated transfer artefact. To do this a gauge block would be used that had itself been calibrated, probably using a mechanical length comparator (see Leach 2009 for a full discussion of the traceability route for gauges blocks). This comparator would be calibrated using a more accurate gauge block that has been calibrated using an optical interferometer with a laser source. This laser source is calibrated against the iodine-stabilised laser that realises the definition of the metre, and an unbroken chain of comparisons has been assured (Leach 2009).

To take an example from the field of surface topography measurement, consider the measurement of surface profile using a stylus instrument (see Sect. 1.5.1). A basic stylus instrument measures the topography of a surface by measuring the displacement of a stylus as it traverses the surface. It is important to ensure that the displacement measurement is calibrated. This calibration may be carried out by measuring a calibrated step height artefact (the transfer artefact). As in the example above, the more accurate instrument measures the displacement of the step using an optical interferometer with a laser source, which is calibrated against the iodine-stabilised laser that realises the definition of the metre. As we move down the chain from the definition of the metre to the stylus instrument that we are calibrating, the accuracy of the measurements usually decreases.

It is important to note the last part of the definition of traceability that states *all having stated uncertainties*. This is an essential part of traceability as it is impossible to usefully compare, and hence calibrate, instruments without a statement of uncertainty. Uncertainty and traceability are inseparable. Note that in practice the calibration of a stylus instrument is more complex than a simple displacement measurement (Leach and Haitjema 2010).

Traceability ensures that measurements are consistent and accurate. Any quality system in manufacturing will require that all measurements are traceable and that there is documented evidence of this traceability. If component parts of a product are to be made by different companies (or different parts of an organisation) it is essential that measurements are traceable so that the components can be assembled and integrated into a product.

4.2 Calibration of Surface Topography Measuring Instruments

Traceability of surface topography measuring instruments can be split into two parts. First, there is the traceability of the instruments and second, the traceability of the analysis algorithms and parameter calculations. Basic instrument traceability is achieved by calibrating the axes of operation of the instrument, usually using calibration artefacts. However, it must be pointed out that calibration of the instrument axes does not necessarily imply that any surface can then be measured and this measurement considered traceable (see Sect. 4.2). Calibration artefacts are available in a range of forms for both profile and areal calibration (see Sect. 4.4), but a primary instrument must calibrate them. Primary instruments are usually kept at the National Measurement Institutes (NMIs) and can be stylus-based (for example, Leach 2000) or optical-based (for example, Wilkening and Koenders 2005). Most primary instrumentation achieves traceability using interferometers that are traceable to the definition of the metre via a laser source.

Traceability of profile measuring instruments has been available now for many years, although it is still common to consider an instrument calibrated when only a single step height artefact has been measured – a dangerous assumption when using the instrument to measure both height and lateral dimensions, or when measuring surface texture parameters. Traceability for areal instruments is still relatively new and there are only a small number of NMIs that can offer an areal traceability service (see Thompsen-Schmidt et al. 2005, Leach et al. 2010).

An important aspect of traceability is the measurement uncertainty of the primary instrument and the instrument being calibrated. Rigorous uncertainty analyses are usually carried out by the NMIs (see for example, Leach 1999, Krüger-Sehm and Krystek 2000, Giusca et al. 2010), but are surprisingly rare in industry for profile measurement using a stylus instrument and almost non-existent for areal measurements, especially when using an optical instrument.

Traceability for parameter calculations can be carried out using calibrated artefacts that have associated parameters, for example the type PDG artefacts used for calibrating profile measuring instruments (see Sect. 4.4). However, the parameter calculations themselves should be verified using software measurement standards (Jung et al. 2004, Bui and Vorburger 2006, Blunt et al. 2008), and for the calibrated artefact an uncertainty calculation has to be made by those institutions that can calibrate these standards.

The calibration of profile measuring instruments has been covered in detail elsewhere (Leach 2001). This book will concentrate on the calibration of optical areal surface topography measuring instruments (the calibration of stylus instruments when used to measure areal surface topography is covered elsewhere (Giusca and Leach 2011)).

4.3 Can an Optical Instrument Be Calibrated?

The calibration of stylus instruments is covered in ISO 25178 part 701 (2010). Currently there is only a draft ISO specification standard on the calibration of optical instruments for measuring areal surface topography and it is expected that this standard (likely to be part 702) will cover all optical instruments that use the areal topography method (see Sect. 1.5). This includes all the instruments described in this book with the exception of those described in Chap. 12 (these are area-integrating methods) that will be covered in a later standard. However, currently the draft of part 702 only covers the calibration of the geometrical characteristics of an instrument (covered in Sect. 4.5). But a question remains: if the geometrical characteristics of an instrument have been calibrated (or at least performance verified), can the instrument measure a rough surface using the geometrical calibration data? The answer to this question is no.

A rough surface is often made up of a complex distribution of heights and spatial wavelengths – such a surface has a spatial frequency bandwidth. To understand how an optical instrument will interact with the rough surface, it is necessary to know the transfer function of the instrument. The transfer function is defined as the response of the instrument (the quotient of input to output) in the frequency (Fourier) domain, i.e. how will the instrument respond to the various spatial frequencies that make up the surface. Note that the use of the transfer function to understand the behaviour of an optical instrument is only applicable for the case of weakly scattering objects where the object causes a small perturbation in the illuminating field (i.e. the interaction is linear) (Coupland and Loberro 2008). With this assumption, each imaging system is characterised in the space domain by its point spread function (PSF) (its response to a point object), and the imaging process can be considered to be a 3D filtering operation. For areal topography measurements of surface topography the assumption of weak scattering can be justified provided the field is returned to the instrument by a single scattering event and this assumption is usually valid for smooth surfaces with gradients that have a magnitude that is less than the numerical aperture of the objective.

In the linear regime, if the PSF can be measured, then the response of the instrument can be calibrated. Indeed, if the PSF can be measured, it is also possible to correct for many of the systematic effects that cause problems when using optical instruments (see Leach 2009 and Chap. 3) by solving the inverse problem (see Palodhi et al. 2010). However, measurement of the PSF is far from trivial and research in this area is still in its infancy. To determine the PSF, Yashchuk et al. (2008) use pseudo-random gratings, Fujii et al. (2010) use chirped sinusoidal gratings and Palodhi et al. (2010) use spheres with a diameter that is smaller than the field of view of the objective. Palodhi et al. (2010) have used a sphere to calibrate the objectives on a commercial coherence scanning interferometer and are developing inverse methods to correct for effects such as optical artefacts (Gao et al. 2008), aberrations and defocus. For now, instrument users should follow the guidelines for measuring the geometrical characteristics – this is not the full story, but is a good starting position when trying to understand and quantify the instrument's performance.

4.4 Types of Material Measure

ISO/CD 25178 part 70 (2011) defines material measures used as measurement standards to calibrate areal surface topography measuring instrumentation. The approach chosen in ISO/CD 25178 part 70 is to combine the profile material measures described in ISO 5436 part 1 (2000) and the newer material measures that can be used to calibrate areal surface topography measuring instruments. Eventually, part 70 will supersede ISO 5436 part 1. A minor inconvenience is that some of the areal and profile material measures are known under different names (Leach 2009). In ISO/CD 25178 part 70 the material measures are separated into two main types: profile and areal. Tab. 4.1 presents the profile material measures including their new and old terminologies and Tab. 4.2 presents the areal material measures.

Table 4.1 Types of profile material measures

Areal Type (ISO/CD 25178: 70)	Previously known types	Name
PPS	[1]C1 and B2	Periodic sinusoidal shape
PPR	[1]C2 and B2	Periodic triangular shape
PGR	-	Periodic rectangular shape
PRO	[1]B2 andC4	Periodic arcuate shape
PRI	[1]A1	Groove, rectangular
PAS	[1]A2	Groove, circular
PDG	[1]D1	Irregular profile
PPS	[1]D2	Circular irregular profile
PPR	[1]E2	Prism
PGR	[1]B3	Razor blade
PRO	[1]C3	Approximated sinusoidal shape
PRI	CS	Contour standard
PAS	ER1	Double groove

[1]ISO 5436: 1 types.

Table 4.2 Type of areal material measures

Areal Type (ISO/CD 25178: 70)	Previously known types	Name
AGP	ER2	Grooves - perpendicular
AGC	ER3	Groove - circular
ASP	[1]E1	Sphere
APS	ES	Plane - sphere
ACG	CG1 and CG2	Cross grating
ACS	-	Cross sinusoidal
ARS	-	Radial sinusoidal
ASG	-	Star-shape grooves
ADT	-	Deterministic

[1]ISO5436: 1 types.

With the exception of types ACS and ASG the material measures listed in Tab. 4.1 and Tab. 4.2 are described in detail elsewhere (Leach 2001, Leach 2009). Type ACS, radial sinusoidal structures are composed of a sinusoidal wavelength along the x axis and a sinusoidal wavelength along y axis. Type ACS measures can be used for overall calibration of the horizontal axis of the instrument and verification of the vertical axis. The measurands are the arithmetic mean height of the surface Sa and root mean height of the surface Sq, but the mean pitches along the x and y axes can also be used for calibration.

Type ASG, star shape grooves (see Sect. 4.5.4) consist of a series of constant height grooves with triangular profiles in the xy plane. Type ASG measures are used for verification of the instrument spatial height resolution (ISO/DIS 25178 part 603 (2011)). The measurand is the depth as a function of pitch on a circular profile extracted concentric to the apex of the artefact.

In some cases, the design of the artefacts has to account for particularities of instruments. For example, focus variation instruments (Chap. 7) are incapable of measuring highly smooth surfaces (those with *Ra* or *Sa* below 10 nm) and specially designed material measures have to be used (Danzl et al. 2008).

With the exception of types PDG, PPS and ADT, material measures are mainly used for the calibration of the instrument scales. However, not all of the artefacts are always required because a large number of the profile artefacts are designed for contact stylus instruments and other artefacts can achieve the same measurement goal.

4.5 Calibration of Instrument Scales

As discussed in Sect. 4.2, basic instrument traceability is achieved by calibrating the axes of operation of the instrument or the instrument scales. According to ISO 25178 part 601 (2010), the instrument scales of the areal topography measuring instrument should be aligned to the axes of a right handed Cartesian co-ordinate system (note that part 601 is for stylus instruments, but the co-ordinate system will be the same for optical instruments). Practically, the scales are realised by various components that are part of the metrological loop of the instrument such that the 3D map of a surface is made up of a set of points measured along the three orthogonal axes.

Some of the instrument components provide a reference surface with respect to which the instrument measures the surface topography and other components provide the vertical axis of the instrument. The quality and the mutual position of these components influence the quality of the areal topography measurements. The areal measurements are also affected by other factors (quantities) such as ambient temperature, mechanical and electrical noise, the quality and type of the instrument's optical components, the mathematical algorithms that are used to process the height information and so on. All these quantities are known as influence factors (ISO/DIS 25178: 603 2011).

One way to estimate the effect of the influence factors on an areal measurement is to establish a meaningful measurement model that links the influence factors to the length measurements along the instrument scales. It is very difficult, and at the same time unnecessary to construct a mathematical model that isolates the effect of each influence factor. Instead, a simple input–output measurement model that is based on a limited number of measurable input quantities can be used. The input quantities are also called metrological characteristics in ISO/DIS 25173 part 603 (2011) and are listed in Tab. 4.3.

The metrological characteristics incorporate the effect of the influence factors and more importantly they can be determined, usually by measuring a feature of a material measure. For example, in the case of a phase shifting interferometer (Chap. 8) the effective wavelength of the light, the bandwidth of the light used for measurement and the state of polarization of the light impinging on the measured surface (all three are influence factors in ISO/DIS 25178 part 603) influence, among other metrological characteristics, the amplification coefficient of the *z* axis scale. The amplification coefficient of the *z* axis scale can be directly measured using a series of calibrated step heights (PRI type), hence, the combined affect of the three influence quantities.

Table 4.3 Metrological characteristics

Metrological characteristic	Symbol	Error along
Amplification coefficient	αx, αy, αz	*x*, *y*, *z*
Linearity	*l*x, *l*y, *l*z	*x*, *y*, *z*
Residual flatness	FLT-Z	*z*
Measurement noise	Nm	*z*
Spatial height resolution	W_R	*z*
Squareness	PER	*x*, *y*

Calibration of the instrument scales consists of measuring the metrological characteristics of the instrument. Some of the metrological characteristics can be affected by the size of the primary extracted surface and the sampling distance, i.e. the spatial measurement bandwidth. Setting the measurement bandwidth is application-dependent and often requires a priori knowledge about the surface to be measured. There are situations in which the surface to be measured is well understood, such as with a milled or turned surface, so the measurement bandwidth can be relatively simple to set. However, there are applications in which the surface, or its functionality, is not known well. In this situation the measurement bandwidth is selected according to a priori theoretical knowledge about the nature of the surface or from prior experimental work, or a combination of both. Unless specified, the spatial measurement bandwidth can be established in accordance with ISO/DIS 25178 part 3 (2011) that recommends some basic rules on how to set the size of the primary surface and the sampling distance. These options are mainly based on the choice of S-filters and L-filters/F-operators, each having a range of preset values called nesting indexes (see Leach 2009, Muralikrishnan and Raja 2010 for descriptions of areal filtering). The S-filter nesting index determines the maximum sampling distance and the L-filter or the F-operator determines the size of primary surface.

Along with the measurement bandwidth, all other calibration conditions should be set in such a manner that all measurement conditions are replicated, including the environmental conditions. Sometimes, especially during research activities, the full instrument calibration could be a very time-consuming task because it is difficult to cover all the measurement conditions in which the instrument could be used. The situation is further complicated due to the number of software settings that are available on the commercial instruments. Fortunately, the often short measuring time of the optical instruments and a careful design of the experiments can partially compensate for the number of measurements required.

The calibration process consists in a series of relatively simple tasks that evaluate the uncertainty associated with the metrological characteristics. The following sections present methods of calibrating the scales and axes of optical areal topography measuring instruments based on calibrated material measures that were presented in Sect. 4.4.

4.5.1 Noise

Two kinds of noise can be measured in practice, static noise and measurement noise (or dynamic noise see ISO 25178 part 601 (2010)) depending on the type of areal surface texture measuring instrument being calibrated. The static noise test is relevant to instruments that operate an *xy* scanning stage and investigates electrical noise and environmental vibration in the absence of the effects of the stage dynamics. The static noise test consists of recording the probe signal, without scanning laterally, at frequencies that correspond to a certain measurement spatial bandwidth. The static noise is not significant for the calibration of the instrument scales because the measurement noise incorporates the effects of static noise.

Measurement noise limits the capability of an instrument to measure high spatial frequencies on the surface of a sample. The measurement noise test requires an optical flat with a deviation from flatness of less than 30 nm. The surface roughness of the optical flat can be as high, and in some instances higher than the measurement noise of the instrument, so that the measurement noise has to be separated from the underlying roughness and the flatness deviation of the optical flat. There are a two noise separation techniques that can achieve such a seperation. One method is based on a subtraction technique (VDI/VDE 2617 2004) and one is based on an averaging technique (Haitjema and Morrel 2005), but both methods should provide similar results.

The subtraction technique requires two measurements at the same position on the sample. The two topography data sets are subtracted from each other such that the form and the underlying roughness of the optical flat are eliminated. The measurement noise Sq_{noise} is estimated using

$$Sq_{noise} = \frac{Sq}{\sqrt{2}}\,. \tag{4.1}$$

The averaging method is based on the assumption that the noise contribution decreases when averaging multiple measurements of the surface topography at the same location on a sample. The measured Sq on the surface of a flat is a function of the instrument noise Sq_{noise} and the roughness of the flat Sq_{flat} as shown in equation (4.2)

$$Sq = \sqrt{Sq_{flat}^2 + Sq_{noise}^2}\,. \tag{4.2}$$

After n repeated measurements at the same location on the surface of the flat, the contribution of the instrument noise into the root mean square height of the averaged surface topography Sq_n is decreased by square root of n while the flat contribution is preserved,

$$Sq_n = \sqrt{Sq_{flat}^2 + \frac{1}{n} Sq_{noise}^2} \,. \tag{4.3}$$

The instrument noise can be extracted from equations (4.2) and (4.3)

$$Sq_{noise} = \sqrt{\frac{n}{n-1}\left(Sq^2 - Sq_n^2\right)} \,. \tag{4.4}$$

Measurement noise tests carried out on a transparent glass flat using a coherence scanning interferometer (CSI) (see Chap. 9) showed that, when using the averaging method, two repeated measurements ($n = 2$) is enough to determine the measurement noise. Tab. 4.4 shows the results for a 50x magnification (0.85 numerical aperture) objective and a one megapixel CCD camera configuration of the instrument that was measuring a primary surface of 0.35 mm by 0.35 mm.

Table 4.4 Measured noise of a CSI that used a 50× magnification objective on a transparent glass flat

Number of repeated measurements	Sq_{noise} / nm
2	0.665 0
4	0.665 0
8	0.665 5
16	0.665 0

Sq_{noise} did not differ significantly (less than 0.1 %) with the increase in the number of repeated measurements. The value of Sq_{noise} after two measurements would have been valid even for a difference of 10 % from the value obtained after sixteen repeated measurements.

The subtraction method results are similar to the addition method results. So, in this case both addition and subtraction methods can be used with the same effect and it is down to the user to determine which of the two methods is more efficient.

Both methods work well in some cases but they have their own shortfalls that are not always easy to predict. Measurement noise can change magnitude from one surface to another so that it is higher on transparent flat surfaces than on those that are highly reflective, or it is different on surfaces that contain features of the same lateral spatial bandwidth but of different amplitudes. It is also difficult to apply either method when the instruments have noise that is non-stationary in a statistical sense (Whitehouse 1976) so that averaging repeated measurements cannot isolate the noise from the residual flatness and flat topography. In the case of non-stationary noise-specific cut-offs and filters that isolate the measurement noise from residual flatness have to be investigated.

4.5.2 Residual Flatness

One important feature of any areal topography measuring instrument is the quality of its areal reference against which the surface topography is measured. The vast majority of areal surface topography measuring instruments use a nominally flat surface as the areal reference and any deviation from it will translate into an error along the z axis measurement direction. In the case of optical instruments the residual flatness can be caused by optical aberrations.

Similar to the measurement noise test, the residual flatness test is performed on a flat surface; however, the parameter that quantifies the magnitude of its effect is the maximum height of the scale limited surface, *Sz*. Unlike *Sq*, the value of *Sz* is affected by the sample local height variation such as scratches or contamination. The difficulty is to completely separate the contribution of the instrument from that of the flat and other spurious measurement data. One way to overcome this difficulty is to measure the topography of the flat at different locations (VDI/VDE 2617 2004) without changing the instrument setup and to average the height measurement of each point of the topography. The contribution of the flat and any spurious data should diminish, whereas the quality of the areal reference should be preserved.

An example of flatness measurement results on a CSI instrument is presented in Tab. 4.5.

Table 4.5 Measured flatness of a CSI that used a 50× magnification objective on a transparent glass flat

Number of repeated measurements	*Sz* / nm
2	6.26
4	3.81
6	3.40
8	3.28
10	3.26

It is difficult to recommend the exact number of repeated measurements because it depends on the rate at which the value of *Sz* stabilises. Instead, the measurements could be repeated until the value of *Sz* becomes stable or is less than the target uncertainty, ideally less than 17 % of the measurement uncertainty. Note that 17 % mark is chosen such that the contribution of the residual flatness to the measurement uncertainty can be ignored but it is rarely achievable especially for uncertainties of the order of nanometres. A small number of repeated measurements will only overestimate the magnitude of the residual flatness which may be acceptable from the point of view of uncertainty estimation.

In the case of instruments that are known to have a residual flatness larger than the *Sz* of the flat surface, one measurement will suffice. The downfall of this method of estimating the residual flatness is that the flat surfaces are not always calibrated in the spatial measurement bandwidth of the instrument and this makes the traceability of the instrument difficult to demonstrate.

In Fig. 4.1 *Sz* is 3.26 nm after averaging ten measurements and it is determined by the virtual scratch present in the areal reference of the instrument. The areal reference is inherited from the quality of flat used during the flat adjustment of the instrument. A better quality flat adjustment would have produced a better quality areal reference. From a practical point of view, if the measurement uncertainty along the *z* direction does not need to be better than 20 nm, four repeated measurements are adequate.

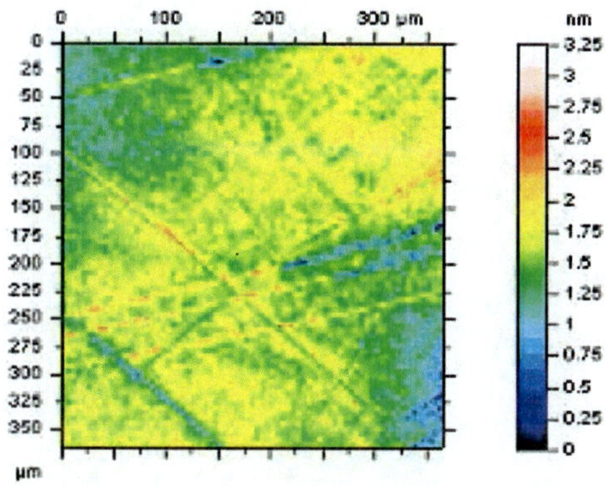

Fig. 4.1 Flatness of a CSI that used a 50× magnification objective to measure a transparent glass flat - result after ten repeated and averaged measurements

For those residual flatness tests that are affected by spurious data of high amplitude, threshold methods have to be investigated and employed in order to achieve effective measurement of *Sz* (Ismail et al. 2010).

4.5.3 Amplification, Linearity and Squareness of the Scales

Amplification and linearity tests establish the relationship between the ideal response curve and the instrument response curve on each of the *x*, *y* and *z* scales. Fig. 4.2 shows a typical example of a linear scale response curve. The linearity of the axes is given by the maximum deviation of the instrument response curve from the linear curve, where the slope is the amplification coefficient.

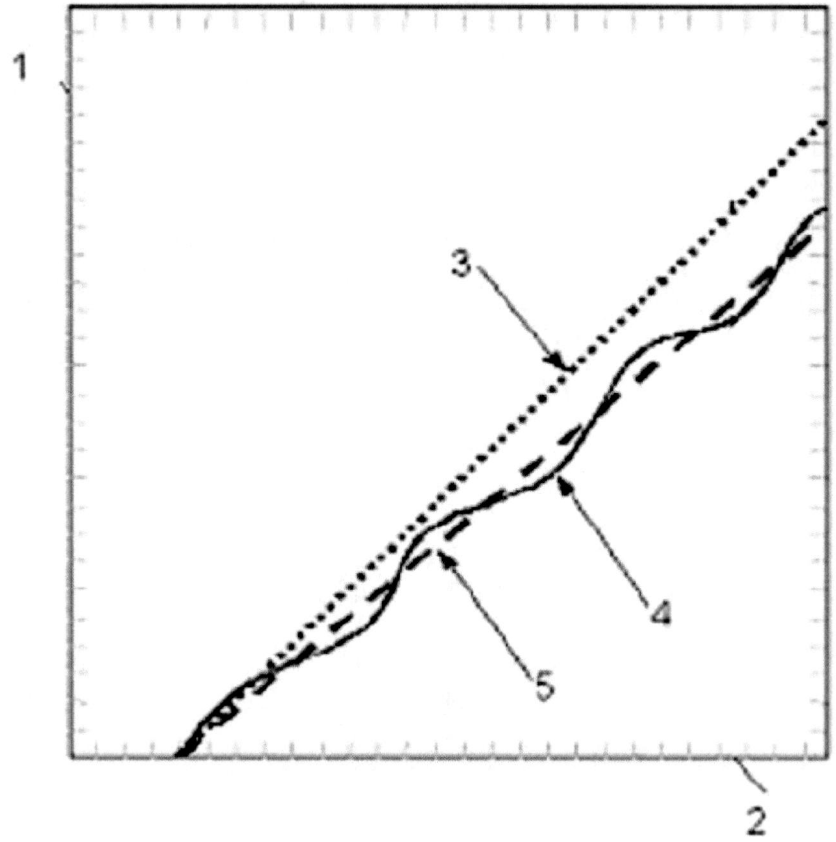

Fig. 4.2 Example of an instrument response curve, where: 1 measured quantities, 2 input quantities, 3 ideal response curve, 4 response curve, 5 linear curve whose slope is the amplification coefficient (from ISO 25178-601 2010)

A thorough amplification and linearity test requires a calibrated artefact capable of providing multiple values uniformly distributed within the instrument range. However, this is a prerequisite rarely fulfilled in practice, especially during the test that provides information about the z axis linearity and amplification coefficient. Often the z axis calibration is performed using a single-step artefact that is also used to adjust the software of the instrument. This situation is potentially very dangerous because it could only shift the response curve in such a way that it crosses the ideal response curve only at that measured point. Unless the response curve is perfectly linear or the instrument is used as a comparator, one calibration artefact does not provide sufficient information about the quality of the z axis scale of the instrument and it will underestimate its contribution into the uncertainty calculations.

The amplification and linearity of the z axis scale should be tested with a series of different step height artefacts with various heights. The step height artefacts should cover the entire range of the z axis scale or at least they should range from the minimum to the maximum height of interest. The linearity and amplification coefficient can be extracted from the measurement results of multiple calibrated step heights of different values by simply fitting a straight line to the data. Example results of the z axis scale calibration with four step heights (19 nm, 300 nm, 3 μm and 17 μm) are presented in Fig. 4.3. The continuous line represents a linear fit set to intersect the axes at zero. The amplification coefficient of the z axis scale is correlated with the slope of the line and the residuals represent the scale non-linearities.

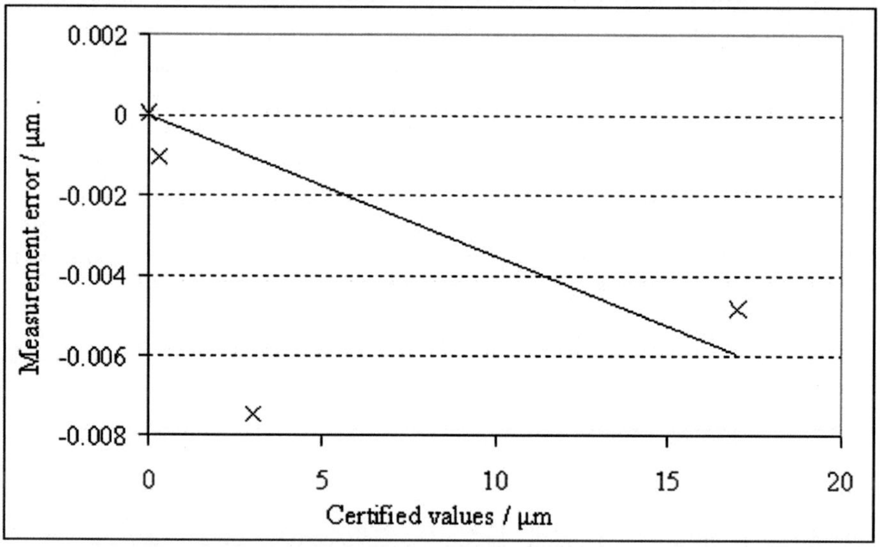

Fig. 4.3 Amplification and linearity of the z axis scale

In the example presented in Fig. 4.3 the value of the largest step height is just a fraction of the z axis range of the instrument. The step heights have to be measured at different positions in the instrument range (25 %, 50% and 75% of the z axis range) in order to certify the whole z axis scale for height measurements. This calibration step is a reproducibility test and it does not certify the instrument for height measurements that exceed the value of the largest step height used during the calibration.

The squareness between the areal reference and the z axis can be determined by measuring the pitch of a periodic structure mounted at different angles relative to the instrument's areal reference (Xu et al. 2008). If the z axis scale is not perpendicular to the areal reference the pitch will change according to the angle between the areal reference and the step height surface. On the other hand, the calibration of the z axis scale with multiple step height artefacts can correct for the squareness errors. The cosine error that is introduced by the z axis scale squareness behaves as an amplification error.

Unlike the z axis scale case, the calibration of the lateral axes uses a calibrated artefact capable of providing multiple values uniformly distributed within the instrument range. Traditionally, the calibration of the lateral axes is performed using a grating with a known pitch. The pitch measuring technique can be applied to calibrate instruments that measure areal topography but the analysis has to be performed in profile mode. The drawback of pitch measurements is that they only estimate the local characteristic of the instrument's scale and do not give information about the instrument response curve. Areal material measures such as cross gratings (Leach et al. 2006) or pyramidal structures (Ritter et al. 2007) are better suited for calibration of the lateral scales of areal topography measuring instruments.

The x and y axes amplification, linearity and the squareness can be measured using a calibrated cross grating artefact. By measuring the positions of the centre of gravity of the cross grating's squares with a traceable areal surface topography measuring instrument (for example Leach et al. 2009, Thomsen-Schmidt and Krüger-Sehm 2008) allows for stable and traceable length measurements along the x and y axes. An example of amplification and linearity measurements along one of the axes is presented in Fig. 4.4 before correcting for amplification errors and in Fig. 4.5 after correcting for amplification errors.

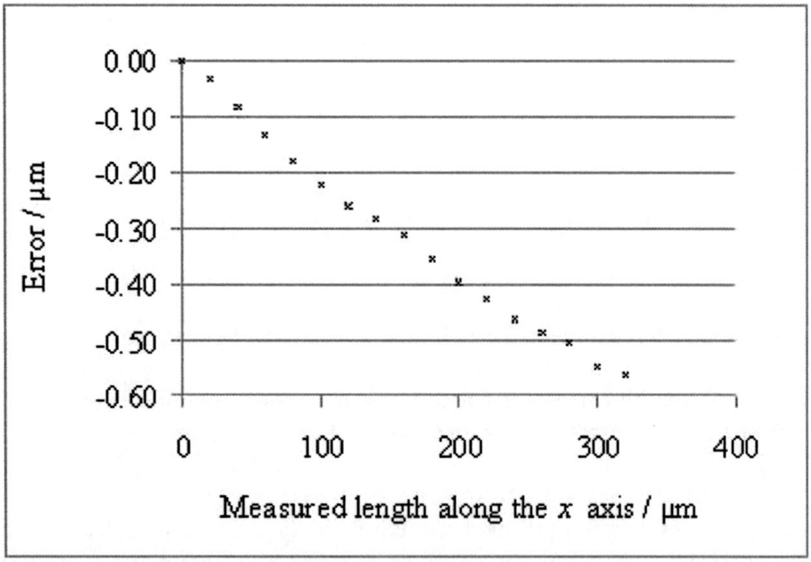

Fig. 4.4 Amplification and linearity of the lateral scale

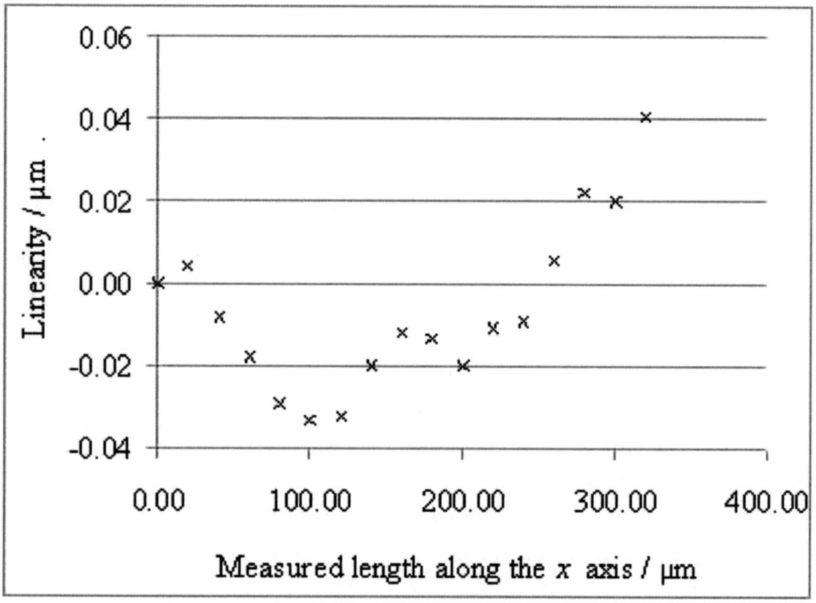

Fig. 4.5 Linearity of the lateral scale

There are two ways to account for the contribution of the amplification of the scales to the measurement uncertainty. The first way is to split the contribution of the amplification coefficient from the contribution of the linearity. The first way can be inconvenient, as it requires uncertainty propagation using the equation of the line fit. The second way is to consider the combined effect of the amplification coefficient and scale linearity. In this case the uncertainty contribution of the magnification of the scales will be given by the largest error that is measured against a calibrated line scale.

The squareness of the x and y axes can be measured by measuring the angle between two nominally orthogonal rows of square holes whose squareness is known. The orientation of each row of squares can be calculated by fitting a line though the centre of gravity of the corresponding squares.

4.5.4 Resolution

It is difficult to design material measures suitable for testing the z axis resolution. In almost all situations the z axis resolution is small compared to other contributors to the uncertainty such as scale linearity, amplification errors and noise.

The resolution of the lateral scales is defined as the smallest lateral separation between two points that can be distinguished. This is a useful definition of the lateral resolution for a 2D microscope or when making lateral measurements. However, for areal surface topography measurements, where the distance between two adjacent points could affect their relative height difference, this definition becomes impracticable. The width limit for full height transmission has been defined (ISO 25178-601 2010) to overcome the problem with the lateral resolution definition but is still under debate. Experimentally, the width limit for full height transmission is measured on gratings or crossed gratings with the pitch value close to the resolution of the instrument. ASG type material measures (Leach et al. 2006, Weckenman et al. 2009) can be used to find the approximate value of the width limit for full height transmission (see Fig. 4.6 and Fig. 4.7) before measuring a grating with a predefined pitch. In Fig. 4.6 the height of the circular profile that is extracted concentric to the apex of the star shaped grooves is not affected by the 32 μm pitch of the square wave, whereas in Fig. 4.7 the measured height drops to 70 % for a square wave of 3 μm pitch.

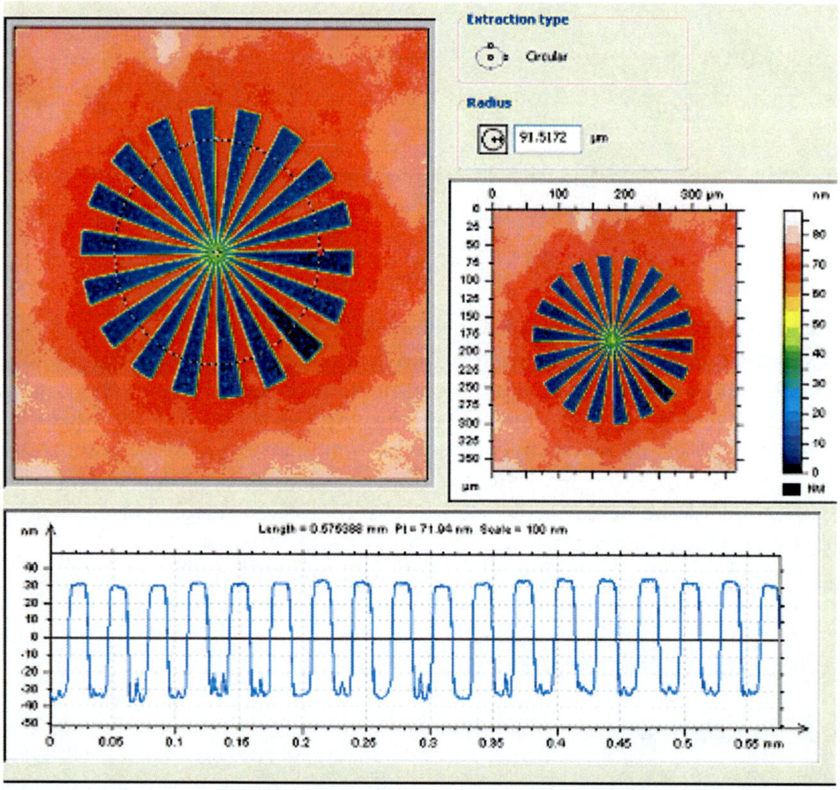

Fig. 4.6 Example of circular profile of 91.5 μm radius extracted from the measured topography of a 3D star pattern. The pitch of the features of the profile does not affect the height of the profile

Other definitions of lateral resolution could be used where the width limit for full height transmission is not appropriate for a specific application. ISO TC 213 WG16 is looking to introduce an umbrella term, spatial height resolution, defined as the ability of a surface topography measuring instrument to distinguish closely spaced surface features (ISO/DIS 25178-603 2011). Currently, under spatial height resolution are six terms: lateral period limit, lateral resolution, width limit for full height transmission, Rayleigh criterion applied to topography measurement, Sparrow criterion applied to topography measurement and point spread function. It is likely that the list of definitions for lateral resolution will change which makes the standardisation process very difficult. Irrespective of the definition, the resolution has to be tested experimentally using artefacts.

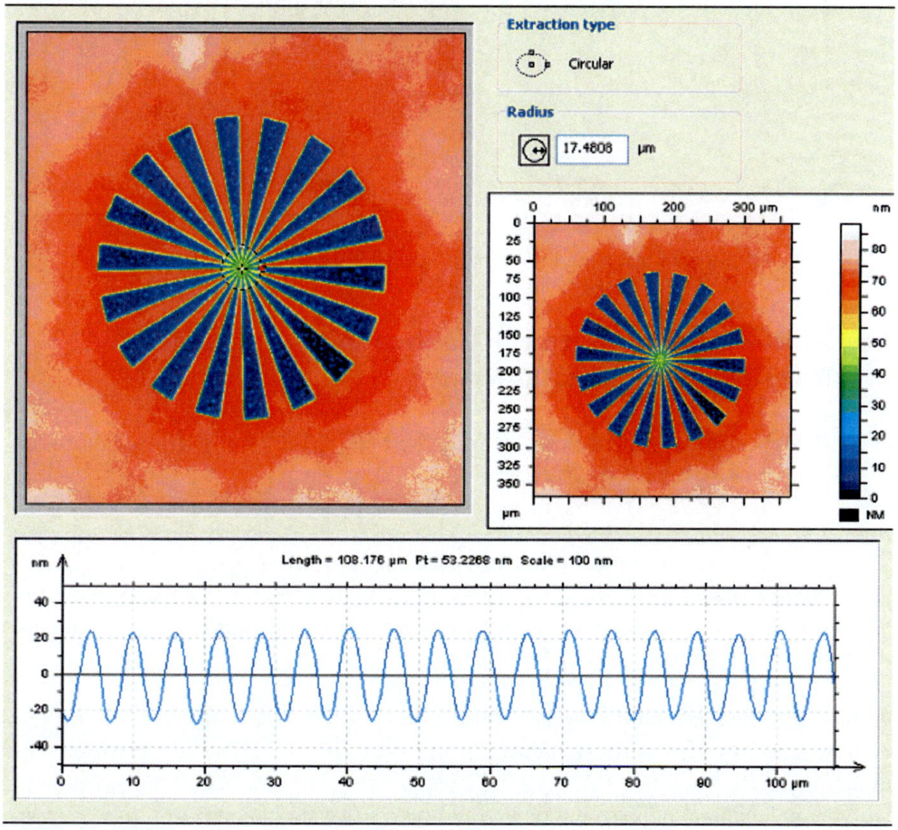

Fig. 4.7 Example of circular profile of 17.5 μm radius extracted from the measured topography of a 3D star pattern. The pitch of the features of the profile affects the height of the profile

As discussed in Sect. 4.4, the design of the resolution artefacts plays an important role in the quality of the resolution measurements. For example, the aspect ratio of the square wave will influence the value of the resolution as well as other errors such as the batwing effect (Gao et al. 2008) that could increase the apparent height of the profile (see Fig. 4.8 – the measured mean depth of the profile is approximately three times higher that its actual height of 200 nm). These complications could be overcome by calibrating the transfer function of the instrument (see Sect. 4.3).

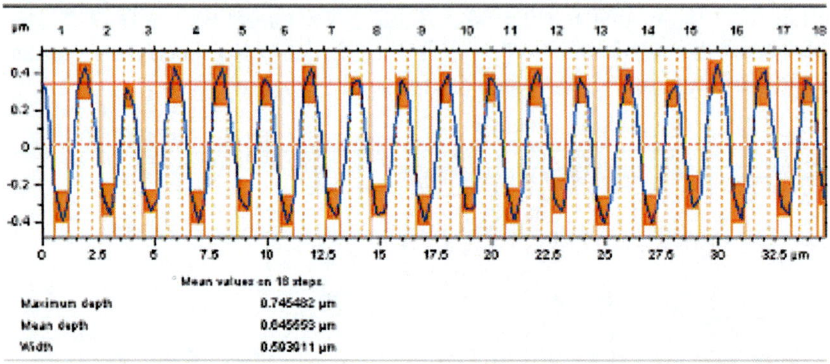

Fig. 4.8 Batwing affecting the height of a square wave of amplitude 200 nm

4.6 Relationship between the Calibration, Adjustment and Measurement Uncertainty

The BIPM *Vocabulary in Metrology* document (BIPM 2008a) defines calibration as

operation that, under specified conditions, in a first step, establishes a relation between the quantity values with measurement uncertainties provided by measurement standards and corresponding indications with associated measurement uncertainties and, in a second step, uses this information to establish a relation for obtaining a measurement result from an indication.

Commonly, the term calibration is used in place of the term adjustment and this leads to confusion in understanding the meaning of the calibration process. The adjustment process tunes some changeable parameters of a metrological tool (it can be a mechanical adjustment or it could be a software change) to provide an indication that is closer to a known value. The adjustment process does not provide any information about measurement uncertainty. Similar results to the ones that the adjustment process produces could be obtained by correcting the measurement result using information that comes from a calibration process without instrument adjustment. Ultimately a meaningful measurement result can be obtained without adjustment but the result should always have an associated uncertainty.

A measurement of areal surface topography consists of the relatively simple task of measuring a set of points across the surface of a component. The difficulty lies in the assessment of the effect of a multitude of influence factors that contribute to the measurement. First, it is necessary to establish the measurement conditions, that is to say the measurement bandwidth and environmental conditions. The following step is to choose appropriate settings that ensure optimal usage of the instrument (magnification, sampling distance, etc.). As the instrument provides a set of data from which the topography data can be extracted, specialised software is required for data processing (i.e. filtering, parameter calculation and visual representation). All these factors will influence the measurement result and their contribution to the measurement uncertainty has to be acknowledged.

If measurement uncertainties are to be estimated using the *Guide to Uncertainty in Measurement* (BIPM 2008b) guidelines an input-output measurement model that allows the output quantity uncertainty to be calculated from the values of the input quantities and their associated probability distributions has to be produced. In this context, the output quantity could be an areal parameter and the input quantities are the metrological characteristics of the instrument. The construction of the measurement model is by far the most complex part of the calculation of the measurement uncertainty because it requires an in depth understanding of all the steps of the measurement process such as those of the filter on the measurement data and the equations used to calculate areal parameters from the 3D data set.

For example, the CSI instrument can accurately measure very small step heights because the mathematical algorithm used to estimate the step value significantly reduces the effect of noise and flatness errors. The depth calculation of a step height standard (type PRI) 100 μm wide has the effect of an S-filter of approximately 30 μm that decreases the measurement noise contribution from 0.67 nm to 0.03 nm and residual flatness contribution from 3.3 nm to 1.4 nm. Assuming a normal distribution $N(0, (0.03\ \text{nm})^2)$ for the measurement noise contribution and a rectangular distribution R(-0.4 nm, 0.4 nm) for the residual flatness contribution their combined effect to the depth standard measurement uncertainty will be ± 0.4 nm. This allows the CSI to measure accurately step heights of the order of nanometres.

4.7 Summary

Assuring traceability for parameter calculations and calibrating the scales of the axes of operation can achieve a restricted calibration of optical surface topography measuring instruments because it is difficult to understand how an optical instrument will interact with the rough surface. The transfer function should explain the way that the instrument responds to the various spatial frequencies that make up the surface; however, it is only applicable for the case of weakly scattering objects and research in this area is still in its infancy. Calibrating the scales of the axes of operation is a good starting position when trying to understand and quantify the instrument's performance.

Calibration of the instrument scales consists of a series of tests aimed at measuring the metrological characteristics of the instrument: measurement noise, residual flatness, amplification coefficient and linearity of the scales of the instrument, spatial height resolution and squareness of the axes. The calibration of the scales has to be performed for well-defined spatial measurement bandwidths and it relies on a limited number of calibrated material measures that can be selected from a wide range of designs, some unidirectional and some bidirectional.

Current measurement noise and residual flatness tests aim to isolate and measure the effect of these two metrological characteristics based on the assumption that the noise is stationary in a statistical sense and the residual flatness is a systematic error that can be easily compensated for. With these assumptions unchallenged by the instrument's characteristics both measurement noise and residual flatness can be accurately isolated and measured using an optical flat. When the measurement noise is non-stationary, the flatness errors are not systematic or the tests require a large number of repeated measurements the current methods of estimating the measurement noise and residual flatness become impractical and other test techniques need to be developed.

The tests for amplification, linearity and squareness of the scales of the instruments require calibrated material measures capable of providing multiple values uniformly distributed within the instrument range, for example calibrated step heights in the case of the z scale and calibrated cross gratings in the case of the x and y scales. Applying a linear fit to the measurement results along each of the three scales, the amplification coefficients and linearity errors of the response curves can be measured. The z axis squareness to the xy plane is incorporated into the amplification of the z scale. The squareness of the x and y scale can be measured separately using the cross grating.

Spatial height resolution has been recently introduced by ISO TC 213 WG16 in order to define the ability of a surface topography measuring instrument to distinguish closely spaced surface features. The measurement tests are still a matter of debate; however, the resolution has to be tested experimentally using artefacts such as gratings or crossed gratings with the pitch value close to the resolution of the instrument. The approximate value of the spatial height resolution can be measured using ASG type material measures.

If measurement uncertainties are to be estimated for areal parameters the effects of the metrological characteristics of the instrument have to be propagated using an input-output measurement model. The measurement model is by far the most complex part of the calculation of the measurement uncertainty. More research is needed to understand the effect of filters and the equations used to calculate areal parameters from the 3D dataset.

Acknowledgements

The authors would like to thank Mr Franck Helary (formally ENSAM) and Mr Tadas Gutauskas (formally Imperial College London) for help in determining the measurement protocols in this chapter.

References

BIPM, IEC, IFCC, ISO, IUPAC, IUPAP and OIML, International Vocabulary of Metrology – Basic and General Concepts and Associated Terms VIM. In: JCGM 200, 3rd edn. (2008a)

BIPM, IEC, IFCC, ISO, IUPAC, IUPAP and OIML, Guide to the expression of uncertainty in measurement, Bureau International des Poids et Mesures. In: JCGM 100 (2008b)

Blunt, L., Jiang, X., Leach, R.K., Harris, P.M., Scott, P.: The development of user-friendly software measurement standards for surface topography software assessment. Wear 264, 389–393 (2008)

Bui, S., Vorburger, T.V.: Surface metrology algorithm testing system. Precision Engineerin 31, 218–225 (2006)

Coupland, J.M., Lobera, J.: Holography, tomography and 3D microscopy as linear filtering operations. Meas. Sci. Technol. 19, 074012 (2008)

Danzl, R., Helmli, F., Rubert, P., Prantl, M.: Optical roughness measurements on specially designed roughness standards. In: Proc. SPIE, vol. 7102, p. 71020M (2008)

Fujii, A., Suzuki, H., Yanagi, K.: A study on response properties of surface texture measuring instruments in terms of surface wavelength. In: Proc. ASPE Summer Topical Meeting on Precision Interferometric Metrology, Ashville, USA (2010)

Gao, F., Leach, R.K., Petzing, J., Coupland, J.M.: Surface measurement errors using commercial scanning white light interferometers. Meas. Sci. Technol. 19, 015303 (2008)

Giusca, C., Forbes, A.B., Leach, R.K.: A virtual machine-based uncertainty evaluation for a traceable areal surface texture measuring instrument. Measurement (accepted for publication 2011)

Giusca, C., Leach, R.K.: The calibration of stylus instruments for measuring areal surface texture. NPL Good practice guide. National Physical Laboratory (2011)

Haitjema, H., Morel, M.A.A.: Noise bias removal in profile measurements. Measurement 38, 21–29 (2005)

ISO 5436-1, Geometrical product specifications (GPS) - Surface texture: Profile method; Measurement standards – Part 1: Material measures. International Organization for Standardization (2000)

ISO 25178-601, Geometrical product specifications (GPS) - Surface texture: Areal - Part 601: Nominal characteristics of contact (stylus) instruments. International Organization for Standardization (2010)

ISO/FDIS 25178-3, Geometrical product specifications (GPS) - Surface texture: Areal - Part 3: Specification operators. International Organization for Standardization (2011)

ISO/DIS 25178-603, Geometrical product specifications (GPS) - Surface texture: Areal - Part 603: Nominal characteristics of non-contact (phase shifting interferometric microscopy) instruments. International Organization for Standardization (2011)

ISO/CD 25178-70, Geometrical product specifications (GPS) - Surface texture: Areal - Part 70: Material measures. International Organization for Standardization (2011)

Ismail, M.F., Yanagi, K., Fujii, A.: An outlier correction procedure and its application to areal surface data measured by optical instruments. Meas. Sci. Technol. 21, 105105 (2010)

Jung, L., Spranger, B., Krüger-Sehm, R., Krystek, M.: Reference software for roughness analysis – features and results. In: Proc. XI Int. Colloq. Surfaces, Chemnitz, Germany, vol. 170, pp. 164–170 (February 2004)

Krüger-Sehm, R., Krystek, M.: Uncertainty analysis of roughness measurement. In: Proc. X Int. Colloq. Surfaces, Chemnitz, Germany (January/February 2000) (in additional papers)

Leach, R.K.: Calibration, traceability and uncertainty issues in surface texture metrology. NPL Report CLM 7 (1999)

Leach, R.K.: Traceable measurement of surface texture at the National Physical Laboratory using NanoSurf IV. Meas. Sci. Technol. 11, 1162–1172 (2000)

Leach, R.K.: The measurement of surface texture using stylus instruments. NPL Good practice guide No. 37. National Physical Laboratory (2001)

Leach, R.K., Chetwynd, D., Blunt, L., Haycocks, J., Harris, P., Jackson, K., Oldfied, S., Reilly, D.: Recent advances in traceable nanoscale dimension and force metrology in the UK. Meas. Sci. Technol. 17, 467–476 (2006)

Leach, R.K.: Fundamental principles of engineering nanometrology. Elsevier, Amsterdam (2009)

Leach, R.K., Giusca, G., Naoi, K.: Development and characterization of a new instrument for the traceable measurement of areal surface texture. Meas. Sci. Technol. 20, 125102 (2009)

Leach, R.K., Haitjema, H.: Limitations and comparisons of surface texture measuring instruments. Meas. Sci. Technol. 21, 032001 (2010)

Muralikrishnan, B., Raja, J.: Computational surface and roundness metrology. Springer, Heidelberg (2008)

Palodhi, K., Coupland, J.M., Leach, R.K.: A linear model of fringe generation and analysis in coherence scanning interferometry. In: Proc. ASPE Summer Topical Meeting on Precision Interferometric Metrology, Ashville, USA (2010)

Ritter, M., Dziomba, T., Kranzmann, A., Koenders, L.: A landmark-based 3D calibration strategy for SPM. Meas. Sci. Technol. 18, 404–414 (2007)

Thompsen-Schmidt, P., Krüger-Sehm, R., Wolff, H.: Development of a new stylus contacting system for roughness measurement. In: Proc. XI Int. Colloq. Surfaces, Chemnitz, Germany, February 2004, pp. 79–86 (2004)

Thomsen-Schmidt, P., Krüger-Sehm, R.: Calibration of an electromagnetic force compensation system. In: Proc. XII Int. Colloq. Surf., Chemnitz, Germany, January 28-29, pp. 323–327 (2008)

VDI/VDE 2617 Part 6.2. Accuracy of coordinate measuring machines. Characteristics and their testing. Guideline for the application of DIN EN ISO 10360 to coordinate measuring machines with optical distance sensors (October 2004)

Whitehouse, D.J.: Some theoretical aspects of error separation techniques in surface metrology. J. Phys. E: Sci. Instrum. 9, 531 (1976)

Weckenmann, A., Tan, Ö., Hoffmann, J., Sun, Z.: Practice-oriented evaluation of lateral resolution for micro- and nanometre measurement techniques. Meas. Sci. Technol. 20, 65103 (2009)

Wilkening, G., Koenders, L.: Nanoscale calibration standards and methods. Wiley-VCH, Chichester (2005)

Yashchuk, V.V., McKinney, W.R., Takacs, P.Z.: Binary psuedorandom grating standard for calibration of surface profilometers. Opt. Eng. 47, 073602 (2008)

Xu, M., Dziomba, T., Dai, G., Koenders, L.: Self-calibration of scanning probe microscope: mapping the errors of the instrument. Meas. Sci. Technol. 19, 025105 (2008)

5 Chromatic Confocal Microscopy

François Blateyron

Digital Surf sarl
16 rue Lavoisier, F-25000 Besançon
France

Abstract. Chromatic confocal probes are single point optical sensors built around a confocal coaxial setting that use chromatic dispersion and decoding to obtain the surface distance. Such sensors are usually installed on scanning stages of surface texture measuring instruments, roundness instruments or coordinate measuring machines. Chromatic confocal probes can measure on, and through, transparent material, detect several interfaces between materials and, therefore, calculate thickness. The metrological characteristics of chromatic confocal probes are close to those of stylus probes and they are often used as a non-contact substitute on stylus profilometers.

5.1 Basic Theory

Chromatic confocal microscopy is a measuring technique comprising a chromatic confocal probe and a lateral scanning system. The chromatic confocal probe senses each point of the workpiece surface and in turn extracts its height (topography) and the associated light intensity. The lateral scanning system allows the measurement of a line profile (with an x axis stage), an areal surface (with x and y axis stages), or any geometrical configuration based on one or several linear or rotary stages.

As a single-point measuring technique, the measurement process is very similar to that of a stylus profilometer (see Sect. 1.5.1) and, therefore, provides the same advantages such as the ability to setup the length of scanned profiles (limited only by the x axis stage) or the ability to measure circular profiles.

As a confocal probe, chromatic confocal microscopy is insensitive to ambient light and stray reflection from the surface, but contrary to traditional imaging confocal microscopy (see Chap. 11), chromatic confocal microscopy does not require any vertical scanning to sense the surface height. This characteristic makes the optical head entirely static, without any spurious vibration generated by an internal mechanism.

Chromatic confocal probes have been integrated into optical profilometers for more than 10 years for surface texture measurement, into coordinate measuring machines (CMMs) for dimensional measurement, and into several other types of machines where an accurate height measurement is required.

Chromatic confocal probe instruments are included in the classification of surface texture instruments (ISO 25178-6 2010) and described in a recently published ISO document (ISO 25178-602 2010).

5.1.1 Confocal Setting

Figure 5.1 shows the classical setup of a single point confocal microscope. The optical path from the light source to the workpiece surface is the same length as the path from the workpiece surface to the photo-detector.

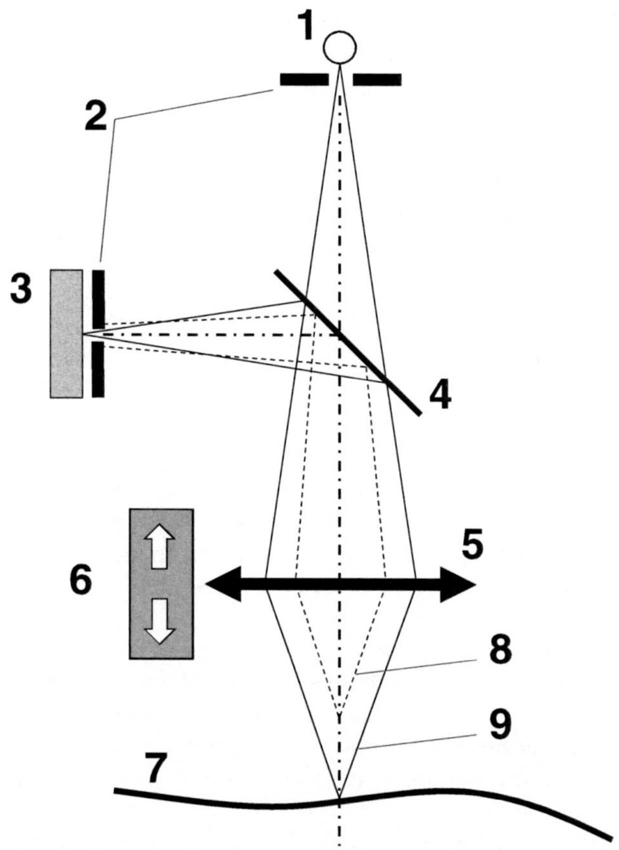

Fig. 5.1 Classical confocal setting

When the focal point is above or below the surface (8), or in other words when the probe is out of focus, the reflected light does not pass through the detector pinhole and the detected intensity is close to zero. When the focal point lies exactly on the surface (9), i.e., when the probe is focused on to the surface, the reflected light is focused back on to the detector and passes through the pinhole, therefore, leading to an intensity peak on the photo-detector.

On a classical imaging confocal probe, the workpiece is mounted on a vertical scanning system that allows the focal point to cross the vertical range from below to above the surface. The photo-detector signal is monitored and the intensity value reaches a maximum at the focused point (see Fig. 5.2). It is, therefore, possible to correlate the intensity curve with the height of the surface.

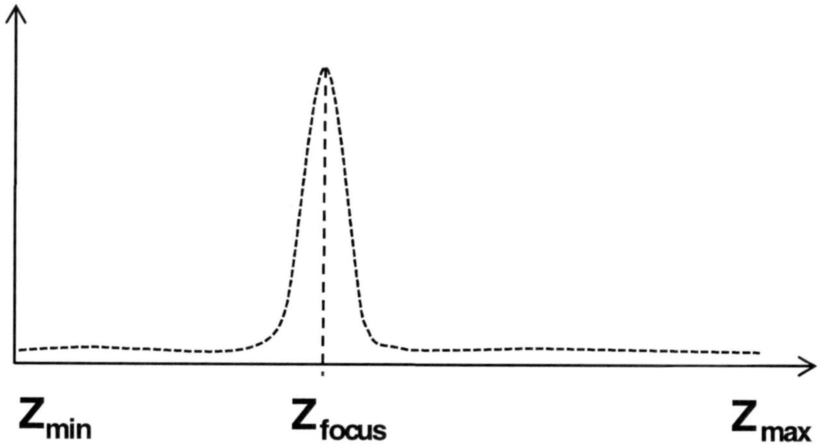

Fig. 5.2 Intensity curve registered by the photo-detector during the vertical scan

The chromatic confocal probe differs from an imaging confocal microscope in that it replaces the objective by a chromatic objective (to create chromatic dispersion), replaces the photo-detector by a spectrometer and does not require the vertical scanning system.

5.1.2 Axial Chromatic Dispersion

Chromatism is a physical property of almost all optical components, refractive, diffractive or gradient-index, and is usually something to avoid when designing optical instruments (see Sect. 2.2). Chromatism generates different focal points for different wavelengths, either along the optical axis (axial chromatism), out-of-axis on the focus plane, or both.

In a chromatic confocal probe, the axial chromatism is used as a space-coding method, by associating a different wavelength to each point of the optical axis within the vertical range, providing a mathematical relationship between the surface height (position of the focal point along the vertical axis) and the wavelength which is focused on the surface (see Fig. 5.3). This operation realises a spectral encoding of the measurement space (Perrin 1994, Cohen-Sabban 2001a).

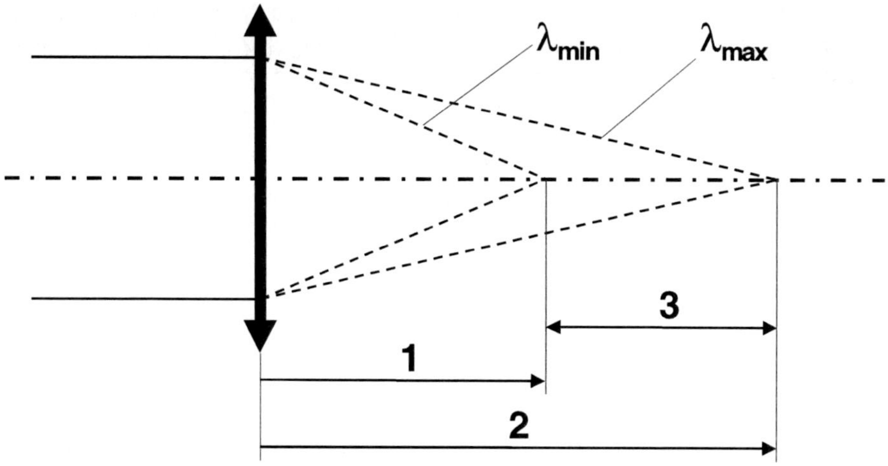

Fig. 5.3 Axial chromatic dispersion: each wavelength is focused at a different point along the optical axis

With reference to Fig. 5.3, short wavelengths, e.g., the blue around 400 nm, have short focusing distances and are focused closer (1) to the objective than long wavelengths, e.g., red around 800 nm (2).

The vertical range of the chromatic probe (in air) is defined by the distance between focal points of extreme wavelengths,

$$\text{vertical range} = \text{focal distance } (\lambda_{max}) - \text{focal distance } (\lambda_{min}). \tag{5.1}$$

The vertical range may vary depending on the acquisition frequency; usually it decreases with the frequency because less light is received by the detector at high frequencies.

The vertical scanning system required by a classical imaging confocal microscope is replaced by the axial chromatic dispersion. The maximum of intensity will occur at the focused wavelength, which can be detected using a spectrometer (instead of a photo-detector).

Spectral encoding was used in an optical profilometer for the first time in 1984 (Molesini 1984) on a low resolution system. It was later integrated into confocal single point probes to achieve high resolution in the z axis (Tiziani 1994).

5.1.3 Spectral Decoding

As described in Sect. 5.1.2, the reflected light is focused back on to a detector through a pinhole. If the focal point of the probe is in focus, the reflected light passes through the pinhole and is detected as an intensity peak. With chromatic dispersion, the white light is spread into individual wavelengths along the optical axis (Fig. 5.3), therefore, a single wavelength is in focus on the surface at a time. All wavelengths that are out of focus, above or below the surface, will be blocked by the detector pinhole and will not contribute towards generating intensity on the detector. On the contrary, the focused wavelength will pass through the detector pinhole and will generate an intensity peak. The detector must, therefore, be able to detect not only the intensity but also the wavelength at which the maximum of intensity occurs. This is usually carried out by a spectrometer.

The wavelength at which the maximum of intensity occurs can be associated with the surface height through calibration (see Fig. 5.4).

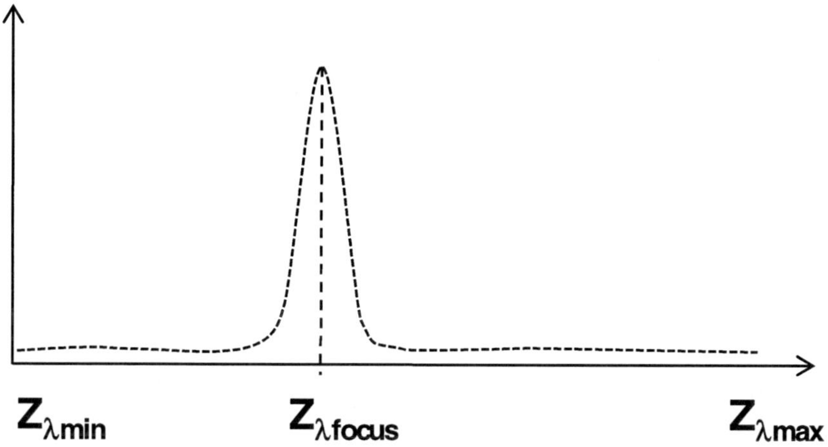

Fig. 5.4 Intensity curve registered by the spectrometer

The intensity curve is analysed for each measured point along the profile. When measuring a step artefact, e.g., the focal point switches alternatively between the top surface and the bottom surface of the grooves. Fig. 5.5 shows the measured spectrum and the corresponding profile. Each line of the spectrum image corresponds to the intensity curve seen on the spectrometer.

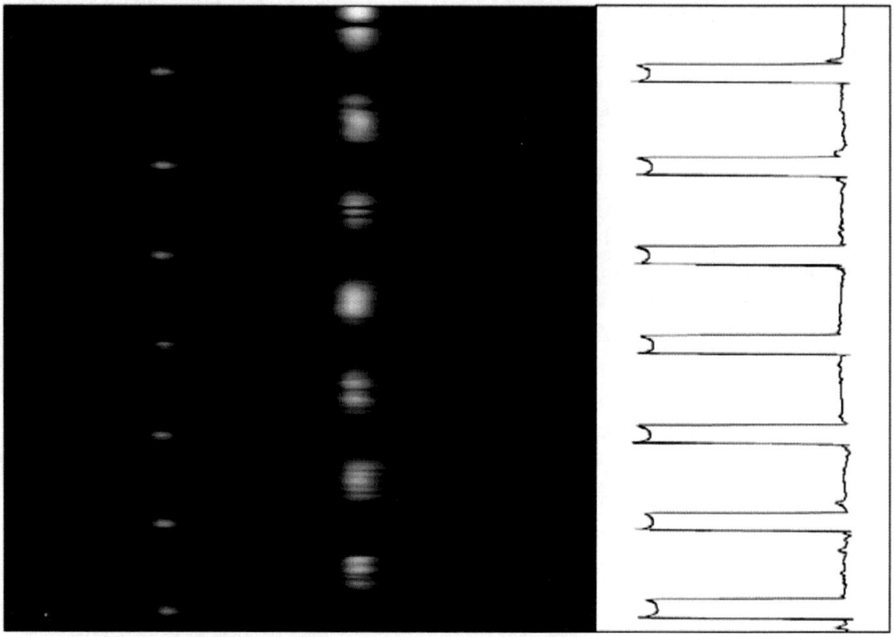

Fig. 5.5 Spectrum (left) and measured profile (right) of a step artefact

5.1.4 Height Detection

The spectrometer provides a curve of intensity with respect to the wavelength. The maximum of intensity of that curve allows the detection of the focused wavelength and, therefore, the surface height. A simple method to find the maximum intensity involves searching for the maximum point of the curve. This method may be used in real time systems where the measurement frequency is very high but it does not provide very high accuracy.

A more accurate method involves fitting a mathematical shape to the intensity peak in order to calculate the maximum of the fitted curve and provide sub-pixel accuracy on the peak abscissa. The mathematical shape can be a parabola or a Gaussian curve, or another type of curve. This method provides high accuracy but requires good computational resources which may not be available in the embedded electronics of commercial probes.

A good compromise consists in calculating the barycentre of the peak area, e.g., the portion above 50 % of the peak intensity. This method provides high accuracy and sub-pixel detection and yet requires low computational resources.

5.1.5 Metrological Characteristics

5.1.5.1 Spot Size

The spot size of the chromatic confocal probe has a direct influence on the lateral resolution of the measurement. It depends on several design characteristics:

- the numerical aperture of the objective;
- the magnification used in the optical head;
- the size of the pinhole (core diameter of the optical fibre); and
- the mean focal distance of the objective.

Typical spot sizes found on commercial probes are between 5 and 10 μm for vertical ranges smaller than 1 mm, and between 10 and 30 μm for larger vertical ranges. The intensity spread function inside a spot is close to a bell shape (Fig. 5.6) and can be described by a Bessel function.

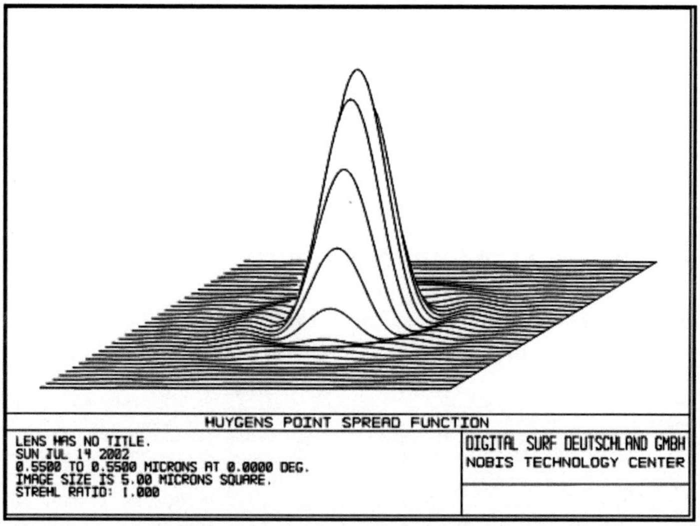

Fig. 5.6 Intensity spread function at focal point

When using a chromatic confocal probe, the user may think that the spot size is larger than specified because the visible spot seems much larger (around several tenths of a millimetre). This is due to an optical phenomenon called caustic, created by the adjunction of multiple light beams focused at different locations (see Fig. 5.7)

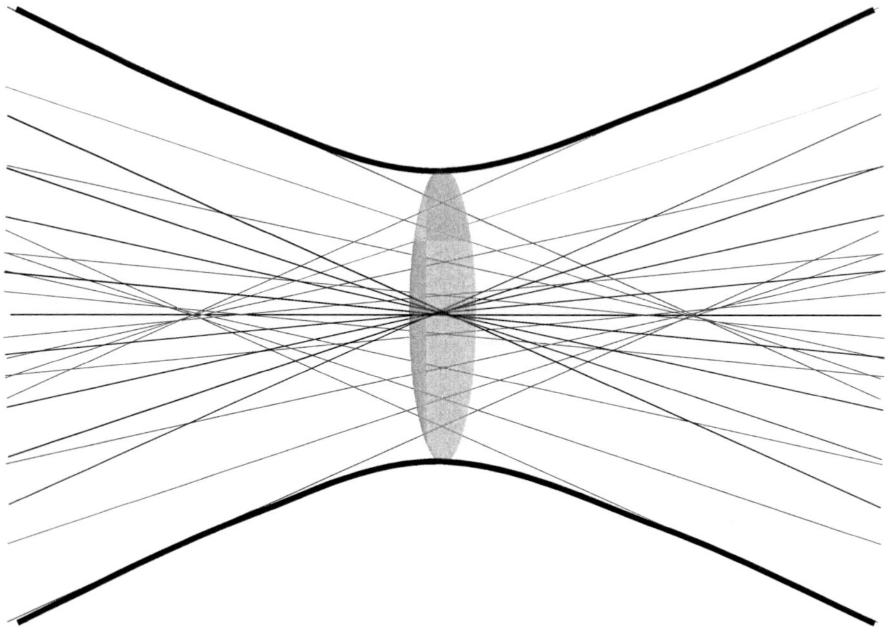

Fig. 5.7 Caustic created by all colour beams focused along the optical axis

5.2 Instrumentation

5.2.1 Lateral Scanning Configurations

As a single point measuring technique, the chromatic confocal probe can be installed on various scanning systems to measure 2D profiles, 3D surfaces, roundness profiles and freeform surfaces.

5.2.1.1 Profile Measurement

Chromatic confocal probes can be used on 2D profilometers to measure surface texture or step height. In such a configuration (see Fig. 5.8), an angled head is convenient (beam reflected at 90°) and could replace a stylus sensor on stylus profilometers. The angled head is compact and lightweight and is linked to the controller by an optical fibre.

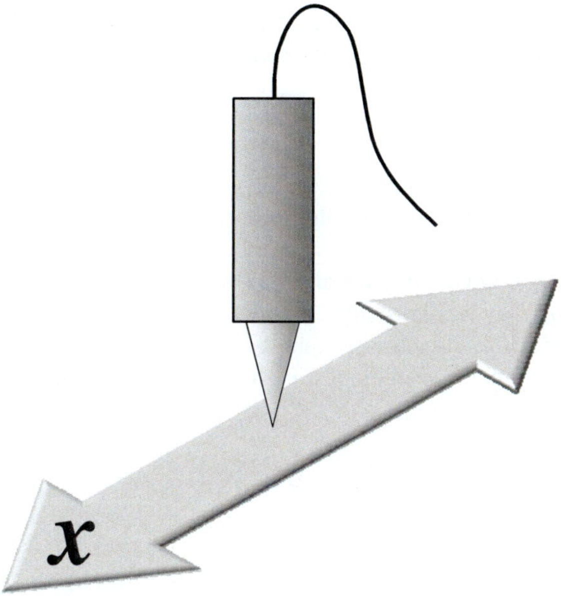

Fig. 5.8 Chromatic confocal probe and a linear scanning system

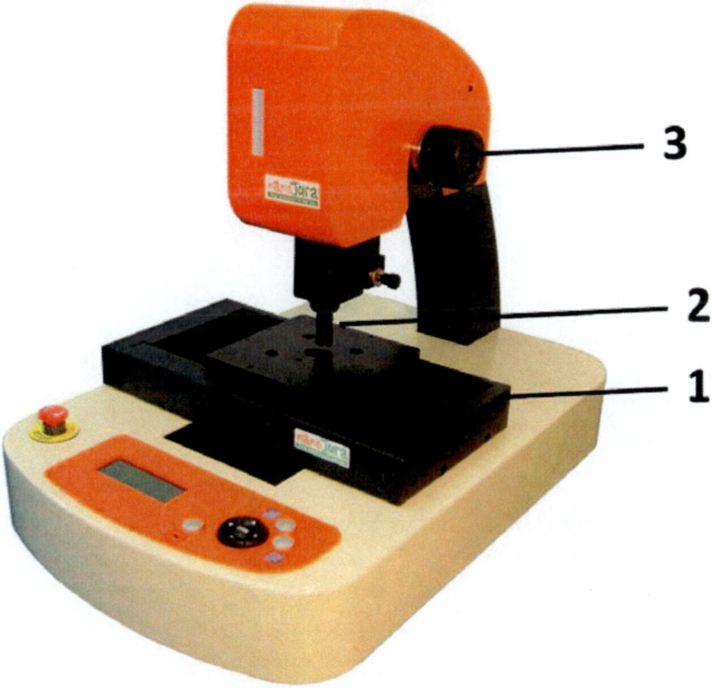

Fig. 5.9 Example of a 2D non-contact profilometer with a chromatic confocal probe. Source nanoJura (www.nanojura.com)

Figure 5.9 shows a simple non-contact profilometer with an x axis stage (1) of 100 mm range, a chromatic confocal probe (2) and a manual z axis stage.

5.2.1.2 Areal Measurement

Chromatic confocal probes can be used for areal measurement of surface texture (see Fig. 5.10). In some areal configurations, the probe is moved in the x axis and the workpiece is moved in the y axis. In other configurations, the probe is fixed and the workpiece is moved in the x and y axes.

Areal scanning may be bidirectional, with one line scanned in one direction and the next in the opposite direction, so that the scan is faster because the probe does not have to return to the beginning of each line before scanning.

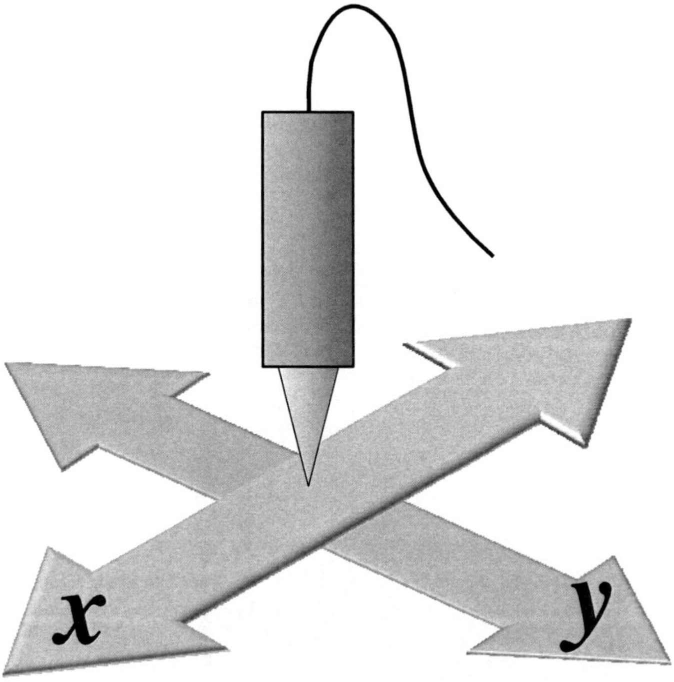

Fig. 5.10 Chromatic confocal probe and *xy* stage

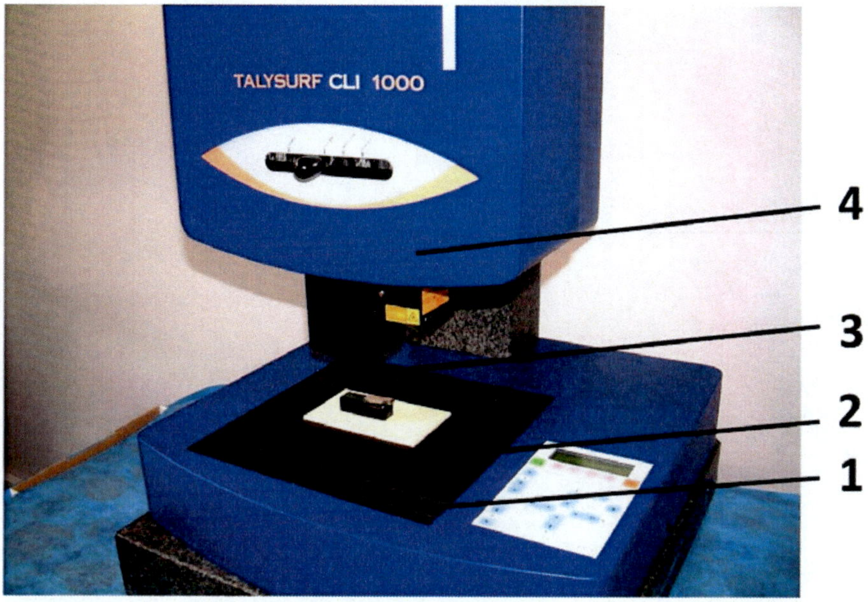

Fig. 5.11 Example of an areal contact and non-contact profilometer with several chromatic confocal probes. Source Taylor Hobson (www.taylor-hobson.com)

Figure 5.11 shows an example of an areal profilometer that offers several contact and non-contact probes. The sample is moved by an x axis stage (2) and a y axis stage. The selected probe (3) is brought into focus with the z axis stage that is installed inside the cover (4).

5.2.2 Optoelectronic Controller

Figure 5.12 is a schema of a typical commercial chromatic confocal probe, with an optical head (1) connected to an optoelectronic controller (3) via an optical fibre (2). The controller contains a light source (5), a beam-splitter, or optical coupler (4), and a spectrometer (6).

It has been demonstrated (Sandoz 1993) that the two conjugate pinholes can be replaced by an optical fibre connecting the light source at one end to the chromatic objective at the other. The small diameter of the optical fibre (typically 50 μm) makes a good pinhole and makes it possible to mount the light source remotely from the measured surface. As white light sources usually generate high temperatures, they need to be cooled using a fan that generates vibration and must be separated from the measurement part of the instrument.

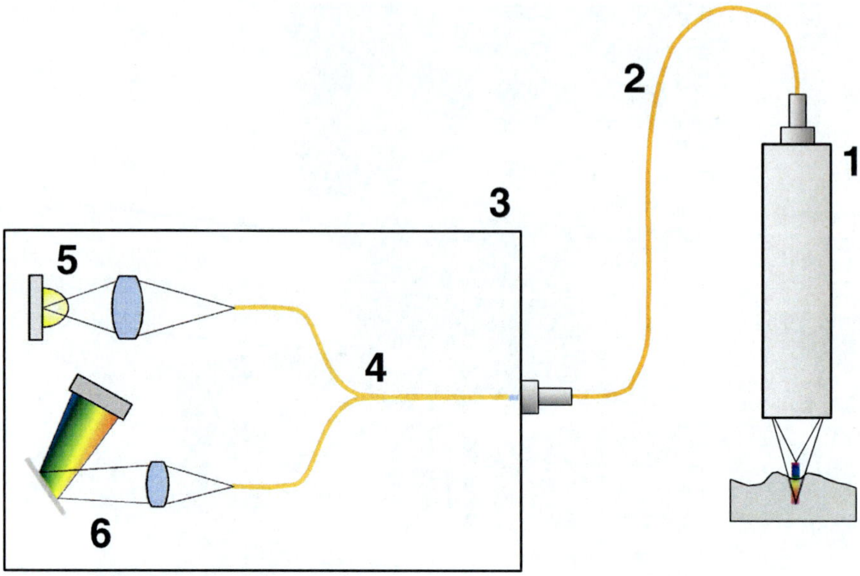

Fig. 5.12 Schema of a chromatic confocal probe

The light emitted by the light source is transferred to the other end of the optical fibre, forming a point source that is then focused onto the surface through a chromatic objective. The reflected light is then focused back onto the optical fibre creating an object pinhole. The reflected light is then sent to a spectrometer through a beam-splitter.

Current chromatic confocal probes are typically made of two main components:

- an optical head which is mounted on a fixture or associated with a lateral scanning system; and
- an optoelectronic controller.

The optoelectronic controller is connected to a PC for post-processing and imaging.

Figure 5.13 shows an example of an optoelectronic controller. It is a module that can be plugged into Volcanyon® controllers along with modules for stage drivers and other probe acquisition electronics. The Nobis® controller provides two inputs and has a white LED source that can easily be exchanged from the outside. The electronic boards contain the optical coupler, the spectrometer and the digital signal processing (DSP) board that contains the firmware for the detection of intensity peaks. This module is used by several instrument manufacturers.

Fig. 5.13 Nobis® opto-electronic module from Digital Surf (www.digitalsurf.com)

5.2.3 Optical Head

The optical head on a chromatic confocal probe projects the source pinhole to the workpiece surface and images the spot.

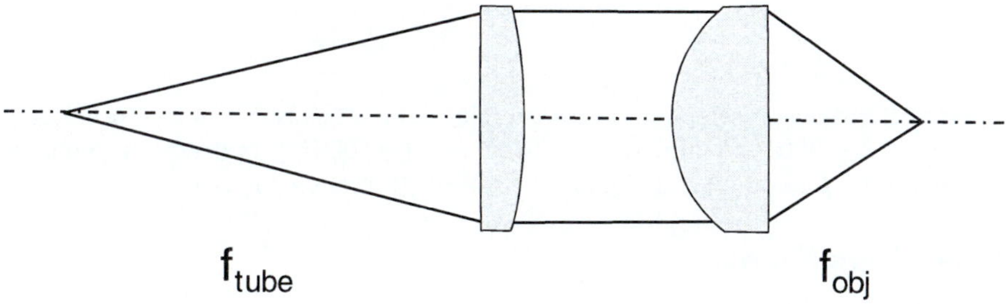

Fig. 5.14 Schema of the optical head

Depending on the manufacturer, the head may contain two or several lenses (Fig. 5.14), but in principle, there is a tube lens whose focal point is positioned on the optical fibre and whose role is to create a parallel beam, and an objective lens that generates chromatic dispersion along the optical axis with a focal point at different distances depending on the wavelength.

Fig. 5.15 Examples of optical heads

Fig. 5.15 shows a selection of commercial optical heads. From top to bottom: Nobis® CLA3000 (3.5 mm range), Nobis® CLA1000 (1 mm range), Nobis® CLA400 (400 μm range) and Nobis® CLA400A (90° angled head).

5.2.4 Light Source

As chromatic dispersion is required, the light source must provide the widest possible bandwidth between wavelengths where the chromatic objective creates the largest dispersion. The intensity of each wavelength must be as constant as possible to allow good linearity in the detection.

Technologies of light sources include:

- halogen bulbs;
- xenon bulbs;
- white LEDs;
- multi monochromatic LEDs; and
- supercontinuum light (Shi 2004).

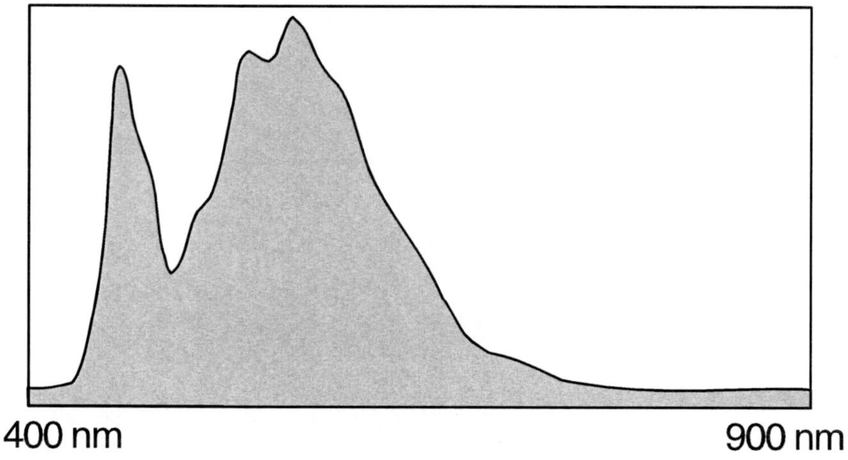

Fig. 5.16 Spectrum of a white LED between 400 nm and 900 nm of wavelength. The spectrum displays a peak at 450 nm (blue)

Halogen and xenon sources produce high intensity and have a wide bandwidth but their bulb requires cooling devices, such as a fan, to reduce their high operating temperature.

White LEDs are used in modern instruments as they provide much longer lifetimes and are cheaper, but their spectrum is not as regular as halogen bulbs (Fig. 5.16).

5.2.5 Chromatic Objective

The aim of the chromatic objective is to generate the chromatic dispersion of white light along the optical axis, in order to have focal points of each wavelength spread at different distances from the lens. Commercial probes usually use an aspheric lens, either a standard off-the-shelf lens or custom-designed.

In principle, any dispersive device could be used as the chromatic objective, such as diffractive lenses (Dobson 1997, Lin 1998) or dynamic wavelength tuning systems. However, the quality of the dispersion and the cost of non-conventional objectives rules out their use in commercial systems. The dispersion depends also on the material used for the lens.

5.2.6 Spectrometer

The space decoding is usually carried out by a spectrometer (see Fig. 5.17) in commercial chromatic confocal probes. The spectrometer contains a lens (2) focused on to the optical fibre connector (1), a grating (3) that will deviate wavelengths laterally, and a spherical mirror (4) to concentrate the light on a linear CCD sensor (5). A simple prism could also be used as a spectrometer but would be less convenient and less compact.

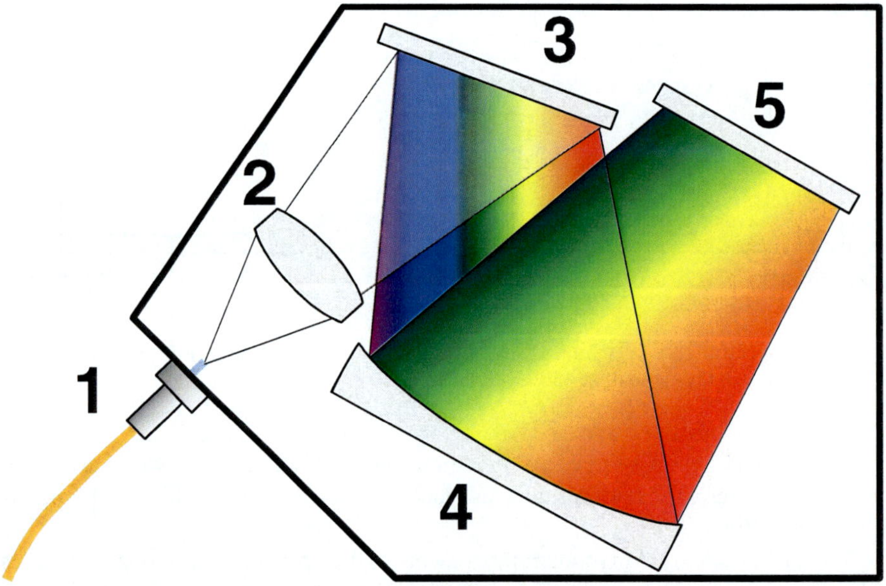

Fig. 5.17 Schema of a spectrometer

The linear CCD collects the light of the dispersed wavelengths, usually from blue to red. Each pixel provides the intensity for a given wavelength. The peak corresponds to the focused wavelength (see Fig. 5.18).

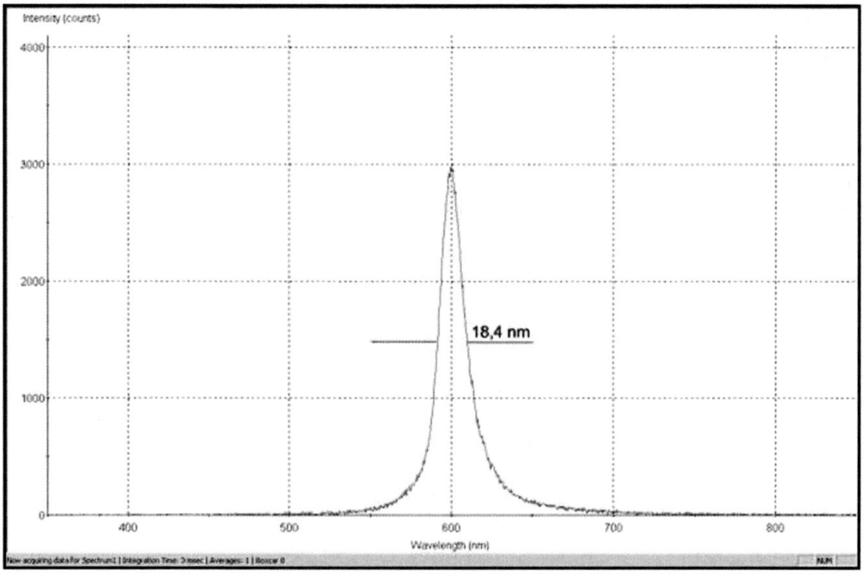

Fig. 5.18 Peak observed on the spectrometer

5.2.7 Optical Fibre Cord

Fig. 5.12 shows that the probe contains at least one external optical fibre (2) and one optical coupler (4). The optical coupler conducts the light from the source to the head, and the reflected light back to the photo-detector. Some implementations use an X-coupler instead of a Y-coupler, to allow the connection of two different heads at the same time, providing that only one is used at a time.

As explained in Sect. 5.1.6.1, the spot size is directly influenced by the size of the pinhole, which is determined by the core diameter of the optical fibre. Common multimode fibres have a core diameter of 50 μm. It is possible to achieve a smaller spot size by using an optical fibre with a core diameter of 10 μm, with the consequence that the probe receives a lower light intensity.

5.3 Instrument Use and Good Practice

5.3.1 Calibration

5.3.1.1 Calibration of Dark Level

When the head is blind, that is when no light enters through the objective lens, the detector can still detect a signal. This signal is called the dark level. The dark level is generated by internal stray light reflected off connectors and optical interfaces, by electronic noise of the components, and by the photonic noise of the detector itself. The dark level depends on the ambient temperature and the internal temperature of components (which depends on their clock frequency).

The dark level is composed of all signal values measured on each wavelength, for example, on each pixel of the CCD on the spectrometer. In order to remove this contribution, an initial calibration should be performed, before other calibrations and before measuring any data. The calibration should be performed once the electronics has reached its operating temperature, usually after a few minutes.

Software algorithms capture the dark level and store it in memory so that it can be subtracted during measurements.

5.3.1.2 Linearisation of the Response Curve

The linearisation of a chromatic confocal probe is the process that establishes a linear relationship between the z axis position of the sample and the z axis position calculated from the peak detected on the spectrometer. The response curve R relates these two quantities via

$$z_{measured} = R\,(z_{sample}). \qquad (5.3)$$

Several internal contributions that influence the response curve can be identified:

- the dispersion characteristics of the chromatic objective (which spreads focal points along the optical axis as a function of the wavelength);
- the angle of the optical fibre from the optical axis;
- the geometry of the spectrometer (position of the reflectors, response of the grating, etc.); and
- the response of the photo-detector.

Fig. 5.19 shows an example of a response curve. This curve contains three domains A, B and C. Domains A and C correspond to z axis positions where the probe is not able to detect any peak. The linearisation algorithm will fit a known curve onto the measured points of domain B.

A polynomial of a given order (or any equation) can be fitted onto the points and then inverted in order to linearise the curve.

Fig. 5.20 shows how the three linearisation steps are carried out in practice. On the left, a known mathematical curve F is fitted to the measured points. In practice the curve approximates the repeatable local variations of the response curve. In the middle, the fitted curve is then transposed (F^T) and stored in memory. On the right, the linear curve L is obtained by applying the transposed curve to the response curve points.

$$L = F.F^T. \qquad (5.4)$$

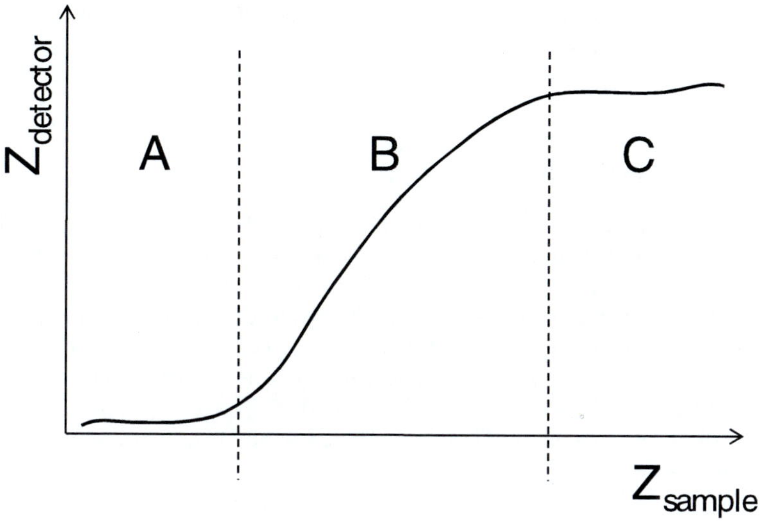

Fig. 5.19 Response curve of the chromatic confocal probe

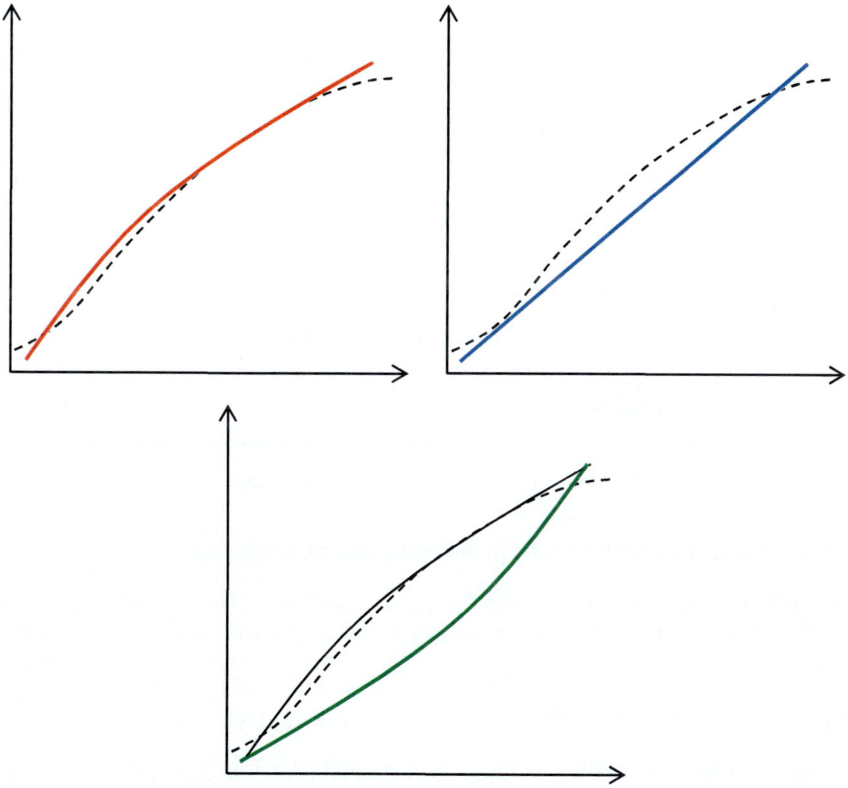

Fig. 5.20 Linearisation process. Left: F curve fitted to the response curve R points. Middle: transposed curve (F^T). Right: linear theoretical response curve L

The following relationship can then be applied onto all measured points in order to get the linearised values

$$z_{linearised} = F^T(R(z_{sample})). \tag{5.5}$$

5.3.1.3 Calibration of the Height Amplification Coefficient

The height amplification coefficient is the slope of the linearised response curve. Calibration of height amplification is carried out in exactly the same way as with stylus probes or another optical single point probes (and see Chap. 4).

Gauge blocks can also be used to create a step of a given height. The upper gauge block (G1) is wrung onto another one (G2). The combined calibrated length (used here as a height) provides a reference step. The use of metal blocks is recommended because their surface reflects most of the incoming light back into the head and the detector (Fig. 5.21). Ceramic gauge blocks are not suitable for this purpose due to their translucency.

The length of block G1 is selected to be a function of the vertical range of the probe, typically one tenth of the range (e.g., 0.1 to 0.2 mm for a 1 mm range head).

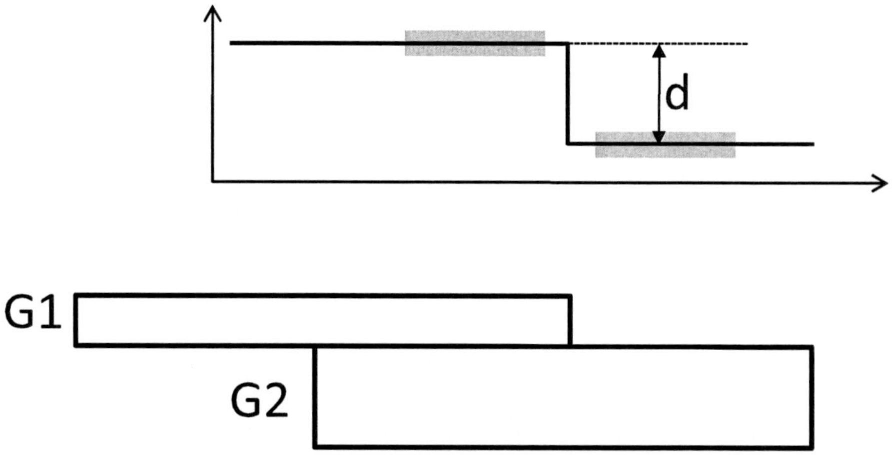

Fig. 5.21 Measurement of a two gauge block step height

5.3.1.4 Calibration of the Lateral Amplification Coefficient

As the lateral amplification coefficient depends only on the lateral scanning system, its calibration can be carried out in the same manner as with a stylus profilometer (see Chap. 4).

5.3.1.5 Calibration of the Hysteresis in Bi-directional Measurement

The measurement process for a chromatic confocal probe consists in capturing the height of a point and moving the probe (or the workpiece) along a profile. The height evaluation consists of a two-step process:

1. Capturing the reflected light on the detector (spectrometer).
2. Calculating the height from the spectrometer curve.

Usually step 2 for point n is carried out at the same time as step 1 for point $n + 1$. Therefore, the height information is given one step later than the light capture. During this period of time, the probe is still moving and this creates a small shift, dx, along the x axis, whose value depends on the acquisition frequency.

When measuring bi-directionally, the shift is doubled as it is generated in the opposite direction, creating a shift between two lines of $2dx$. This value can be corrected automatically by the acquisition program, either after calibration or using predefined values.

5.3.2 Preparation for Measurement

Before starting a measurement with a chromatic confocal probe, the following steps are carried out:

- The z axis column is adjusted to focus the probe on to the workpiece.
- The frequency is adjusted by looking at the spectrometer peak value to ensure enough light is received (or to avoid too much light).
- The z axis column is adjusted to have the signal centred in the range, or in some cases, to have the signal at the top or bottom of the range, depending on the geometry of sample.

5.3.3 Pre-processing

A raw measurement file is often affected by local measurement defects that need to be removed. This process can be thought of as an initial "cleaning" of the surface before metrological analysis.

This cleaning process includes:

- Removing outliers in the data by applying an appropriate filter (for example a median or alternate sequential morphological filter).
- Filling in non-measured points in small non-measured areas (larger areas may be holes or surrounding areas of a non-rectangular sample).

Following the steps above, the metrological pre-processing can be carried out that consists of:

- Levelling (using a least-squares plane, minimum zone, levelling from a selected area, etc.).
- Micro-roughness filtering (λs filter on a profile, S-filter on an areal surface).

5.4 Limitations of the Technique

5.4.1 Local Slopes

The light focused on the workpiece is reflected towards the detector for analysis. However, a flat and smooth surface acting as a mirror may reflect the light outside of the objective lens (specular reflection). In Fig. 5.22, a ray of light coming through zone 2 will be reflected back through the objective lens and to the detector. A ray coming through zone 1 will be reflected outside the objective. The loss in the quantity of light is proportional to the slope of the surface and the numerical aperture of the objective lens.

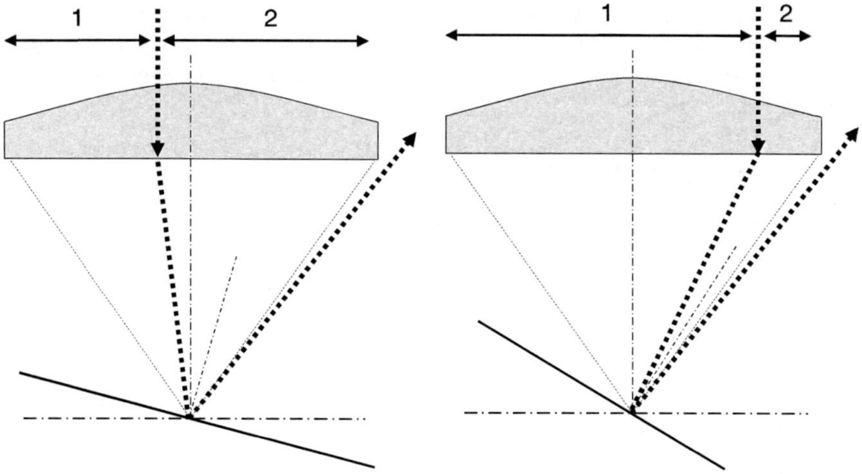

Fig. 5.22 Left: only the light coming through zone 2 is reflected to the detector. Right: almost all light is reflected outside the detector

If the detector does not receive enough light, the detection algorithm will not be able to detect the peak position and will generate non-measured points.

The maximum slope, α_{max}, allowed in the case of specular reflection is equal to half of the aperture angle, which is linked to the numerical aperture of the lens (see Tab. 5.1 and equation (5.6))

$$\alpha_{\max} = \sin^{-1} A_N \,. \tag{5.6}$$

Table 5.1 Maximum measureable slope against numerical aperture

Numerical aperture	Maximum slope (specular case)
0.3	±18°
0.4	±24°
0.5	±30°
0.6	±37°
0.7	±44°

In practice, surfaces usually have roughness, creating a diffuse reflection (see Fig. 5.23). A part of the light will be reflected back to the detector, even if the surface slope exceeds the maximum slope of the specular case. Slopes can be measured up to 80° on diffuse material, such as matt metal, plastics, rubber, etc.

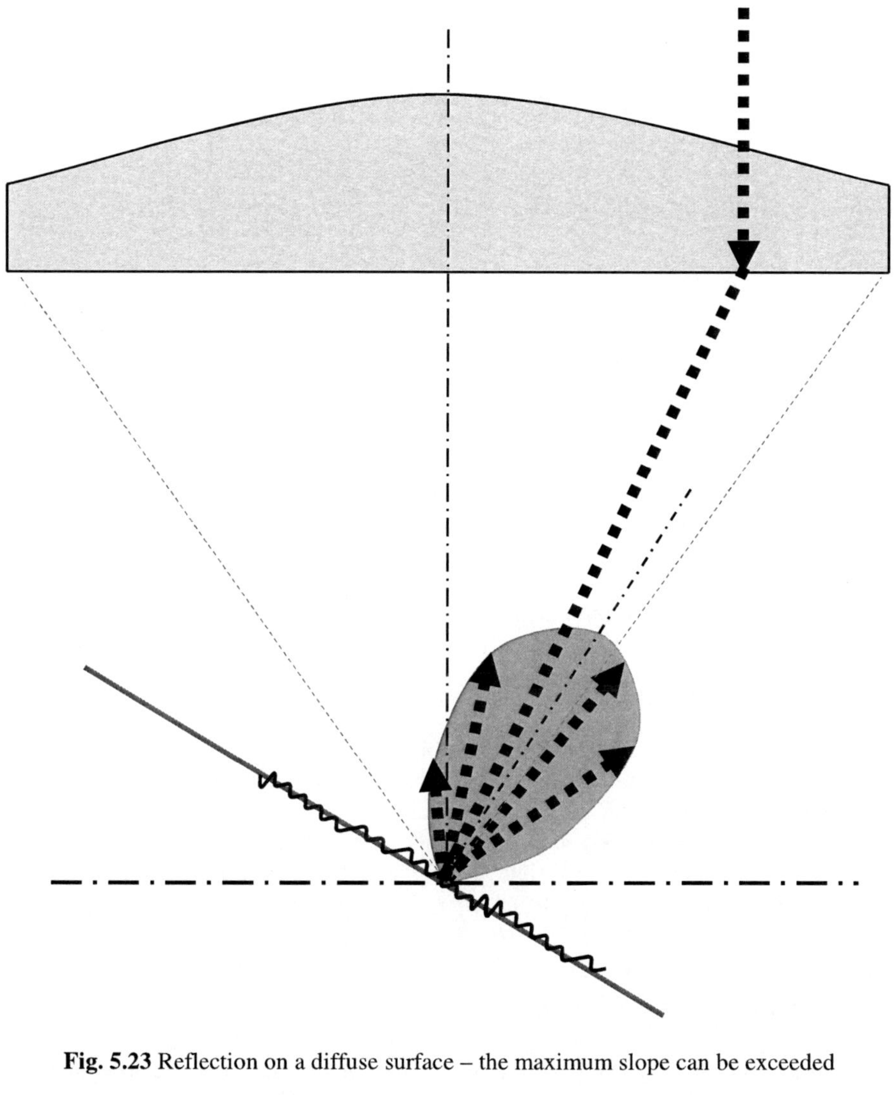

Fig. 5.23 Reflection on a diffuse surface – the maximum slope can be exceeded

Other factors may influence the maximum measurable slope, e.g., the power of the light source, the colour or reflectivity of the workpiece and acquisition frequency.

5.4.2 Scanning Speed

The scanning speed is a parameter which is usually selected by the user, as a function of the capabilities of the *x* axis stage. One speed limitation is a result of the acquisition frequency of the probe f_{probe}. This frequency determines the exposure time during which the detector accumulates light coming back from the sample. The exposure time is the inverse of the acquisition frequency

$$t_{exp} = \frac{1}{f_{probe}}. \tag{5.7}$$

During t_{exp}, the *x* axis stage moves a distance d_{exp} that depends on the speed s_{stage}

$$d_{exp} = s_{stage} t_{exp}. \tag{5.8}$$

For example, with a frequency of 200 Hz and a scanning speed of 1 mm s^{-1}, the distance covered during acquisition of each point is 5 µm.

This distance covered during acquisition has two effects:

- it applies a low-pass filter to the *x* axis, similar to a λs filter or the mechanical filtering of a stylus probe; and
- it limits the lateral spacing and, therefore, the number of points of the profile.

These two effects can limit the lateral resolution of the measurement.

5.4.3 Light Intensity

As with any optical measurement technique that uses a photo-detector, the light intensity received by the detector must be within two limits. Firstly, if not enough light is received, there will be no peak detected above the noise level and, therefore, no *z* axis position calculated (non-measured point – see Sect. 5.4.4). Secondly, if the light level is too bright, the detector may become saturated and the peak will be truncated leading to an erroneous calculation of the *z* axis position (outlier – see Sect. 5.4.5).

5.4.4 Non-measured Points

If the light intensity received on the detector (spectrometer) is too low, surface height cannot be detected. In such cases, the probe delivers an appropriate error code meaning that the current point was not measured. Non-measured points can result from many different causes, some of which are listed in Tab. 5.2.

Table 5.2 Typical causes of non-measured points and potential solutions

Typical causes of non-measured points	Potential solutions
the surface height is above the upper limit (or below the lower limit) of the vertical range, so that the focal point cannot be found	move the z axis stage down (or up) the surface until it comes back inside the vertical range of the probe
the focal point falls into a hole or outside the workpiece	none
the local slope is too steep so that all the reflected light is sent outside the objective	in some cases, increase the light intensity or decrease the measurement frequency
the material is such that, the reflected light intensity is too low	increase the light intensity or decrease the measurement frequency
the material is too shiny so that the reflected light intensity is too high and saturates the detector	decrease the light intensity or increase the measurement frequency

As the surface is measured by scanning profiles, specific post-processing may be applied to the measured profiles to eliminate non-measured points if necessary. Usually, a small number of consecutive non-measured points can be eliminated as they are likely to be due to bad detection by the probe during the scan, although a large number of consecutive non-measured points probably means that the probe was outside the workpiece area (e.g., when measuring a non-rectangular workpiece or a workpiece that contains a hole).

5.4.5 Outliers

In some cases, the intensity curve measured by the spectrometer does not allow correct evaluation of the surface height. If the calculation is not possible at all, the probe delivers a non-measured point. If a peak is high enough above the noise level to be detected, the surface height is then correctly calculated. In all other cases, between no point and a correct point, there are cases where a value for surface height is calculated although it is not correct. These cases create outliers - points that are significantly above or below the surface.

Outliers may be created in the following cases:

- The curve displays no peak but the noise level is above the threshold, so any point above the threshold may be misinterpreted as a peak.
- The peak shape is distorted by spurious light reflections or other optical aberrations, so that the calculation is affected by a significant error.
- Multiple peaks occur due to interference, semi-transparent material or other causes, leading to bad or incorrect detection.

- The local slope is close to the limit, so that the intensity is slightly above the threshold.
- The local curvature creates spurious peaks due to ghost reflections.

Outliers are responsible for sudden changes in the slope measured on the surface (for example, on steps or rectangular grooves).

Outliers may be eliminated using specific filters (such as median filters) applied to each scanned profile. The ISO/TS 16610 part 49 (2006) standard also describes useful filters called alternating symmetrical morphological filters that can eliminate outliers. Such post-processing may be available as an option of the acquisition software or the analysis software.

5.4.6 Interference

When measuring on a transparent material that has thickness which is small compared to the probe range (such as on thin films), interference may be created between the upper and lower interfaces. The result is a modulation of the peaks on the spectrometer and errors in the determination of the z axis position.

5.4.7 Ghost Foci

When the surface of the sample is locally spherical, the focal point may be detected at the centre of the sphere instead of on the surface itself (see Fig. 5.24). The probe generates outlier points at the bottom of circular grooves when the reflection is specular. These outlier points are known as ghost foci.

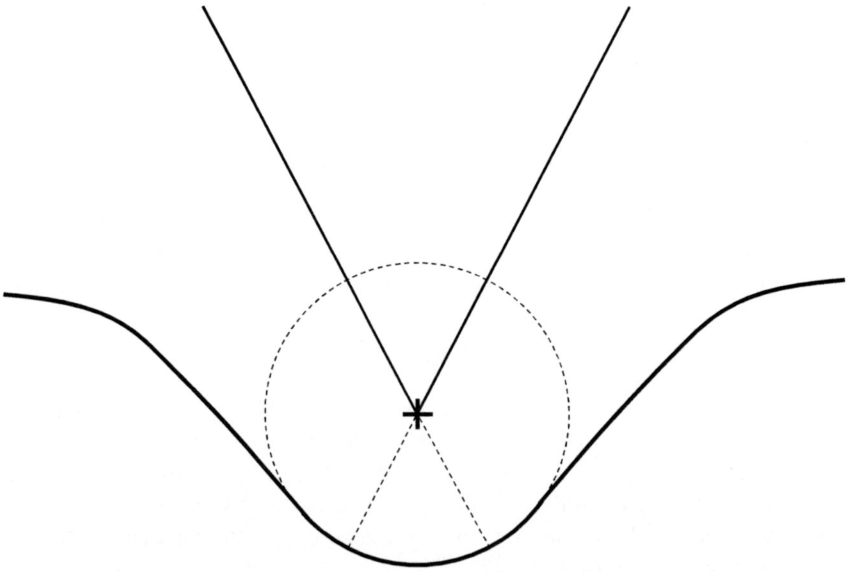

Fig. 5.24 Ghost focus on a circular groove

5.5 Extensions of the Basic Principles

5.5.1 Thickness Measurement

When measuring a surface inside a transparent material, the refractive index of the material must be taken into account when calculating the true depth (see Fig. 5.25).

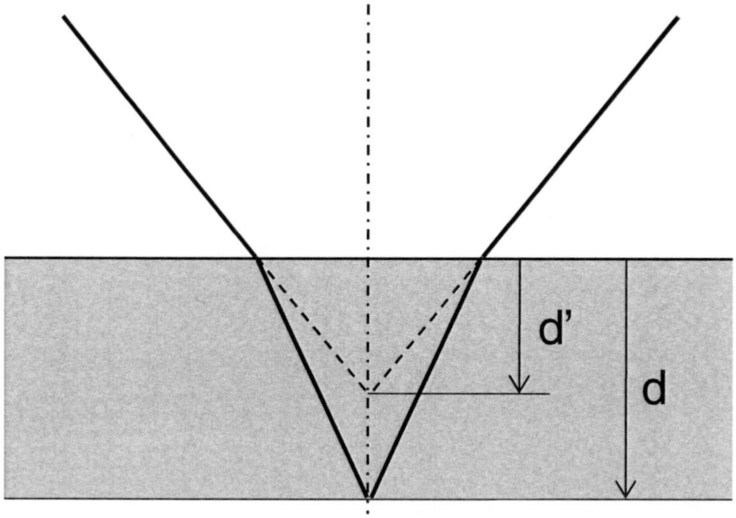

Fig. 5.25 The sensor "sees" a depth *d'* while the true thickness is *d*

The measured depth *d'* must be corrected using the following equation

$$d = \frac{d'n'}{n} \tag{5.9}$$

where n is the refractive index outside the material (typically in air, n is approximately unity) and n' is the refractive index of the transparent material (for example, in polymethyl metacrylate (PMMA) $n' = 1.49$).

In practice, when measuring a transparent material, the incident beam may have two focal points (see Fig. 5.26); one in the upper interfacial surface and one in the lower interfacial surface.

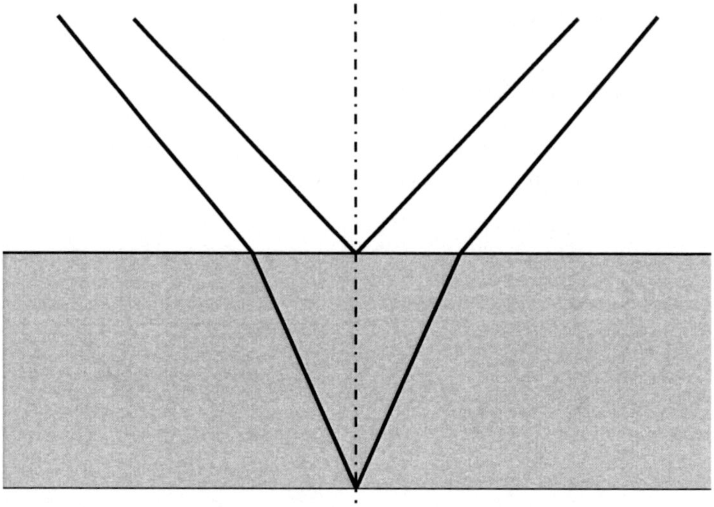

Fig. 5.26 Measurement of the two interfacial surfaces at a time

In this case, the two focused wavelengths will pass through the detector pinhole and will be detected by the spectrometer, producing two peaks in the intensity curve (see Fig. 5.27).

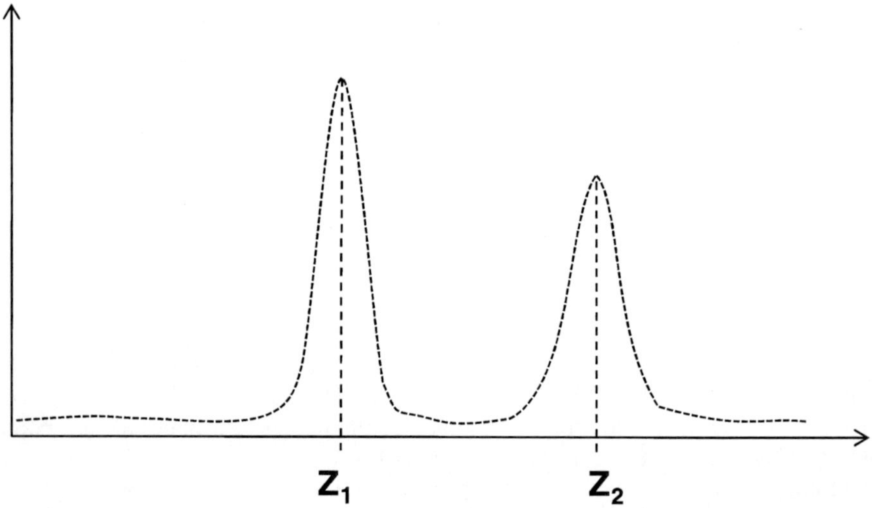

Fig. 5.27 Thickness measurement: Z_1 is the upper surface, Z_2 is the lower surface

The thickness d is given by the following relationship

$$d = \left| Z_1 - Z_2 \frac{n}{n'} \right|. \tag{5.10}$$

5.5.2 *Line and Field Sensors*

Some manufacturers have extended the single point sensor in order to measure lines or even areas (Cha 2000, Ruprecht 2004). Currently, commercial linear or areal probes use oscillating mirrors - they are not really true line or areal sensors as there is a lateral scanning device which is integrated into the optical head in order to provide a compact system that can be mounted onto a coordinate measuring machine or another lateral scanning system.

Line or areal sensors lose two properties of the original chromatic confocal setup:

- they not static because of the oscillating mirror; and
- the beam is not projected onto the surface parallel to the optical axis due to the sweep angle.

5.5.3 *Absolute Reference*

The ability to measure two interfaces through a transparent material leads to an ability to automatically correct the straightness of the lateral scanning system (Cohen 2001b, Vaissière 2010).

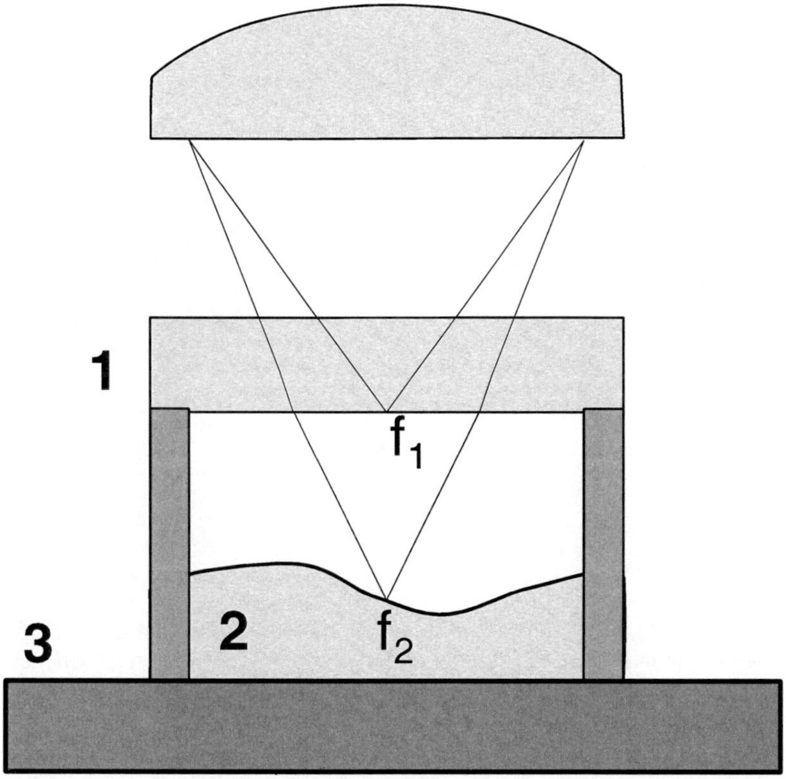

Fig. 5.28 Principle of the absolute reference

A reference glass flat (1) is installed above the sample (2) and attached to the x axis stage (3) (Fig. 5.28). The probe is brought to a position in z axis where a first focus is obtained on the lower interface of a reference glass flat, and a second focus is obtained on the surface of the sample. The two distances are recorded simultaneously and subtracted from each other.

The straightness errors of the x axis stage adds a contribution in z axis on both measurements given by

$$z_{f1} = z_{glassflat} + z_{straightness}, \tag{5.11}$$

$$z_{f2} = z_{sample} + z_{straightness}, \tag{5.12}$$

$$\begin{aligned} z_{measured} &= z_{f2} - z_{f1} \\ z_{measured} &= z_{sample} - z_{glassflat} \end{aligned} \tag{5.13}$$

Subtracting the height measured on the reference glass from the height measured on the sample removes the component due to the straightness errors of the stage (see Fig. 5.29).

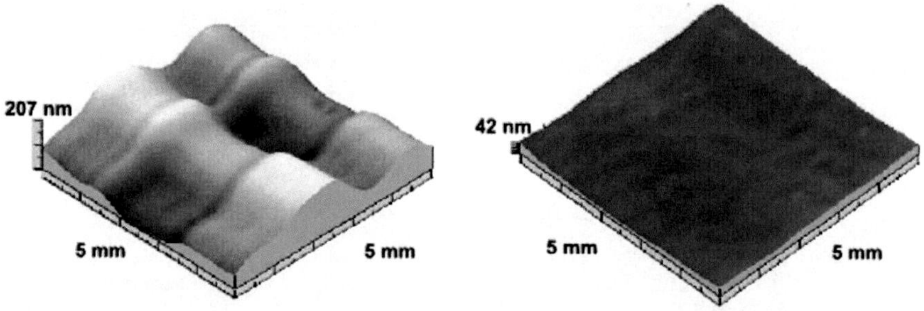

Fig. 5.29 Measurement on a $\lambda/10$ glass flat. Left: without an absolute reference. Right: with an absolute reference (Vaissière 2010)

5.6 Case Studies

The following examples were measured with a Nobis® chromatic confocal probe and various optical heads, installed on 3D scanning profilometers. Images were generated using the analysis software package MountainsMap® version 6.0 (website: www.mountainsmap.com).

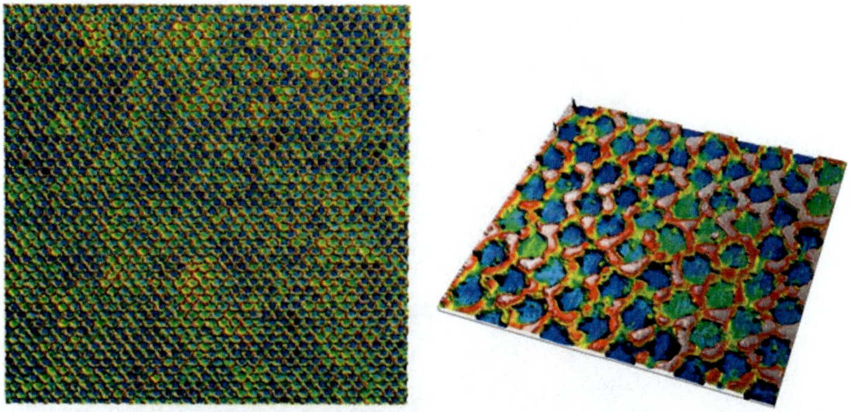

Fig. 5.30 Electron beam textured metal sheet
Measured with a Nobis® CLA400 probe on a silicon rubber imprint
Left: 10 mm by 10 mm. Vertical range: 50 μm
Right: 2 mm by 2 mm. Vertical range: 25 μm

Fig. 5.31 Detail of a brass medal
Measured with a Nobis® CLA3000 probe at 1000 Hz
36 mm by 36 mm. Vertical range: 2 mm

Fig. 5.32 Micro-electro-mechanical system
Measured with a Nobis® CLA1000 at 500 Hz
4.5 mm by 4.5 mm. Vertical range: 15 μm

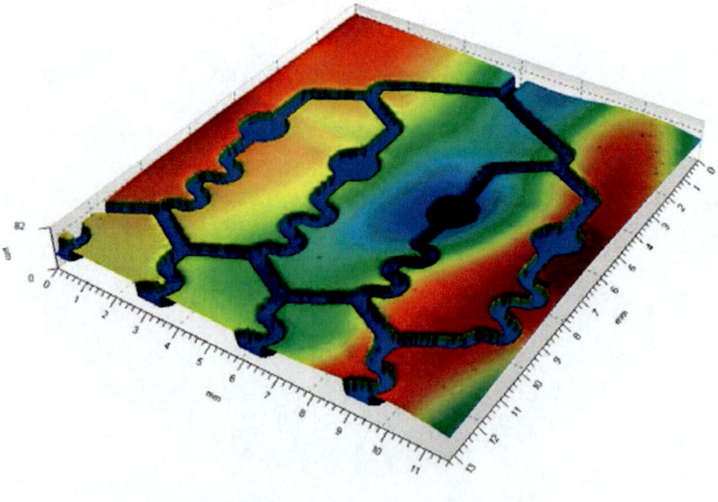

Fig. 5.33 Microfluidics component
Measured with a Nobis® CLA400 probe at 200 Hz
11.7 mm by 13.4 mm. Vertical range: 82 μm

Fig. 5.34 EPROM die
Measured with a Nobis® CLA400 probe
Lateral spacing: 1 μm; original measure contained 3000 lines of 1800 points measured at low speed in 16 hours. Source: nanoJura (www.nanojura.com)

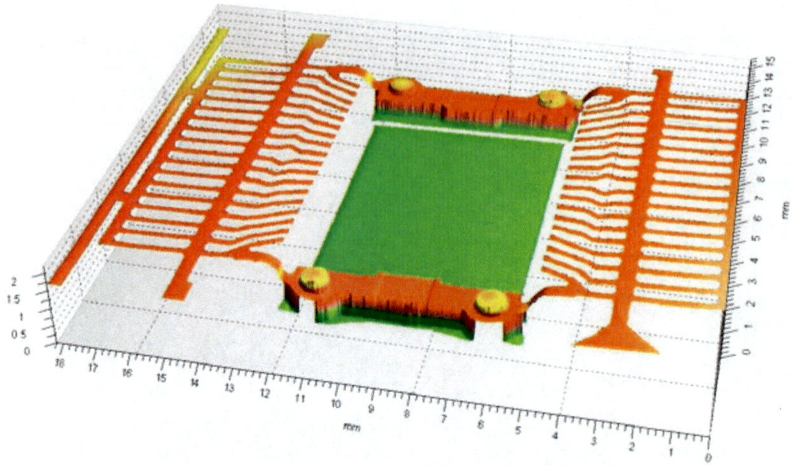

Fig. 5.35 Lead frame around an integrated circuit die
Measured with a Nobis® CLA-3000 probe
18 mm by 15.5 mm. Vertical range: 2.25 mm. Source: nanoJura (www.nanojura.com)

Fig. 5.36 Paper fibers
2 mm by 2 mm. Vertical range: 600 μm

Fig. 5.37 Abrasive paper, grade 180
5 mm by 5 mm. Vertical range: 200 μm

Acknowledgements

The author would like to thank Dr Hans-Joachim Jordan (Digital Surf Deutschland GmbH) for providing data on optical heads, Mrs Isabelle Cauwet, Mr Antony Caulcutt and Mr Benoît Moritz (Digital Surf) for providing examples and illustrations, and the members of the French committee UNM 09 G5 for their contribution to the draft standard ISO 25178 part 602 (2010) during its preparation.

MountainsMap®, Volcanyon® and Nobis® are registered trademarks of Digital Surf, France (www.digitalsurf.com).

References

Cha, S., Lin, P., Zhu, L., Sun, P.-C., Fainman, Y.: Nontranslational three-dimensional profilometry by chromatic confocal microscopy with dynamically configurable micromirror scanning. Appl. Opt. 39, 2605–2613 (2000)

Cohen-Sabban, J., Gaillard-Groleas, J., Crepin, P.-J.: Quasi confocal extended field surface sensing. In: Proc. SPIE, vol. 4449, pp. 178–183 (2001a)

Cohen-Sabban, J., Poizat, D., Vaissière, D.: Colloque de la Société Française d'Optique: Méthodes et techniques optiques pour l'industrie, Trégastel (2001b)

Dobson, S.L., Sun, P.-C., Fainman, Y.: Diffractive lenses for chromatic confocal imaging. Appl. Opt. 36, 4744–4748 (1997)

ISO 5436 part 1, Geometrical product specification (GPS) — Surface texture: Profile method. Measurement standards — Material measures. International Organization for Standardization (2000)

ISO/TS 16610 part 49, Geometrical product specifications (GPS) – Filtration – Part 49: Morphological profile filters: Scale space techniques. International Organization for Standardization (2006)

ISO 25178 part 6, Geometrical product specification (GPS) – Surface texture: Areal – Part 6: Classification of methods for measuring surface texture. International Organization for Standardization (2010)

ISO 25178 part 602 , Geometrical product specification (GPS) – Surface texture: Areal – Part 602: Nominal characteristics of non-contact (chromatic confocal probe) instruments. International Organization for Standardization (2010)

Lin, P.C., Sun, P.C., Zhu, L., Fainman, Y.: Single-shot depth-section imaging through chromatic slit-scan confocal microscopy. Appl. Opt. 37, 6764–6770 (1998)

Molesini, G., Pedrini, G., Poggi, P., Quercioli, F.: Focus-wavelength encoded optical profilometer. Opt. Comm. 49, 229–233 (1984)

Perrin, H., Sandoz, P., Tribillon, G.: Profilometry by spectral encoding of the optical axis. In: Proc. SPIE, vol. 2340, pp. 366–374 (1994)

Ruprecht, A.K., Körner, K., Wiesendanger, T.F., Tiziani, H.-J., Osten, W.: Chromatic confocal detection for high speed micro-topography measurements. In: Proc. SPIE, vol. 5302, pp. 53–60 (2004)

Sandoz, P.: Profilométrie en lumière polychromatique et par microscopie confocale. PhD Thesis, Université de Franche-Comté, Besançon, France (1993)

Shi, K., Li, P., Yin, S., Liu, Z.: Chromatic confocal microscopy using supercontinuum light. Optics Express 12, 2096–2101 (2004)

Tiziani, H.-J., Uhde, H.-M.: Three-dimensional image sensing by chromatic confocal microscopy. Appl. Opt. 33, 1838–1843 (1994)

Tiziani, H.-J., Achi, R., Krämer, R.: Chromatic confocal microscopy with microlenses. J. Mod. Opt. 43, 155–163 (1996)

Vaissière, D.: Métrologie tridimensionnelle des états de surface par microscopie confocale à champ étendu. Ed. Univ. Européennes (2010)

6 Point Autofocus Instruments

Katsuhiro Miura and Atsuko Nose

Mitaka Kohki Co., Ltd.
1-18-8 Nozaki, Mitaka-shi
Tokyo 181-0014, Japan

Abstract. A point autofocus instrument is a non-contact surface texture measuring instrument that consists of an autofocus microscope and a high precision *xy* scanning stage. A point autofocus instrument is capable of measuring various surfaces with different surface properties, such as reflectivity, colour and slope angle, with high resolution and high accuracy. This chapter presents the basic principles of point autofocus instruments, their instrumentation, use and good practice, the limitations of the technique, and extensions of the basic principles, e.g., to allow the measurement of aspherical surfaces with a data stitching method.

6.1 Basic Theory

A point autofocus instrument (PAI) measures surface texture by automatically focusing a laser beam at a point on a specimen surface, moving the specimen surface in a fixed measurement pitch using an *xy* scanning stage, and measuring the specimen surface height at each focused point (ISO/CD 25178-605 2011). This technique enables measurement over a relatively large area with a high degree of accuracy and is very popular for the geometrical measurement of various specimens predominately in the ultra-precision machining field.

For PAIs, there are several different measuring methods, such as the knife-edge method (Fukatsu and Yanagi 2005), astigmatic method, focus detection by critical angle total reflection, and others. However, this chapter focuses specifically on the beam offset method. Fig. 6.1 illustrates a typical PAI operating in beam offset autofocus mode. A laser beam, with high focusing properties is generally used as the light source. The input beam passes through one side of an objective lens and the reflected beam passes through the opposite side of the objective lens after focusing on to a specimen surface at the centre of the optical axis.

An image is formed on the autofocus sensor after passing through an imaging lens. Fig. 6.1 shows the in-focus state. The coordinate value of the focus point is determined by the *xy* scanning stage position and the height is determined from the *z* axis position sensor.

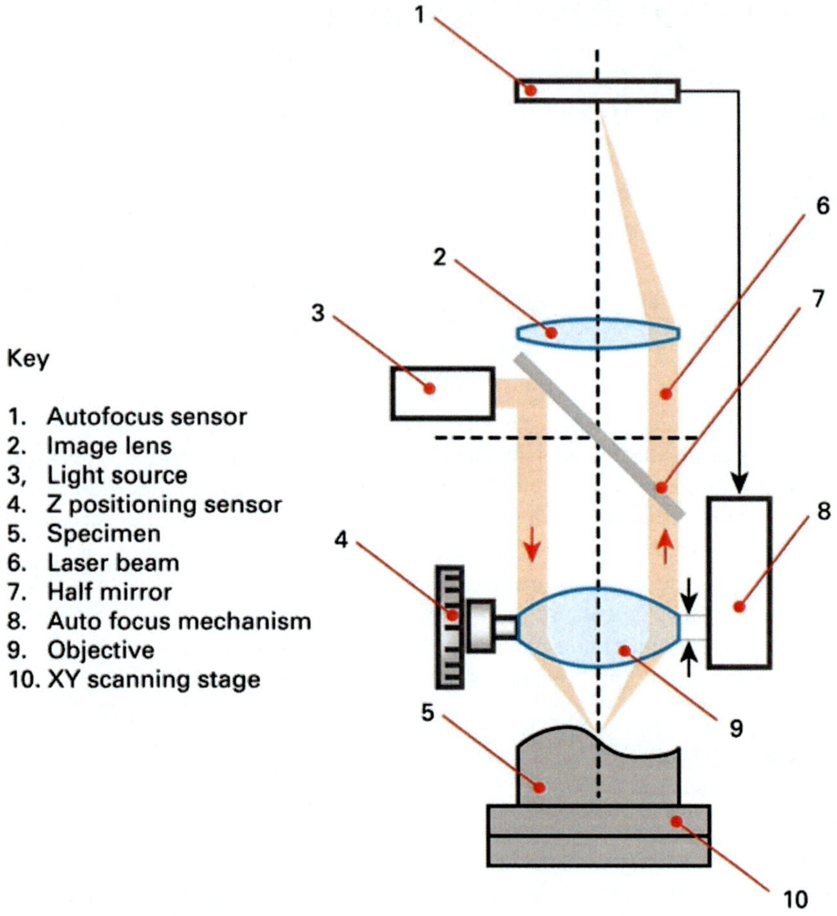

Fig. 6.1 Schema of a point autofocus instrument

Fig 6.2 shows the principle of point autofocus operation. Fig. 6.2A shows the in-focus state where the specimen is in focus and Fig. 6.2B shows the defocus state where the specimen is out of focus. The surface being measured is displaced downward (-Z), and the laser beam position on the autofocus sensor changes accordingly (W). Fig. 6.2C shows the autofocus state where the autofocus sensor detects the laser spot displacement and feeds back the information to the autofocus mechanism in order to adjust the objective back to the in-focus position. The specimen surface height, Z1, is equal to the moving distance of the objective, Z2, and the vertical position sensor (typically a linear scale is used) obtains the height information from the surface being measured (Leach 2009).

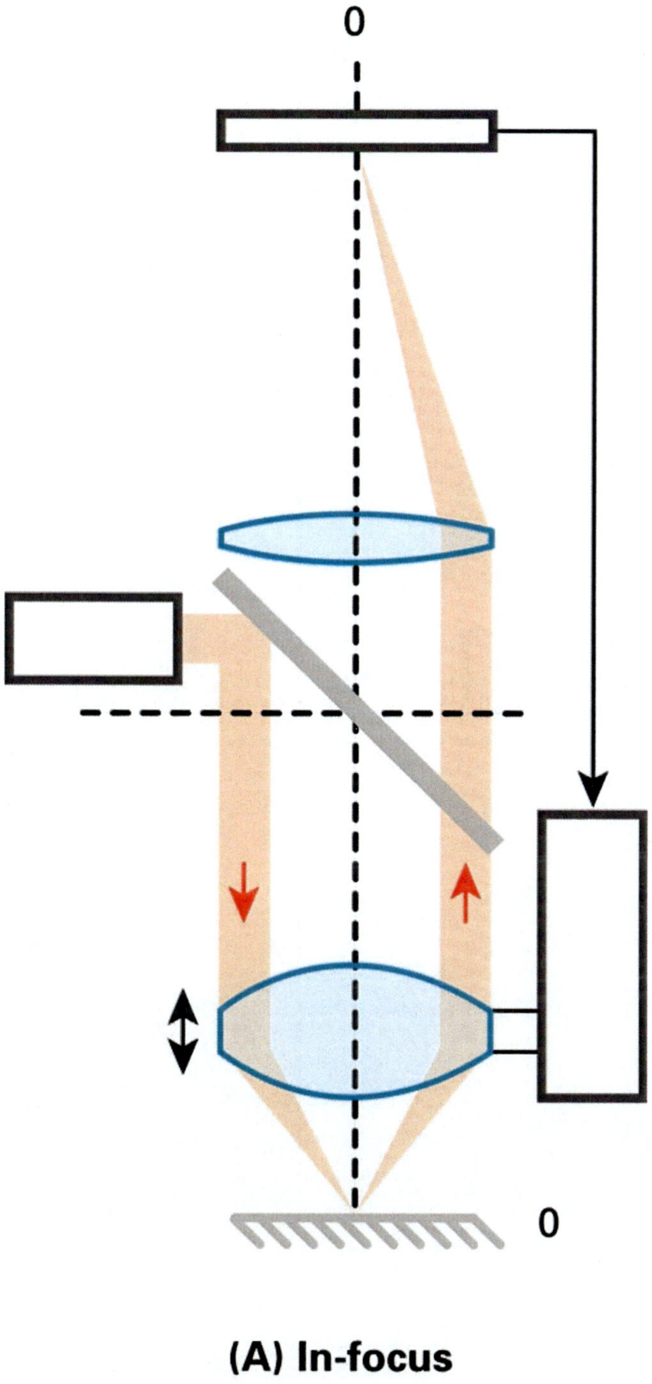

Fig. 6.2 Principle of point autofocus operation

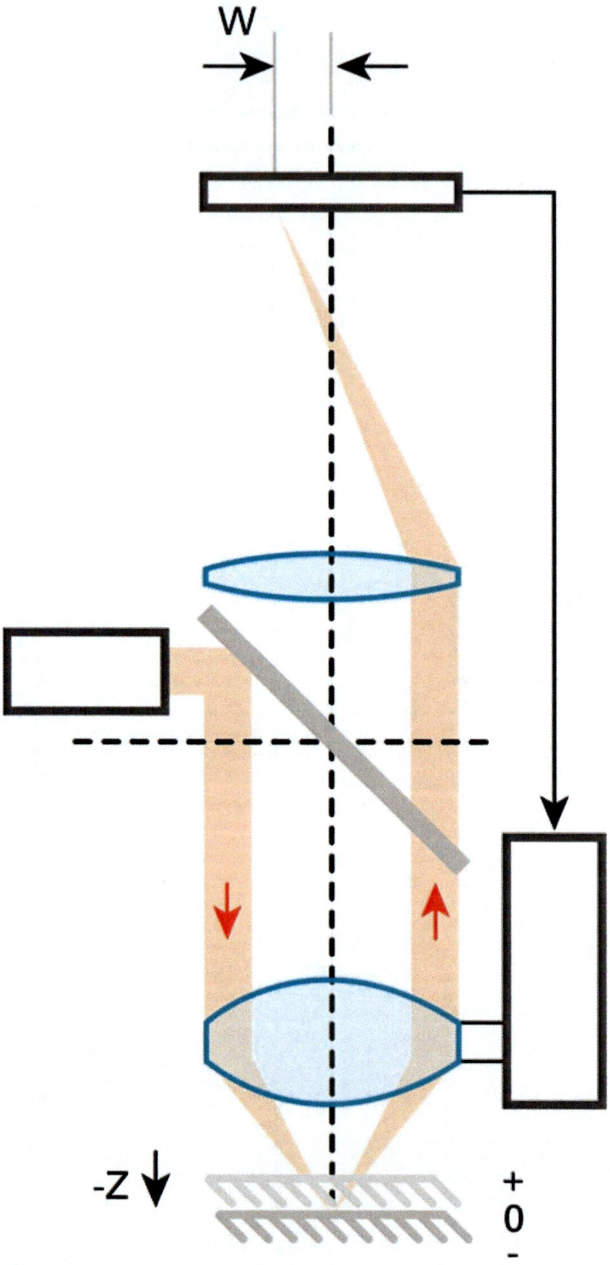

(B) Defocus

Fig. 6.2 (*continued*)

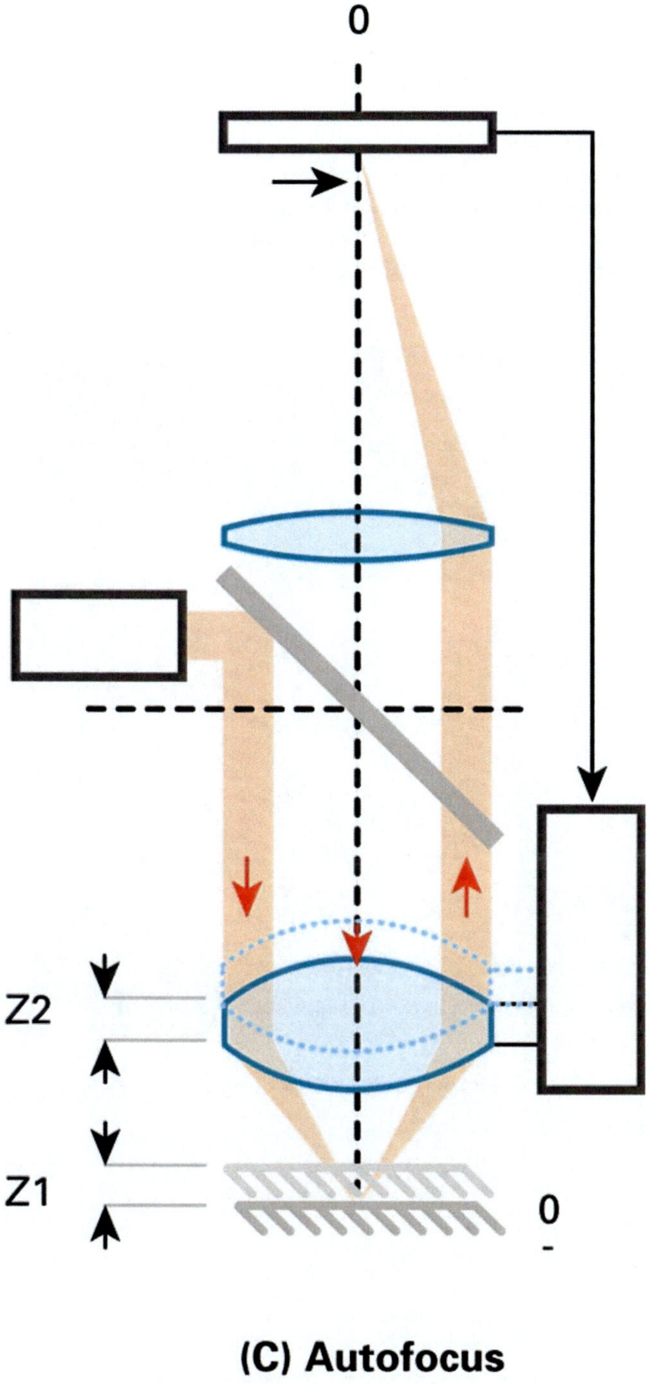

Fig. 6.2 (*continued*)

6.2 Instrumentation

Figure 6.3 shows a typical PAI and Fig. 6.4 shows a PAI system diagram. The laser beam and the illuminating light travel along the same axis to the specimen surface, which is monitored by a CCD camera. The autofocus mechanism moves only the objective in order to obtain a high autofocus response signal and the autofocus mechanism determines the resolution of 1 nm within the range of movement (typically 10 mm). The autofocus microscope mounted on the z axis coordinate stage enables the measurement of specimens up to 120 mm in height.

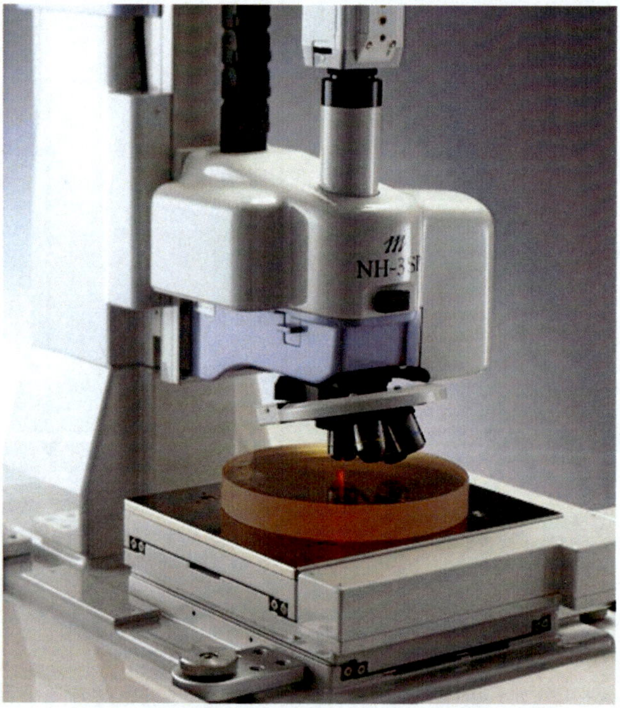

Fig. 6.3 Point autofocus instrument

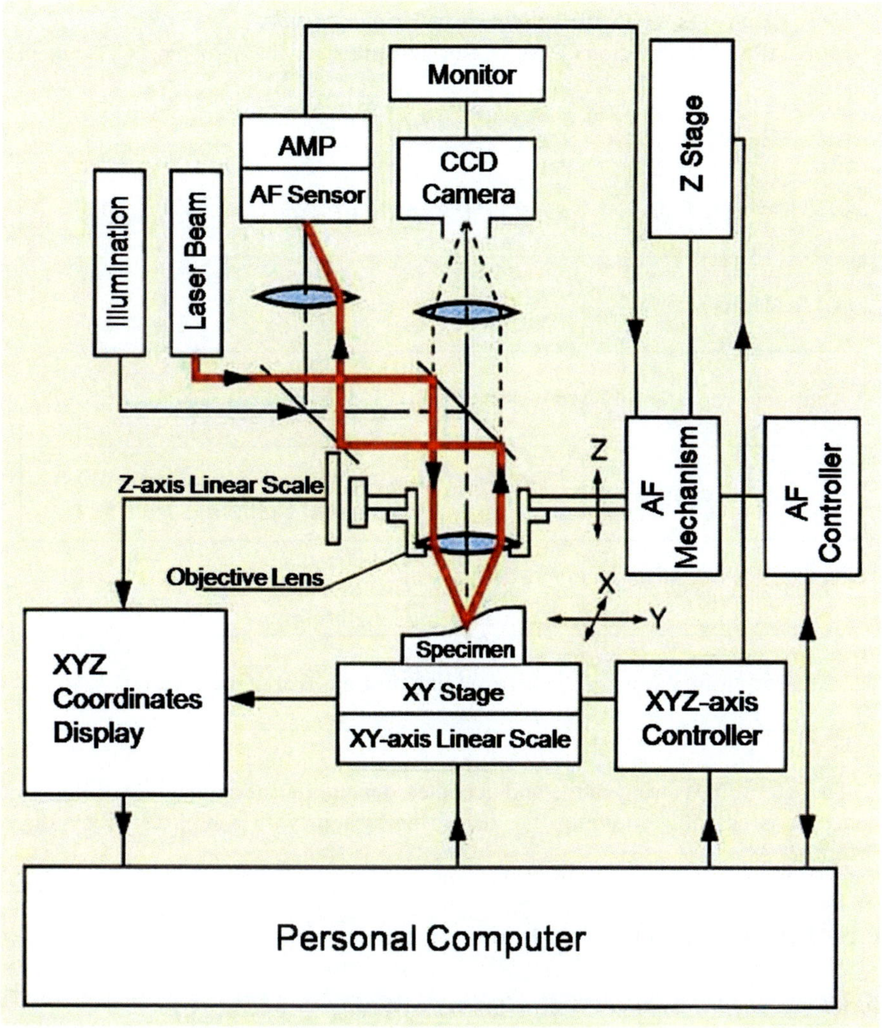

Fig. 6.4 System diagram

The specifications of the PAI instrument shown in Fig. 6.3 are listed in Tab. 6.1, and some features of the PAI are listed below.

- Large measuring range with high resolution.
- Capable of measuring steep angles over 45°.
- High autofocus repeatability (nanometre level).
- Immune to surface reflectance properties (the limit of reflectivity is approximately 1 %).

- Compares well with roughness material measures.
- Allows observation of the measuring surface via the mounted CCD camera.

Table 6.1 Specifications of a typical PAI

Measuring range	$xyz = (150 \times 150 \times 10)$ mm
Scale resolution	$xy = 0.01\,\mu m$, $z = 0.001\,\mu m$
Autofocus repeatability	$\sigma = 0.01$ μm
Laser spot diameter	1 μm (magnification of the objective lens: 100×) 2 μm (magnification of the objective lens: 50×)
Measuring accuracy	$xy = (0.5 + 2.5L/1000)$ μm, $z = (0.1 + 0.3L/10)$ μm (L = measuring length/mm)
Measuring functions	1) Size, 2) line profile, 3) areal surface

The lateral measuring range and accuracy depend on the *xy* stage specification, hence, it is essential to equip the measuring system with a high accuracy stage (Miura 2008).

6.3 Instrument Use and Good Practice

6.3.1 Comparison with Roughness Material Measures

The PAI is a stage scanning instrument that measures over the same range as contact stylus instruments and calculates roughness parameters under the same conditions for comparison (Miura 2004a). Fig. 6.5 shows a PAI measuring result on a material measure that is usually used for the calibration of stylus instruments. The PAI used gave a result for the material measure of 0.971 μm (*Ra*) compared to the certified value of 0.972 μm. The PAI measurement resulted in a pitch of 39.994 μm, whereas the certified value was 40 μm. These measurements are within the expanded uncertainty associated with the material measure. Fig. 6.6 also shows the PAI measuring result for a further material measure. The PAI again measured an *Ra* value that was within the expanded uncertainty associated with the material measure.

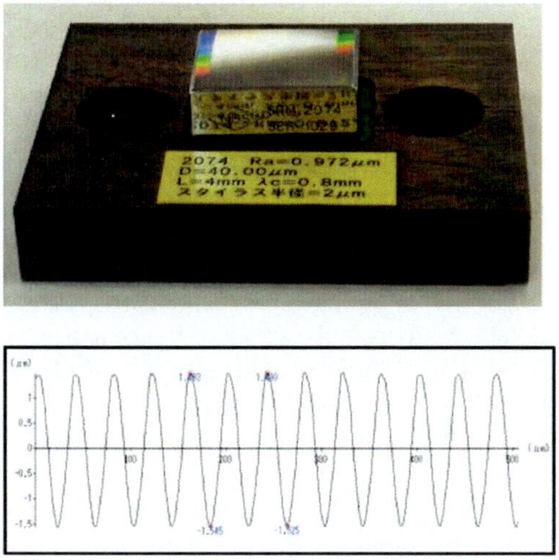

Measuring conditions:
*NIST Calibration certificate of SRM2074 (NIST)
*Stylus tip width = 2 µm
*λc = 0.8 mm, lr = 4 mm.

PAI:
*Objective: 100x (A_N = 0.8)
*Spot diameter = 1 µm
*λc = 0.8 mm, lr = 4 mm

Unit: micrometre

	NIST-SRM 2074	NH-3SP	Difference
Ra	0.972 (±0.025)	0.971	-0.001
RSm	40.00 (-0.02 + 0.06)	39.994	-0.006

Fig. 6.5 Roughness material measure (NIST SRM2074)

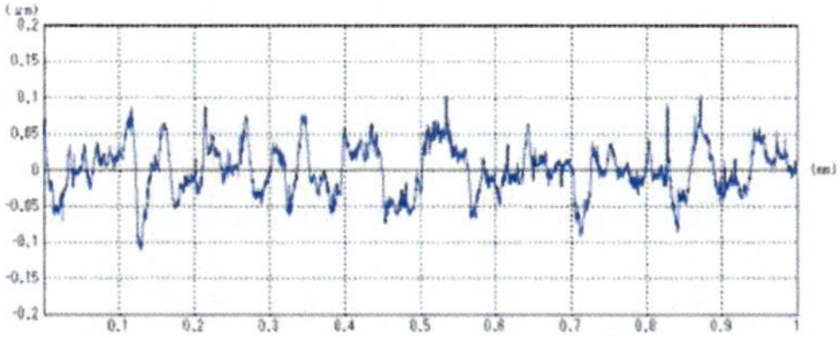

Measuring conditions:
*Type: D2 (PTB)
*Stylus tip diameter = 4 µm
*λc = 0.25 mm, lr = 1.25 mm

PAI:
*Objective: 100x (A_N = 0.8)
*Spot diameter = 1 µm
* λc = 0.25 mm, lr = 1.25 mm

Unit: nanometre

	PTB-Type: D2	NH-3SP	Difference
Ra	26 (±1.3)	27.3	+1.3

Fig. 6.6 Roughness standard (PTB type D2)

The data in Figs. 6.5 and 6.6 were obtained using a 100× objective (spot diameter is 1 µm). The radius of the contact stylus used to calibrate the measure was 2 µm, therefore, it was necessary to apply a 4 µm radius morphological filter to the extracted data for accurate data comparison.

Typical PAIs are equipped with a revolving turret that enables the objective magnifications to be changed. Generally, the laser spot diameter S_d is determined by the objective's numerical aperture and the source laser wavelength λ, and is calculated using

$$S_d = \frac{1.2\lambda}{A_N}. \tag{6.1}$$

However, the spot diameter from the beam offset method does not always correspond with the value from equation (6.1) since it uses only a part of the objective's aperture. The actual measured values of spot size obtained with a typical PAI are approximately 1 µm with 100×, 2 µm with 50×, and 10 µm with 10×.

Figure 6.7 shows the PAI measurement result on a material measure using various laser spot sizes. The horizontal axis represents the maximum roughness *Rt* and the vertical axis represents the value obtained by dividing the measuring result by the material measure's reference value. The PAI has a one to one correlation with a 1 µm laser spot diameter using the 100× objective. The objective magnifications are inversely proportional to the laser spot diameter. For contact stylus instruments, larger stylus diameters effectively filter the high spatial frequency components of a surface, which results in a smaller roughness value. Conversely, for PAI, larger laser spot diameters

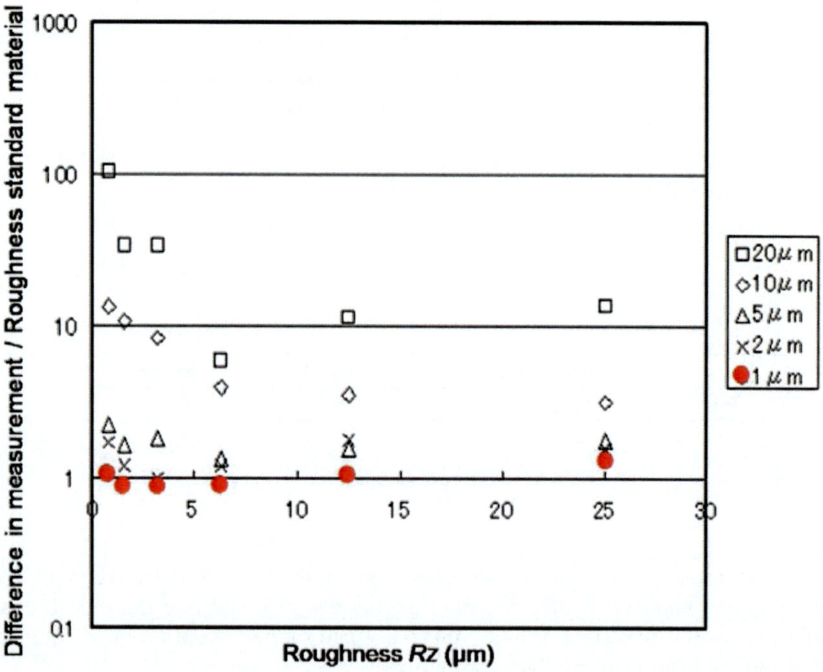

Fig. 6.7 Roughness measurement errors against various laser spot diameters

generate larger measurement errors because of the focus shift, which is produced by speckle (Fukatsu 2006) generated by the surface roughness within the laser spot diameter. It is important for a PAI to use a smaller laser spot diameter for roughness measurement since the speckle noise leads to greater measurement error.

6.3.2 Three-Dimensional Measurement of Grinding Wheel Surface Topography

The PAI enables the measurement of specimens with relatively high asperities and a range of reflectivities. This section describes PAI measurements of a grinding wheel surface, which is a challenging specimen for surface texture evaluation and analysis.

Quantitative evaluation of grinding wheel surface topography is a prerequisite to an understanding of grinding mechanisms, such as the material removal process of the grain cutting edge, the surface integrity of the finished surface and grinding wheel wear. Previously, only the stylus method has been available for the measurement of grinding wheel surface topography, but this method has problems with stylus penetration, which results in measured surface distortion. PAIs provide a potential solution to this problem (Miura et al. 2000).

Figure 6.8 shows the surface topography maps and their corresponding Abbott curves (Leach 2009) before and after grinding of a diamond wheel. Fig 6.8a shows well-protruded diamond grains with nominal diameters of 30 μm. This topography changes into a smoother surface without asperities after grinding due to the piling

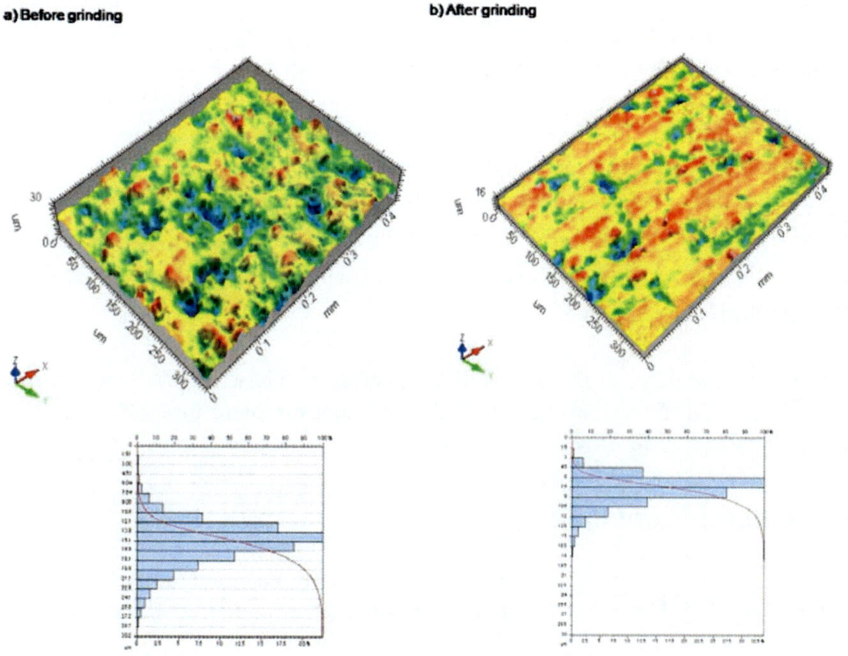

Fig. 6.8 Topography maps and Abbott curves of diamond grinding wheel

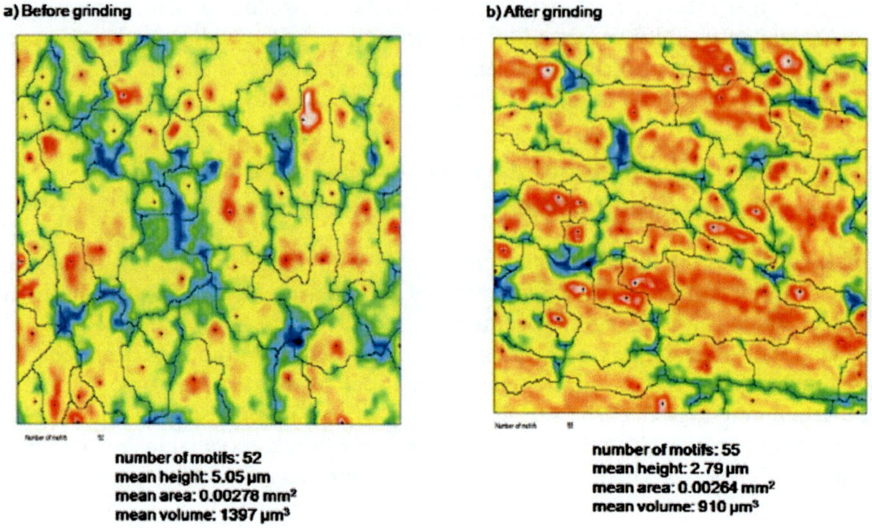

Fig. 6.9 Motif analyses of a diamond grinding wheel

up of debris in chip pockets, troughs of the wheel surface profile and embedding of diamond grain into the resin bond matrix (see Fig. 6.8b). The two Abbott curves show surface heights obtained from the corresponding surface data. It can be confirmed from Fig. 6.8 that the deepest point of the working surface rises up from a depth of 30 μm to a level of 15 μm after grinding. Fig. 6.9 shows the result of a motif analysis of the diamond grinding wheel; the grain cutting edge positions, their values and each height were measured from the bottom of the trough. The motif analysis enables a detailed quantitative evaluation of important factors, such as the grain cutting edge positions, their numbers, the volume of pores, etc.

6.4 Limitations of PAI

There are several limitations of the PAI technique. These limitations are now discussed in detail. Note that the relatively long measurement time may also be a limitation in some applications.

6.4.1 Lateral Resolution

The lateral resolution of a PAI is determined by the optical resolution of the objective as in Sect. 2.5. Fig. 6.10 shows the measurement of a diffraction grating with

1200 lines per millimetre, which was measured over an area *xy* of 20 μm by 20 μm with a sample spacing of 0.1 μm. Fig. 6.10 shows that the fine texture (pitch = 0.8 μm and depth = 0.1 μm) is clearly measured.

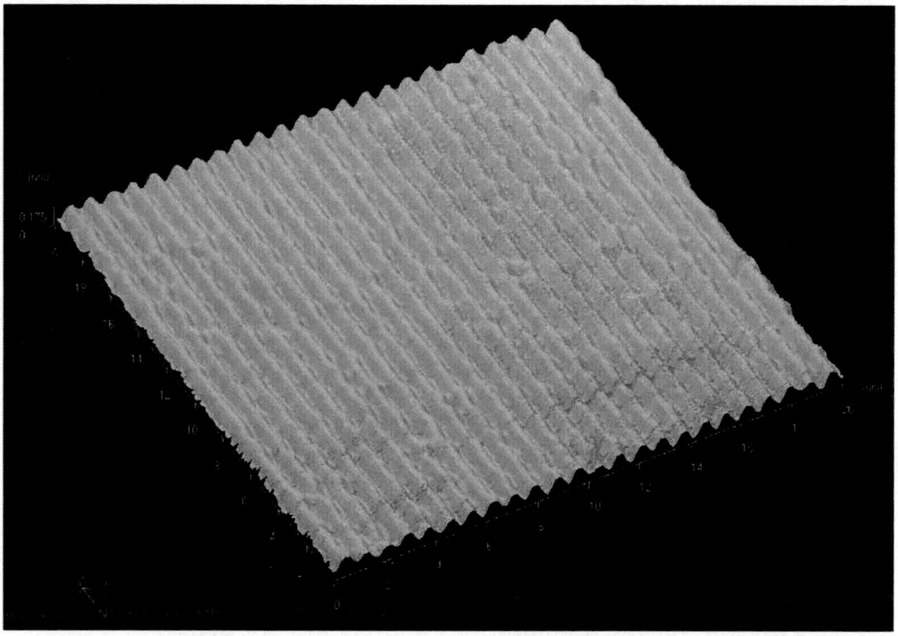

Fig. 6.10 Measurement of a grating (1200 line/mm, α = 8.5°)

6.4.2 Vertical Resolution

The vertical resolution of a typical surface topography measuring instrument based on a microscope is determined by the depth of focus of the objective (see Sect. 2.7). However, the vertical resolution of a PAI is determined by the repeatability of the autofocus mechanism and the linear scale resolution. Fig. 6.11 shows the results of a measurement of the autofocus repeatability. This data were obtained on a mirror specimen with a 100× objective with a numerical aperture of 0.8, and a defocus of ± 50 μm from the focal position with 20 measurement points. The result, 3.5 nm peak to valley, clearly indicates a high repeatability of the point autofocus mechanism.

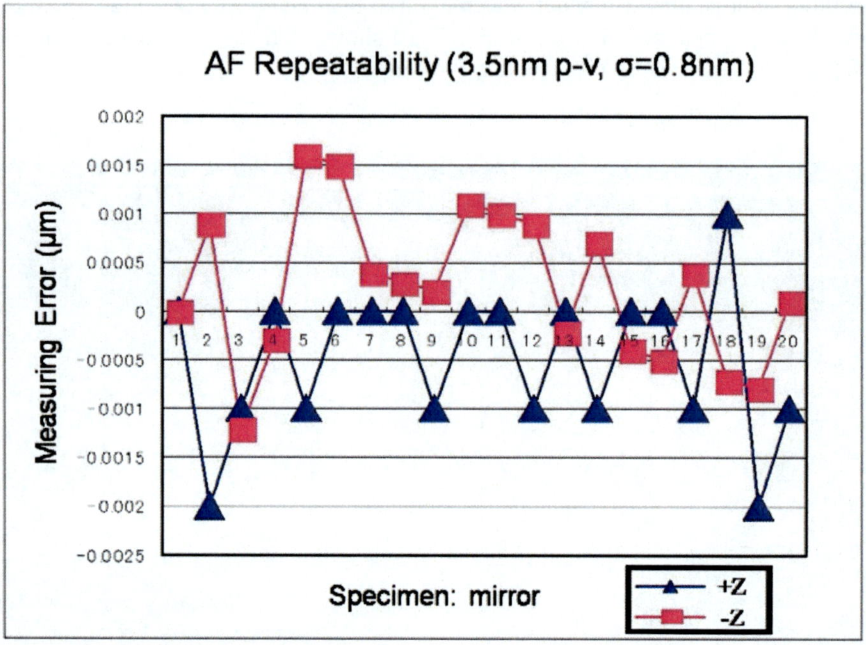

Fig. 6.11 Autofocus repeatability

6.4.3 The Maximum Acceptable Local Surface Slope

The maximum acceptable local surface slope with a PAI varies with the laser beam offset direction. Fig. 6.12 shows the laser beam direction being offset in the y direction. The maximum acceptable local slope parallel to the offset direction is A_1 and A_2 (refer to Fig. 6.12-1). Fig 6.12-2 shows the maximum acceptable local slope A_3 perpendicular to the offset direction. The maximum acceptable local slopes, are given by

$$A_1 < a$$
$$A_2 < (\alpha + b/2 - a)/2 . \tag{6.2}$$
$$A_3 < \frac{\alpha}{2} + \frac{b}{2}$$

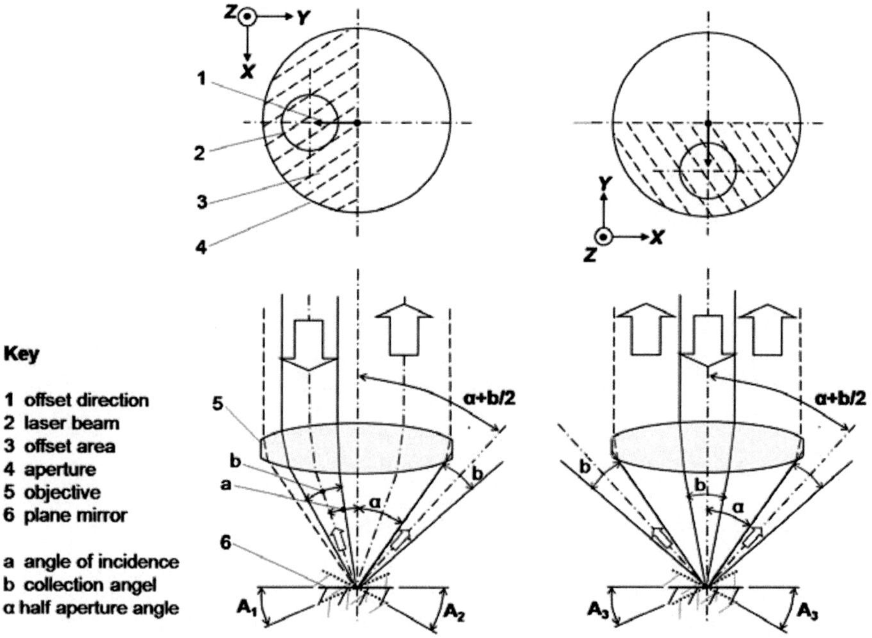

Fig. 6.12 The maximum acceptable local surface slope to the offset direction

Steep slope angles are measurable using scattered light that is caused by surface roughness. Fig. 6.13 shows the measuring result for a steel bearing. The measuring conditions are as follows:

- objective: 100× ($A_N = 0.8$, $\alpha = 53°$);
- direction of measurement is perpendicular to the offset direction; and
- collection angle, $b = 80°$.

The above condition gives $A_3 < 66.5°$, however, the result indicates that the local surface slope of 80° is achieved due to the scattered light generated by the surface roughness of the steel bearing ($Ra = 0.02$ μm).

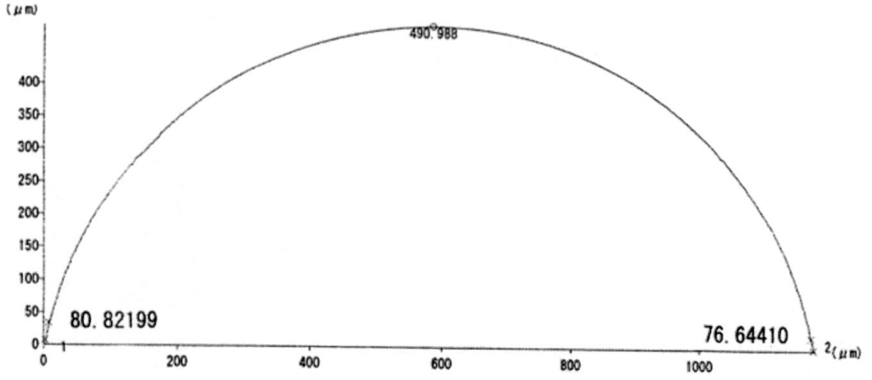

Fig. 6.13 Steel bearing ($Ra = 0.02$ μm)

6.5 Extensions of the Basic Principles

PAIs are a popular technology for measuring dies and moulds for aspherical lenses since they have a wide measuring range and a measuring accuracy at the sub-micrometre level. Using PAIs, steep slope angles in the proximity of 40° can be measured on a surface with 1 % reflectance. However, the maximum slope angle that a PAI is capable of measuring accurately at the sub-micrometre level would be approximately 30° since steeper angles give less reflectivity and generate more noise (Miura 2004b). The stitching measuring method, which uses a rotation stage, is effective for measuring aspherical lenses with steeper angles. Fig. 6.14 is a photograph of the measurement set up for an aspherical lens with a rotation stage, and Fig. 6.15 shows the measuring procedure of the stitching method. The specimen is rotated and has its surface measured within ±25° with the stage scanning method. All the sampling profiles must be partially overlapped with adjacent profiles. The method puts the centre of the rotation stage in to polar coordinates and combines all the overlapping data to calculate the entire surface map.

Axisymmetric aspheric profiles (those having their optical axis parallel to the z axis and their radial axis parallel to the x axis) are expressed with the following equation

$$Z(x) = \frac{Cx^2}{1+\sqrt{1-(K+1)C^2x^2}} + \sum_{i=1}^{n} A_i x^i \qquad (6.3)$$

where C is the curvature at the apex, K is the conic constant and A_i are the aspheric surface coefficients. With $K < -1$ the surface is a hyperbola, for $K = -1$ it is a parabola, for $-1 < K < 0$ it is an ellipsoid with its long axis parallel to the z axis, for $K = 0$ it is spherical and for $K > 0$ it is an ellipsoid with its long axis parallel to the x axis. To calculate the form error, the design coefficients are substituted into equation (6.3) and a best-fit line is fitted.

Fig. 6.14 Aspherical lens measurement using a rotation stage

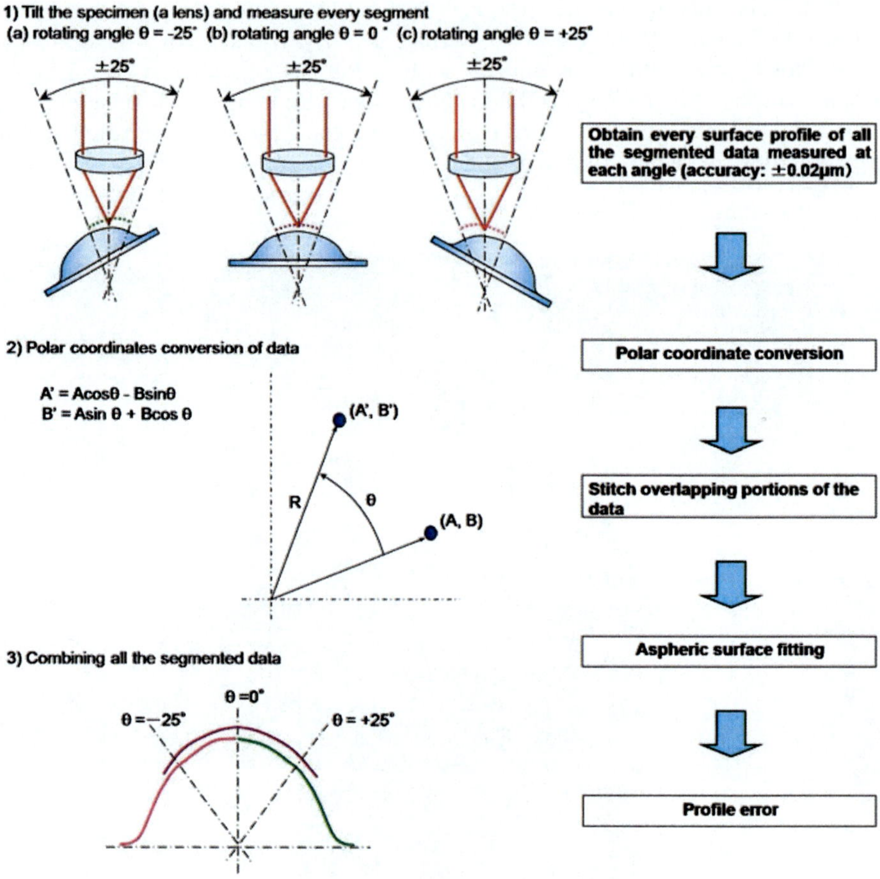

Fig. 6.15 Aspherical surface stitching measurement

Fig. 6.16 shows the measurement result for a standard glass sphere (the certified diameter is 5.198 mm and the roundness is 0.06 μm) within ± 80° for the accuracy validation of the stitching measurement. Fig. 6.16-1 is the basic measurement result. Fig. 6.16-2 shows the best-fit result by stitching seven profiles with 23° pitch. The best-fit result clearly indicates that the standard glass ball was measured to an accuracy of several tens of nanometres within the ± 80° range. Fig. 6.17 shows the measurement result for an aspherical lens used in cellular telephones with cameras. The result indicates a ± 0.15 μm form deviation.

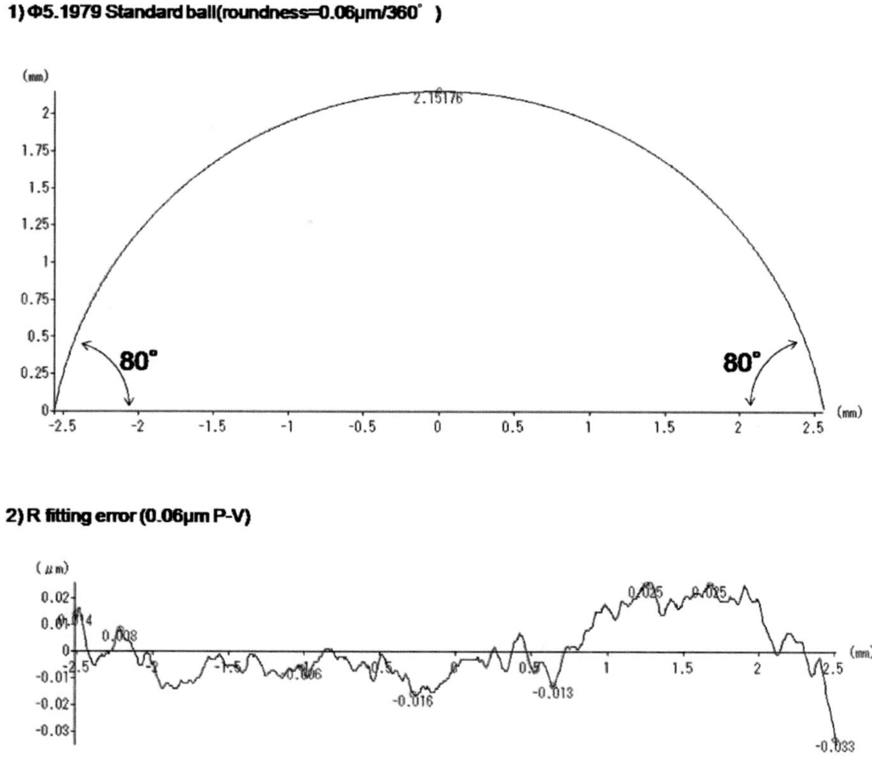

Fig. 6.16 Accuracy validation of the stitching measurement

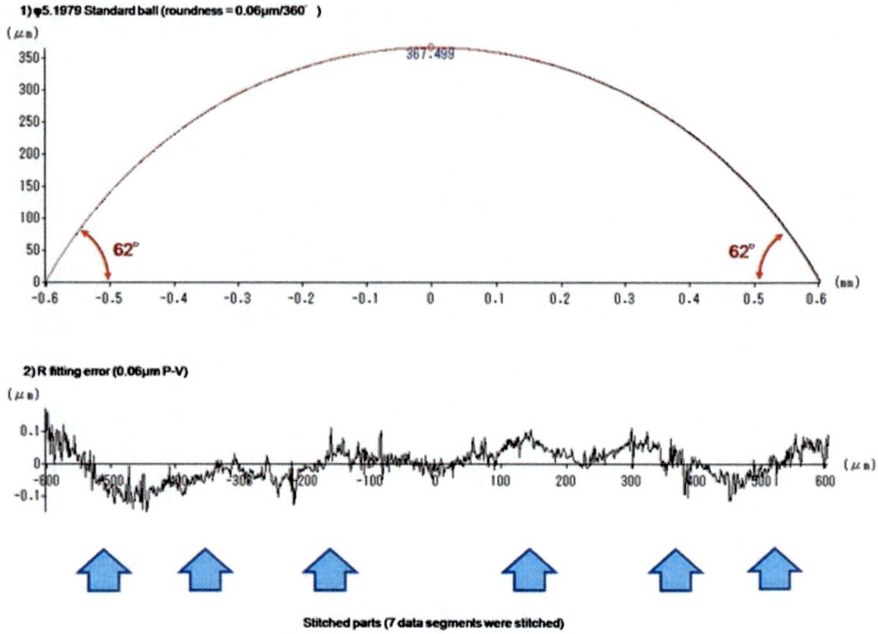

Fig. 6.17 Form deviation of an aspherical lens

6.6 Case Studies

PAIs enable the measurement of a wide range of high precision surfaces and are used in the areas of ultra-precision machining, MEMS, semiconductors, medical fields, and more. Fig. 6.18 shows some examples of surfaces measured using PAI. Fig. 6.18-1 is an areal topography map of a microlens array die (size in the *xy* plane is 3.5 mm by 2.25 mm). It takes approximately 30 minutes to obtain this measuring result because the PAI uses the stage scanning method. The important factors for microlens array measurement are the form deviation and the pitch of each lens pair (Miura and Okada 2003).

Figure 6.18-2 shows the surface map of a 45° vee-groove array for an acrylic LCD light guide panel for camera cellular telephones. LCD light panels require the measurement of form deviation at the micrometre level and roughness with an *Ra* value of less than 5 nm for thin panels with homogenous luminance. Fig. 6.18-2 shows that the 8 μm depth of the vee-grooves with an *Ra* value of 5 nm with < 10 % reflection of the acrylic material is clearly measured.

Figure 6.18-3 shows the surface map of a prosthetic tooth made of ceramic measured with a magnification of 10× and a numerical aperture of 0.3.

Figure 6.18-4 shows the surface texture of a ground linear rail, which clearly shows a 1 μm dent around the six screw holes. These dents were generated by the fluctuation in the depth of cut, which was caused by a grinding resistance change due to the difference in processing surface area. The data in Fig. 6.18-4 can be used for high precision linear scale production.

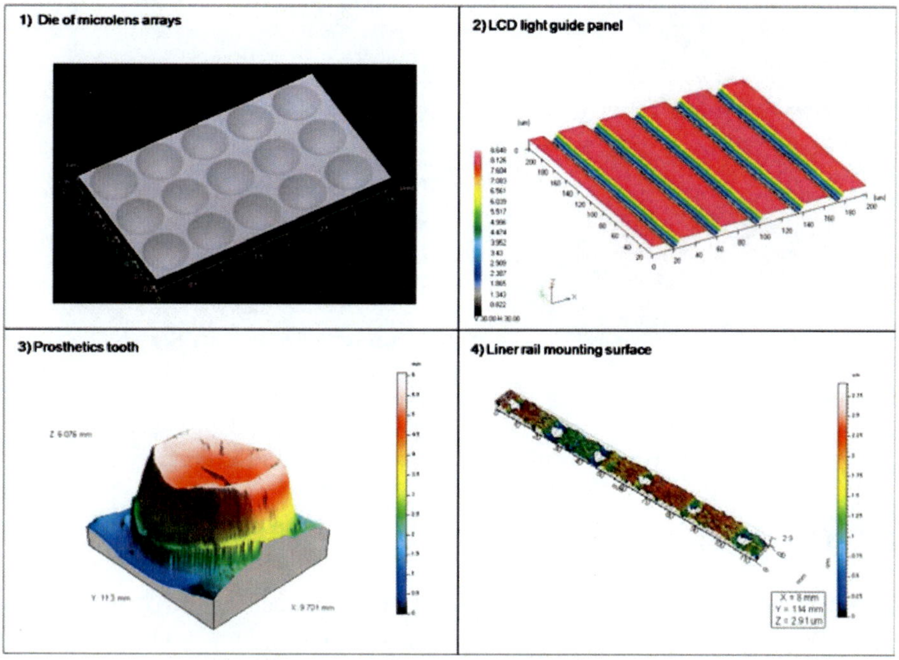

Fig. 6.18 Measurement examples with PAIs

Figure 6.19 shows examples of quantitative evaluations of surface texture, where areal parameters defined in ISO 25178 part 2 (2010) have been calculated following form removal. Fig. 6.19-1 is a ground spur gear. The waviness that generates gear noise has been extracted by removing the form of the tooth flank. Fig. 6.19-2 is the number printed on a 1,000 yen bill. The form removal function enables the evaluation of the amount of ink that is applied to the number by removing low frequency waviness.

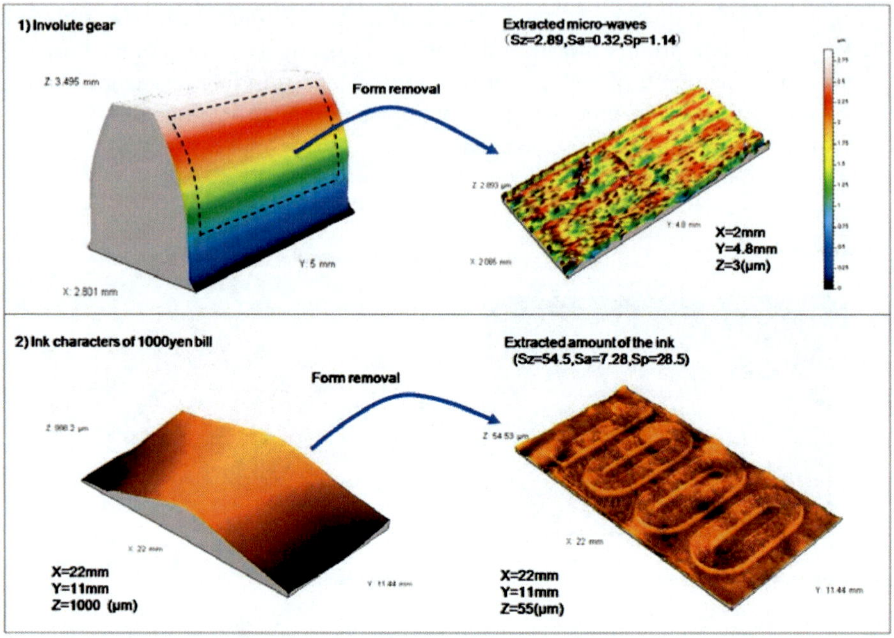

Fig. 6.19 Assessment of quantities from the areal surface texture

6.7 Conclusion

PAIs are widely used for quality control of various products and R&D in technology fields, despite the disadvantage of requiring longer measuring times than many other non-contact measuring methods. The positioning accuracy of the PAI mechanism determines the accuracy of the measurement. Hence, it is crucial for the PAI to have precision mechanisms, temperature drift corrections and vibration isolation systems in order to carry out high precision measurement. However, the main advantages of PAIs are that they can measure various specimens with different reflectivities, steep angles, and have a wide measurement range with high resolution and accuracy.

References

Fukatsu, H., Yanagi, K.: Development of an optical stylus displacement sensor for surface profiling instruments. Microsys. Tech. 11, 582–589 (2005)

Fukatsu, H.: Development of an optical profiling sensor suppressing outliers and able to deal with inclinations. Research achievement paper (2006)

ISO 25178 part 2, Geometrical product specification (GPS) – Surface texture: Areal – Part 2: Terms, definitions and surface texture parameters. International Organization for Standardization (2010)

ISO/CD 25178 part 605, Geometrical product specification (GPS) – Surface texture: Areal – Part 605: Nominal characteristics of non-contact (point autofocus probe) instruments. International Organization for Standardization (2011)

Leach, R.K.: Fundamental principles of engineering nanometrology. Elsevier, Amsterdam (2009)

Miura, K., Okada, M., Tamaki, J.: Three-dimensional measurement of wheel surface topography with a laser beam probe. Advances in Abrasive Technology III, 303–308 (2000)

Miura, K., Okada, M.: Measurement and evaluation of surface texture with optical contact probe. In: Ultra precision process and mass production technology of a micro lens (array). Technical Information Institute Co. Ltd (2003)

Miura, K.: Roughness measurement by optical probe. In: Proc. 303rd Workshop of The Japan Society for Precision Engineering (2004a)

Miura, K.: 2D and 3D surface measurement with a laser probe. In: Diffractive optics technical guide. Technical Information Institute Co (2004b)

Miura, K.: 3D surface texture/roughness measuring instrument with the point autofocus profiling. In: Practical encyclopedia of precision positioning technique. Technical Information Institute Co. Ltd (2008)

7 Focus Variation Instruments

Franz Helmli

Head of R&D, Alicona
Teslastraße, Grambach
Austria

Abstract. The focus variation method uses vertical scanning with limited depth of focus. Because of its ability to measure steep flanks and its robustness in relation to different materials, focus variation enables the measurement of roughness and form at the same time.

7.1 Introduction

This chapter describes the technology of focus variation. The main goal is to describe the basic principle of the technology to measure depth values. Following a description of the basic theory, the main components of a focus variation instrument are described in detail. A good practice guideline is formulated that contains advice and rules to ensure successful measurement results. After listing the limitations of the method, a case study shows the performance of focus variation for measuring the surface texture and form.

The main focus of this chapter is on using focus variation to measure surface texture. The application range of focus variation itself is much larger. Even form measurements on complex geometries with steep slopes are possible. More details can be found in Danzl and Helmli (2007) and Danzl et al. (2007). Moreover, the aim of the chapter is to describe the generation of the depth map, but not the calculation of surface texture or form parameters or the interpretation of these results.

7.2 Basic Theory

Focus variation is a method that allows the measurement of areal surface topography using optics with limited depths of field and vertical scanning. Compared to other methods, focus variation is very new in the field of measuring surface texture although its principle was first published in 1924 (von Helmholtz 1924). In the following sections, focus variation is described in detail.

7.2.1 How Does It Work?

Depth measurement by focus variation (Scherer et al. 2007, ISO/WD 25178-606 2011) is performed by searching the best focus position of an optical element pointing to a sample. This focus position is related to a certain distance from the sample (depth) value. By carrying out this process for many lateral positions, a depth map of the sample is generated. The basic components of a focus variation instrument are:

- an optical system with limited depth of field to detect the best focus;
- the illumination device/source;
- a CCD sensor to detect focus; and
- a driving unit for focus search.

A schematic diagram of a typical focus variation instrument is shown in Fig. 7.1. White light from LED light sources is transmitted through the semi-transparent

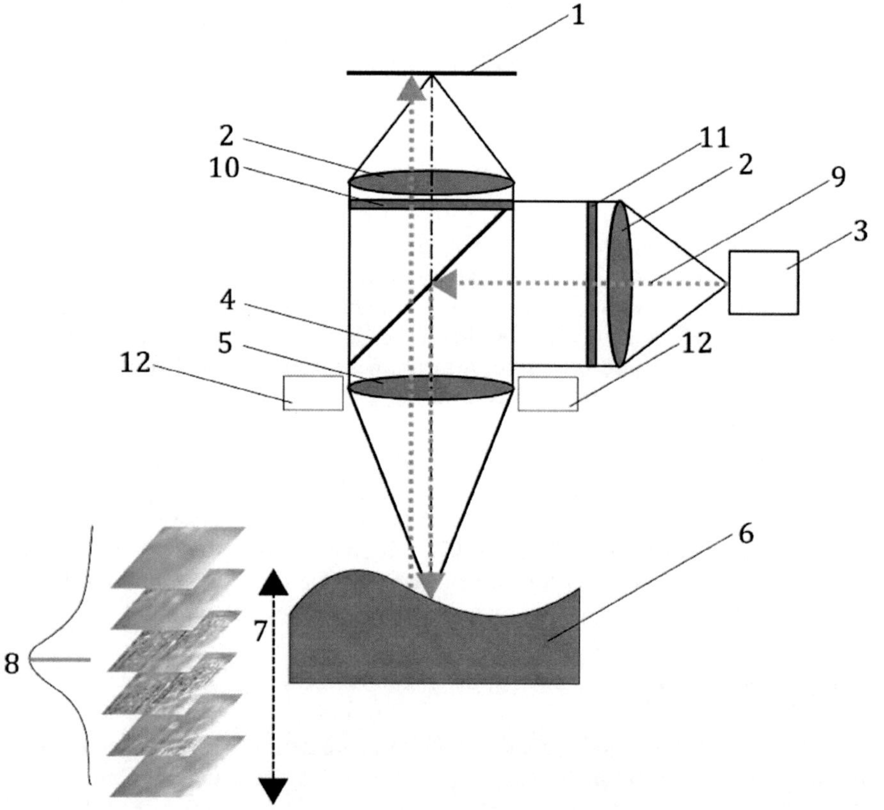

Fig. 7.1 Schematic diagram of a typical measurement device based on focus variation: (1) CCD sensor, (2) lenses, (3) white light source, (4) semi-transparent mirror, (5) objective lens with limited depth of field, (6) sample, (7) vertical movement with driving unit, (8) contrast curve calculated from the local window, (9) light rays from the white light source, (10) optional analyser, (11) optional polarizer and (12) optional ring light

mirror and the objective lens to the sample. Due to variations in the topography and the reflectivity of the sample, the light is reflected in different directions. The reflected light is partly collected by the objective and projected through the semi-transparent mirror and the tube lens to the charge-coupled device (CCD) sensor. Depending on the vertical position of the sample in relation to the objective lens, the light is focused to varying degrees on to the CCD sensor.

7.2.2 Acquisition of Image Data

By moving the sample in the vertical direction in relation to the objective lens, the degree of focus varies from low to high and back to low again. This change of focus is related to a change of contrast on the CCD sensor. By analysing this contrast on the CCD sensor, the position where the sample was in focus can be measured. By repeating this for every lateral position on the CCD sensor, the topography of the sample in the field of view can be measured. In addition to measuring the position where the sample was in focus, the colour of the sample can be determined.

7.2.3 Measurement of 3D Information

In order to measure the contrast in the image on the CCD sensor, a small region around the actual pixel position has to be considered. This also results in a lateral resolution that is less than the size of one CCD element on the sample. Several methods of measuring the contrast in the image of the CCD sensor have been reported. One method to measure the focus is to calculate the standard deviation of the grey values of a small local region.

$$F_z(x, y) = FM(reg_w(I_z, x, y)). \tag{7.1}$$

In equation (7.1), the content of the image I_z at height z is used as input for, what is know as, the region operator $reg_w(I_z, x, y)$, which extracts the information from I_z at a lateral position (x, y) over a certain rectangular region of $w \times w$ pixels. This content is used to calculate the amount of focus F_z with a focus measure FM.

If the focus level is very low (if the specimen is far away from the focus plane), the grey values are almost identical and the standard deviation of the grey values will be very low. In the case of a highly focused specimen, the variation of the grey values in a region is much higher and so the focus measure yields higher values. This is schematically shown in Tab. 7.1. More advanced focus measures can be found elsewhere (Nayar 1989, Niederöst et al 2003).

Table 7.1 Sample grey levels and their standard deviation at different focus positions. Region 1: point of interest for which the focus information is calculated. Region 2: 5×5 neighbourhoods used to calculate the focus information (standard deviation)

Scan position	Surface image	Standard deviation
Out of focus	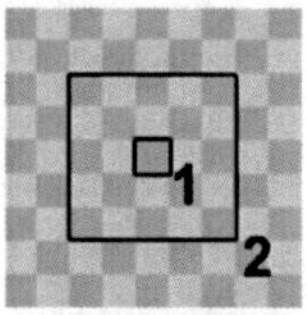	10
Almost in focus	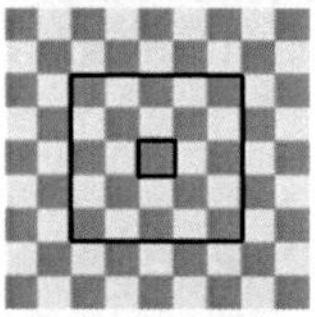	20
In focus	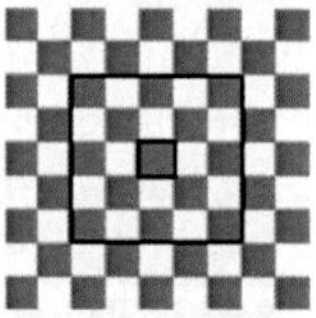	50
Almost in focus	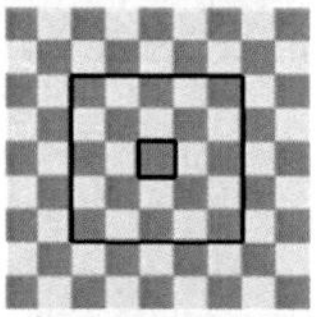	20
Out of focus	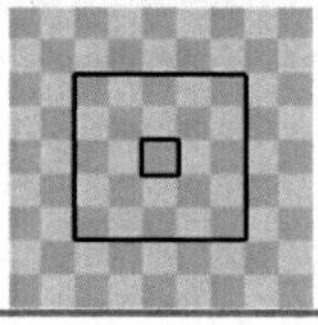	10

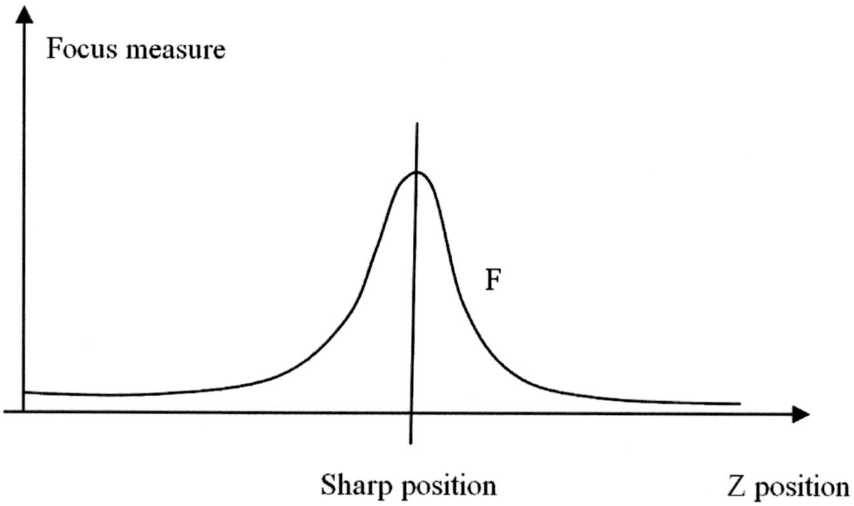

Fig. 7.2 The diagram shows the change of the focus with respect to the z axis position. The peak of the curve is identical with the focused position

The next step in the focus variation method is the calculation of the focus curve and the calculation of its maximum. By calculating the focus at every position in the stack of images I_z, the focus curve F is determined (see Fig. 7.2). This curve contains a peak that corresponds to the most focused position. The detection of this peak can be carried out using one of the following methods:

- maximum point;
- polynomial curve fitting; and
- point spread function curve fitting.

The maximum point is the fastest of the three methods but has the lowest accuracy. In this case, the depth value is calculated out of the index of the largest focus information as shown in equation (7.2) for a focus information curve with n z axis positions between z_1 and z_n.

$$\text{depth} = \arg(\max F_z) \text{ for } z_1 \leq z \leq z_n \,. \tag{7.2}$$

A further method for peak detection is the polynomial curve fitting method (see equation (7.3)). The advantage of this method is the higher resolution of the depth values in relation to the distance between the image planes. Hereby, the left and the right points around the maximum focus point are used to fit a polynomial curve $p(z)$ using a least squares technique (see equation (7.4)).

$$p(z) = az^2 + bz + c \,. \tag{7.3}$$

$$\min_{a,b,c} \sum_{z_1 \leq z \leq z_n} \left(F_z - \left[az^2 + bz + c\right]\right)^2 . \tag{7.4}$$

The coefficients a and b in equation (7.3) can be used to calculate the maximum of the fitted polynomial (equations (7.5) and (7.6))

$$p'(z) = 2az + b = 0 , \tag{7.5}$$

$$z_{maximum} = -\frac{b}{2a} . \tag{7.6}$$

The point spread function curve fitting method is the slowest method, however, it gives the highest accuracy. In this case, the measured focus values F_z are used to fit the point spread function (determined from the optics) and to calculate its maximum to obtain the depth value. This is schematically shown in Fig. 7.3 where a curve has been fitted to all points.

After performing the maximum detection for all lateral positions of the CCD sensor, a depth map is available. In Fig. 7.4 a typical 3D measurement result of a focus variation instrument is provided, showing a part of an angular gear wheel with overlaid true colour image and in pseudo-colours where colours represent different height values.

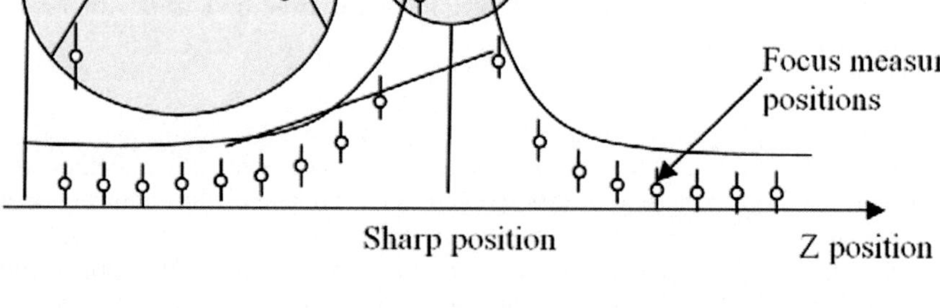

Fig. 7.3 The measured focus value is obtained by analyzing the whole curve. This gives a robust and accurate depth value

Fig. 7.4 3D measurement by focus variation of an angular gear with true (left) and pseudo colour information (right)

7.2.4 Post-processing

Due to non-ideal focus and other optical effects, a post-processing procedure is performed after determining the maxima of the focus curves. The post-processing procedure can include several steps.

One important step is the deletion of non-ideal depth values. A criterion to delete a depth value can be the quality of the fitted curve during the polynomial fitting. Another criterion can be the colour information. Fig. 7.5 illustrates a bad

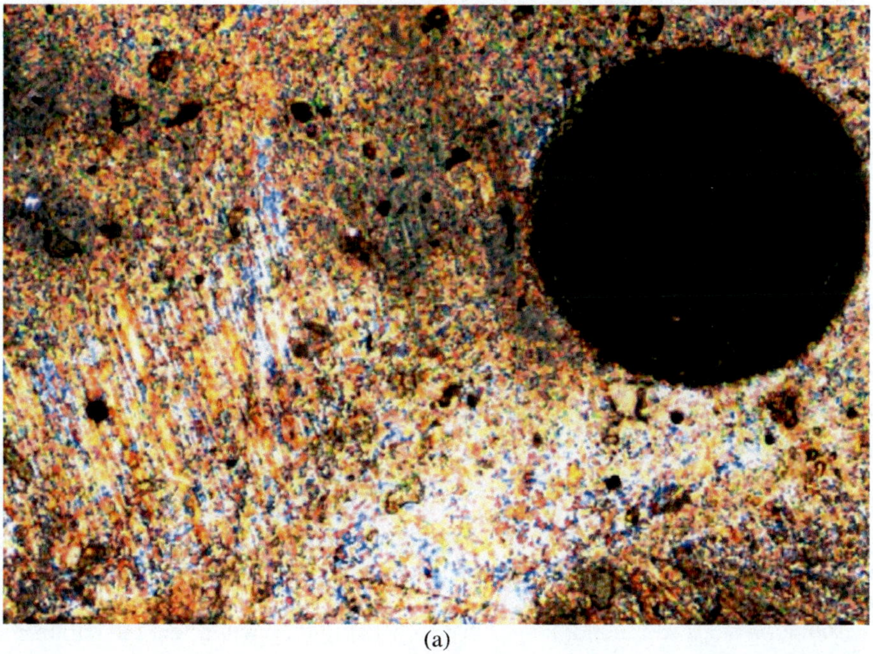

(a)

Fig. 7.5 (a) True colour image of a metallic sample with a dark hole region. (b) 3D measurement of the sample without post-processing, showing insufficient measurements. (c) 3D measurement after automatic post-processing where bad measurement points have been removed

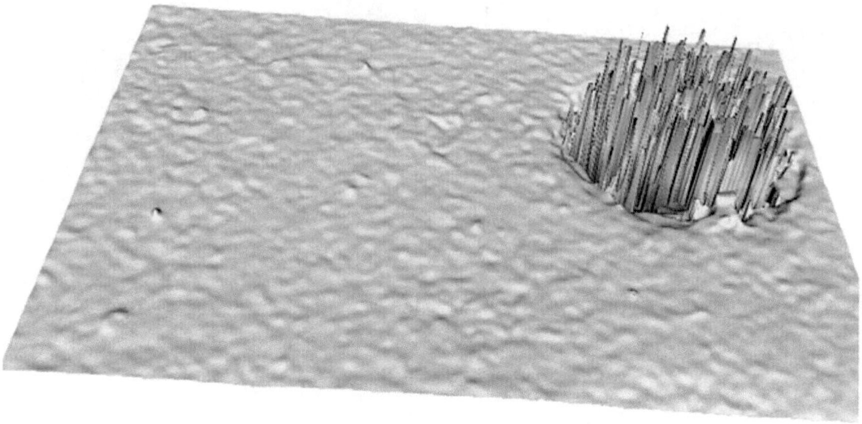

(b)

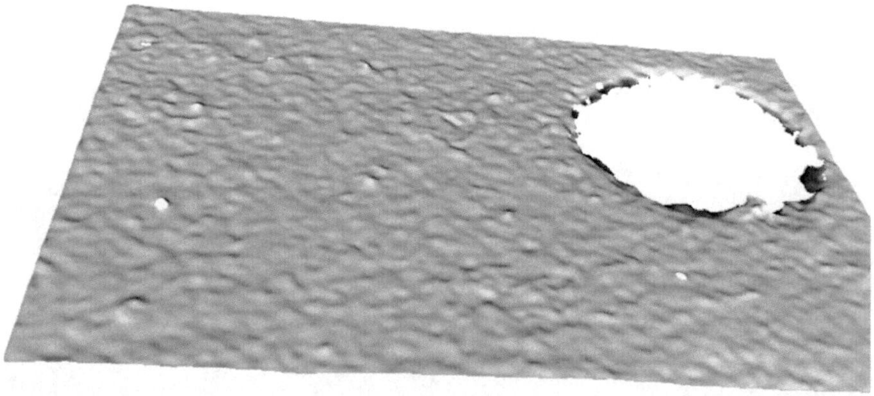

(c)

Fig. 7.5 (*continued*)

measurement result where insufficient data points have been deleted in a post-processing step.

After the deletion of some points, the depth map will contain holes. In order to delete the holes, a filling algorithm can be used. Filling is carried out using the height information of the nearest valid adjacent points. Typically, the points at the border polygon around the hole are used in combination with a spline or NURBS interpolation. If the resulting dataset is used for surface texture calculations, such interpolation steps should not be performed because they will affect the parameter calculation.

7.2.5 Handling of Invalid Points

If measured points have been deleted during the post-processing steps, this information must be stored for subsequent visualisation and the calculation of parameters. This can be carried out with a valid map or using certain depth values to encode the invalid points. A special depth value can be *NAN* (not a number), the maximum float or double precision.

For calculations on partly invalid depth maps, it is very important to be aware of whether a depth value represents a single point or an area. Depending on the representation, the area of invalid points can be small or large (see Fig. 7.6).

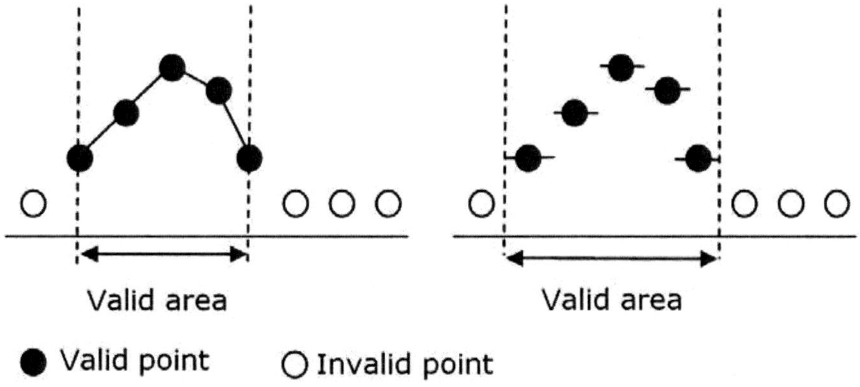

Fig. 7.6 Left: the valid area is shown for the case that a depth value represents a single point. Right: the (larger) valid area is shown for the case that depth values represent a small area

The most important issue when handling invalid points occurs during surface texture calculation. In this case, two steps have to be taken. The first step is the filtering procedure, which can be carried out using a convolution of the depth values. In order to get undistorted results, the weights for this convolution have to be adapted with respect to a valid mask. The second important step for surface texture calculation is the integration step. Many surface texture parameters require an integration of depth values over a certain area. The parameter *Sq*, for example, requires the integration of the squared depth values. If the depth map consists of invalid points, the border of the integration is equal to the border of the valid points, which makes the issue shown in Fig. 7.6 very important.

7.3 Difference to Other Techniques

Since focus variation is relatively new in the field of surface texture measurement, it is sometimes confused with other techniques. This section describes the main differences to imaging confocal microscopy (see Chap. 11) and the point auto focusing method (see Chap. 6).

7.3.1 Difference to Imaging Confocal Microscopy

There are several differences between focus variation and imaging confocal instruments (see Chap. 5 and Chap. 11). Two of the differences are described in more detail in the following section.

The first difference is the width of the curve for maximum sharpness calculation. The focus curve (see Fig. 7.2) has a larger width in the scan direction as compared to the intensity curve of confocal methods. Therefore, for confocal methods there can in principle be an inferior vertical resolution but that there is also a larger range for evaluation of the curve. In practice, this leads to less vibration sensitivity and less influence of high frequency errors of the scanning axis. Because vibration spectra usually have a Gaussian distribution, and that the errors of the scanning axis are well-known, the use of more points typically yields results of a higher quality.

Secondly, confocal instruments only allow the use of coaxial illumination. This can result in a limited maximum measureable slope angle. See Sect. 7.3 for more details about possible illumination sources for focus variation.

7.3.2 Difference to Point Auto Focusing Techniques

One of the major differences between focus variation and the point auto focusing technique (see Chap. 5) is the lateral scanning. Because point auto focusing measures only a single point, the sample needs to be moved laterally in order to get a depth profile or a depth map. Focus variation does not need to move the sample because it is an areal-based method that measures many points with one vertical scan. The second difference is that focus variation is a vertical scanning method while the point auto focusing method tracks the depth of the surface heights.

7.4 Instrumentation

This section lists and describes the components of a focus variation instrument. The following components are a part of a focus variation instrument:

- optical system,
- light source,
- CCD sensor,
- microscope objectives,
- driving units, and
- PC with software.

7.4.1 Optical System

The basic idea behind focus variation is to represent focus as a function over depth and to determine the depth positions with maximum focus. This maximum can only be determined if sharp and blurred regions can be well distinguished. Therefore, high contrast information from the CCD is required. This is achieved by maximising the optical transfer function and minimising the stray light in the illumination and detection path of the optics. Additionally, focus variation often uses colour information. Therefore, chromatically corrected optical elements are needed; a requirement that simple lenses do not fulfil.

The correction of the following aberrations (see also Sect. 2.2) and other optical effects must be performed to some extent in order to maximize the contrast on the CCD and the accuracy of the resulting 3D data:

- axial and longitudinal chromatic aberration,
- barrel distortion,
- stray light,
- optical transfer function,
- point spread function, and
- coma.

7.4.2 CCD Sensor

The CCD is used to detect the light from the specimen. In order to distinguish between in focus and out of focus for low contrast samples, the sensor must have the following characteristics:

- high radiometric resolution; and
- high spatial resolution.

High spatial resolution is needed to minimise the region in which thc focus is calculated (see equation (7.1)). A high radiometric resolution is needed to calculate the focus on samples with low contrast.

The CCD sensor can be a monochrome or a colour sensor. Monochrome sensors have the advantage of a higher spatial resolution; colour sensors have the advantage of a higher radiometric resolution and the possibility of 3D visualisation with real colour. The latter effect makes the interpretation of the 3D data easier for the operator. The advantage of having a true colour image and not just the depth values is demonstrated by example in Fig. 7.7.

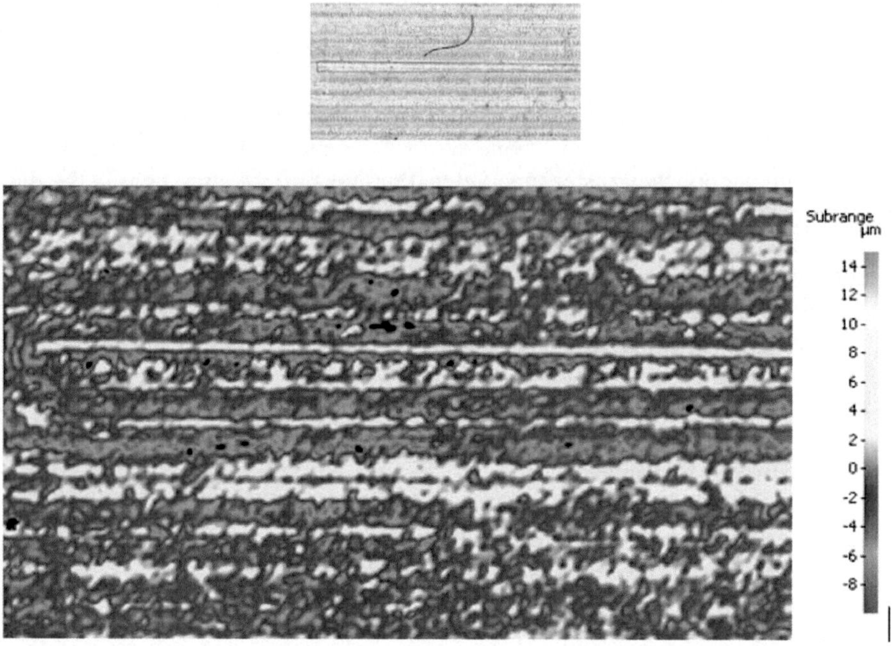

Fig. 7.7 Top: a true colour image (converted to black and white for print) of a milled surface structure with a fibre is shown. Below: a depth map is provided where the fibre is not visible

7.4.3 Light Source

The light source is used to illuminate the sample. The reflected light is used to generate the image on the CCD. Because focus variation needs contrast in the image, the light source is the basis for a high quality measurement. The following list gives the three most important aspects of light sources for focus variation:

- spectrum of the light source,
- polarization of the light, and
- direction of the illumination.

The spectrum of the light source must be adequate for the sample and the CCD. If the CCD is able to detect colour, a white light source should be used. An easy way to produce such light is to use an LED light source. This has the advantage of long lifetime and good stability.

Focus variation allows the use of illumination sources other than the established coaxial illumination. In particular, there are different ways of projecting the light to the sample and different types of light that can be used.

Different ways of projecting light to the sample help to achieve higher contrast in the image of the CCD sensor. In particular, there are options for coaxial illumination, ring light, dark field illumination, diffuse illumination or point light sources as can be seen in Fig. 7.8.

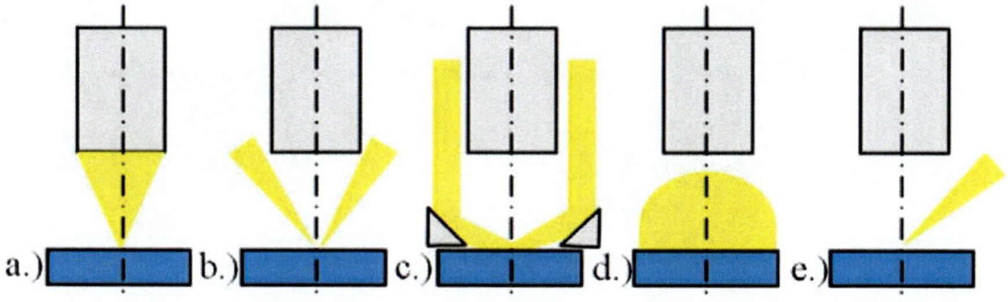

Fig. 7.8 Individual types of illumination for focus variation. a) coaxial illumination, b) ring light, c) dark field, d) diffuse illumination, e) point light source

Polarized light is beneficial when measuring metal samples with high reflectivity. In this case, a polarizer is used to generate polarized light for illumination. The reflected light is then filtered by an analyzer. The angle between the polarizer and the analyzer is called the polarisation angle. Typically, a polarization angle of 90° is used; a state also referred to as cross polarization. This means that only light is projected to the CCD sensor where the polarization is rotated by 90° on the sample surface. On metallic samples, the polarization is changed differently for regions with diffuse and specular reflection. Using polarization, specular components, which produce problematic highlights on the CCD sensor can, therefore, be reduced. In Fig. 7.9 an example of the advantage of polarisation is illustrated.

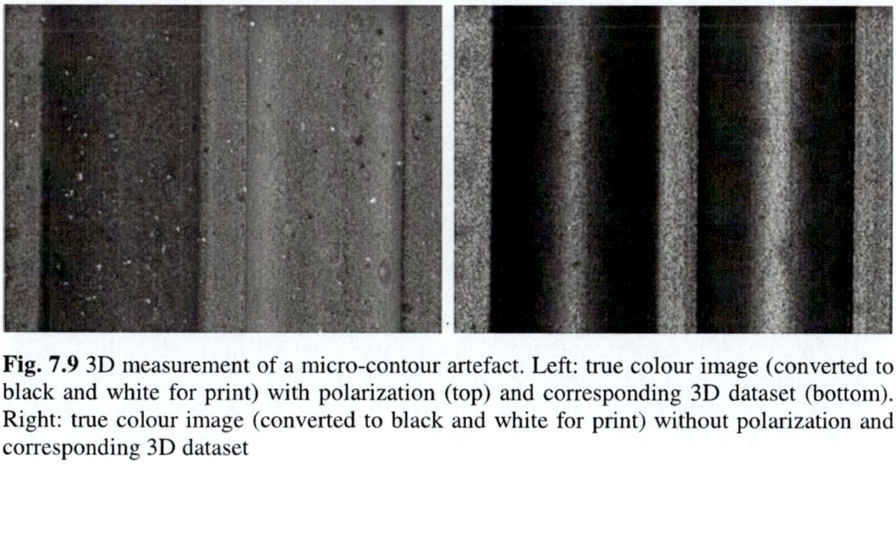

Fig. 7.9 3D measurement of a micro-contour artefact. Left: true colour image (converted to black and white for print) with polarization (top) and corresponding 3D dataset (bottom). Right: true colour image (converted to black and white for print) without polarization and corresponding 3D dataset

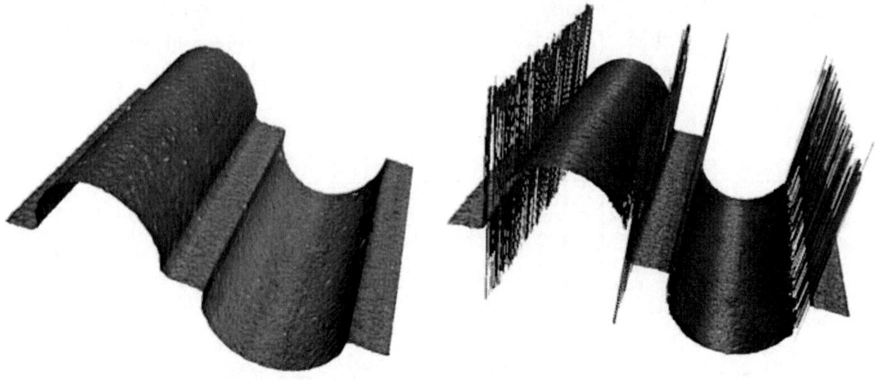

Fig. 7.9 (*continued*)

7.4.4 Microscope Objective

Focus variation works in different magnification ranges. Because of the prerequisite of having limited depth of field, it is reasonable to use microscope objective lenses (see Sect. 2.3). For microscope objectives, the relationship between the numerical aperture and the depth of field is given by equation (2.6) (see Sect. 2.7). This means that an objective lens with high numerical aperture yields a small depth of field, which means a good vertical resolution for focus variation.

7.4.5 Driving Unit

Focus variation needs to vary the degree of focus. Typically, this is carried out by moving the sample in relation to the objective - that is by moving the optics and the objective lens in relation to the sample. The driving or scanning units for a focus variation instrument are very important for speed of measurement and the accuracy of the measured surface data. There are several types of scanning units:

- piezo-electric drives,
- direct drives, and
- spindle drives.

A piezo-electric drive has the highest resolution but a very limited driving range. Direct drives have the highest speed but are difficult to use in the vertical direction. Spindle drives have high resolution, can be used over a large scan range and can drive a high load. So, if form measurement is required, spindle drives are a good solution. It is important that the driving units have a linear scale built in to them to generate traceable results.

If the above components are well implemented in practice, the focus variation instrument allows the measurement of very steep slopes and sample surfaces with a high variation in reflective properties.

7.4.6 Practical Instrument Realisation

At the time of writing, there are very few instruments commercially available that use focus variation. One of these instruments (Fig. 7.10 and 7.11) is described in this section. This instrument consists of a motorized *xy* stage and a sensor for the focus variation. The sensor consists of LED illumination, optics optimized for contrast, a high resolution colour CCD sensor with two million pixel elements and a driving unit that is able to move the optics over a vertical range of 100 mm. The objectives are mounted on a motorised turret.

The system can be used with microscope objectives between 2.5× and 100× magnification. This gives a vertical repeatability of the measured height data of 1 μm for the 2.5× down to 3 nm for the 100× objective.

It is also possible to equip commercial light microscopes with motorized *z*-axis drives and CCD sensors in order to perform focus variation measurements (Fig. 7.12). However, such systems usually have a lower accuracy than specially designed focus variation instruments, and moreover their calibration is typically rather complex.

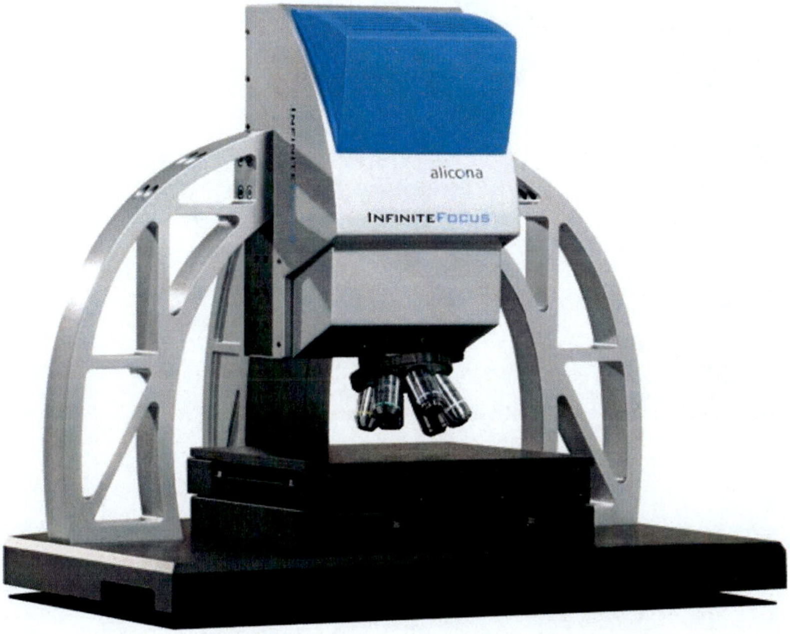

Fig. 7.10 A commercial laboratory system. From top to bottom: the sensor for focus variation, the turret with the objectives, the motorized *xy* stage and the passive vibration isolation system

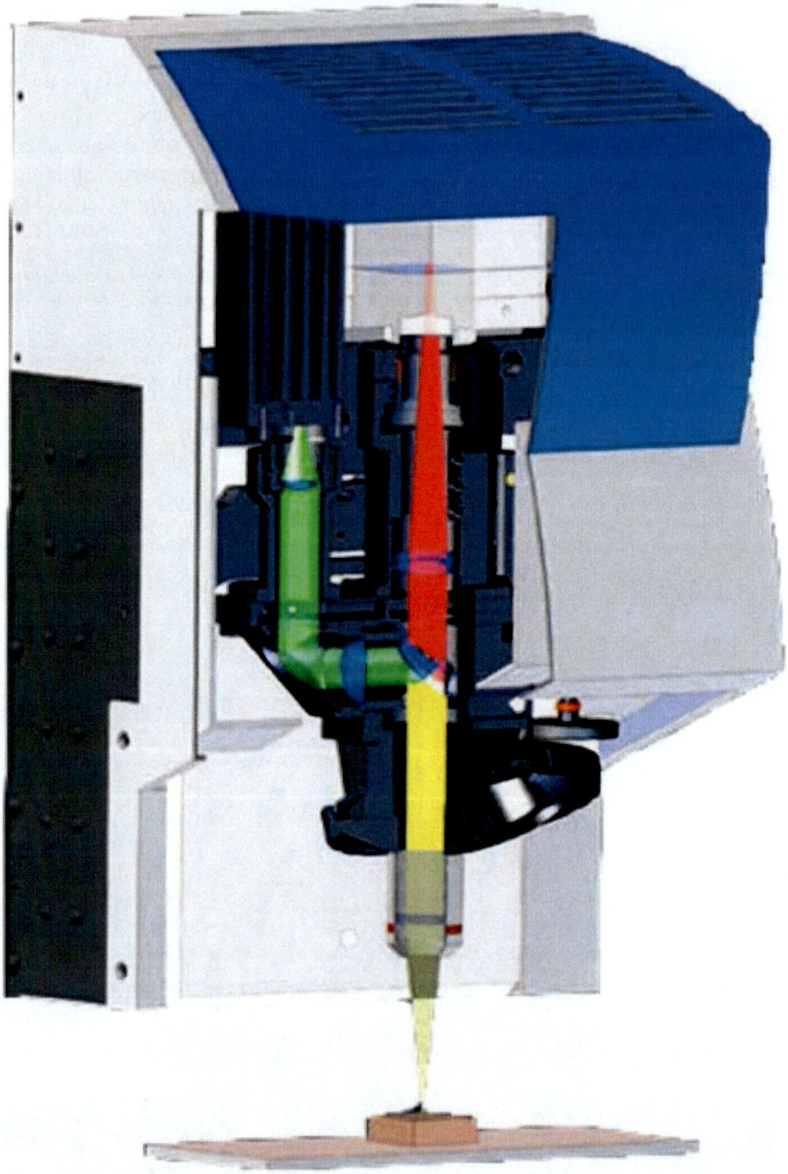

Fig. 7.11 Schematic of the sensor in Fig. 7.10. The illumination (green and yellow) and the detection path (red and yellow) are shown

Fig. 7.12 A commercial light microscope that is equipped with a motorized z axis drive and a CDD camera. This setup allows the generation of 3D datasets with focus variation. The drawback is that this instrument is not designed for metrology and so the accuracy may be low

7.5 Instrument Use and Good Practice

Optical surface texture measurement is a relatively new alternative to traditional contact stylus instruments (see Chap. 1). However, optical instruments require that many important instrument parameters be adjusted in an application-dependant manner. The following setup sequence for the relevant instrument parameters can help in achieving a good measurement result. For focus variation, it is important to choose the correct settings before a measurement is carried out. Some of the steps can be carried out automatically depending on the instrument or the software mode used.

Choose the correct objective lens. The objective lens will influence the measurement area, the possible lateral and vertical resolution and the illumination condition (different objectives have different numerical apertures for detection and illumination). The use of higher magnification objectives will yield results with higher quality. If an application requires a certain objective, it is better to use this objective and stitching methods (see 2D alignment in Sect. 7.6) in contrast to the use of a lower magnification objective without stitching.

Light source. The light source needs to be adjusted so that the illumination is appropriate for the sample material and geometry. On flat samples and samples with little curvature, non-polarized coaxial illumination is sufficient. If the sample has high slope angles or a higher material contrast, the aperture of the illumination should be increased (by using a ring light) or polarized light should be used. The live image can contain useful information in order to evaluate the quality of the illumination. Sometimes the instrument provides information as to whether the settings are acceptable or not. It should be borne in mind that focus variation uses scanning, therefore, the settings should be optimized at different vertical positions.

Adjust the settings of the CCD sensor so that the contrast is maximized. For this task, the settings exposure time and gamma of the CCD sensor are often available. The contrast is optimized if there are no black or white areas in the image. To decide whether the image has good contrast, the histogram of the images of the CCD sensor can be used (see Fig. 7.13 and Fig. 7.14).

Adjusting the scan range. It is important that the scan range is a little higher than the vertical range of the sample. This is needed because focus variation uses a certain range before and after the maximum peak in the focus curve to calculate the depth value. The additional range is dependent on the depth of field. Ideally, adjustment of the scan range is automatically carried out by the instrument.

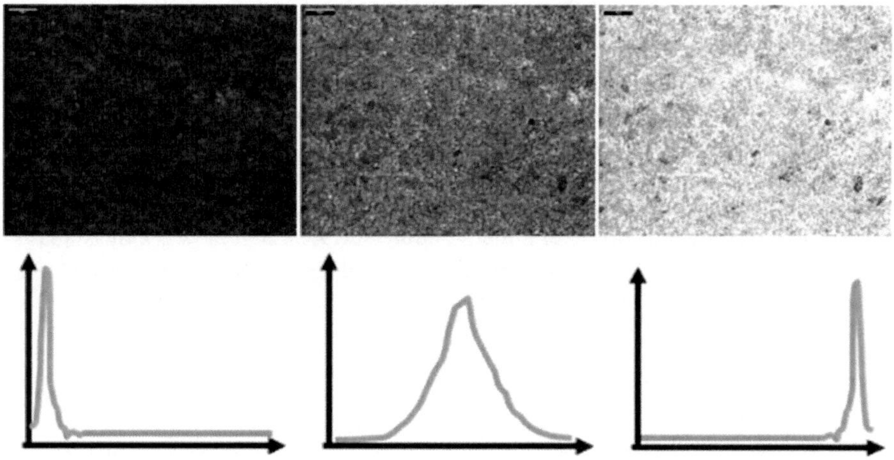

Fig. 7.13 The left image and the left histogram show an image that is too dark. The right image and histogram show an image that is too bright. Only the centre image and the corresponding histogram show no over and under saturation

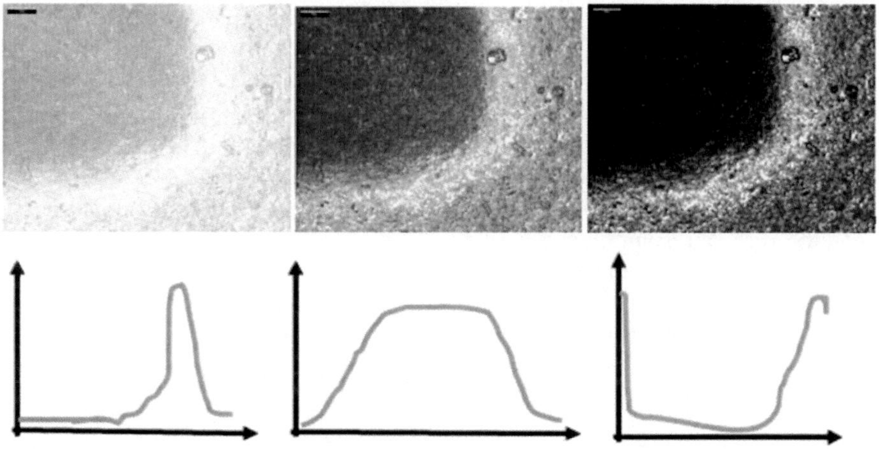

Fig. 7.14 The left image and the left histogram have too low a contrast. The right image and the right histogram have too high a contrast. The centre image has good contrast

Adjust the lateral and vertical resolution depending on the application. For roughness measurements, the lateral resolution should be set to a value smaller than the required λs filter value (Leach 2009). The vertical resolution should also be adjusted to the needs of the measurement. If roughness measurements are carried out, the vertical resolution should be adjusted depending on the estimated roughness. If it is assumed that the roughness value *Rq* is approximate, the vertical resolution value, that is the minimum requirement for an appropriate measurement, can be estimated. If it is assumed that the surface roughness is Gaussian distributed with a standard deviation σ_s, then *Rq* will be approximately σ_s. The instrument will add some instrument noise to the measured surface roughness with a standard deviation σ_i. This yields into an *Rq* value of σ_d given by

$$\sigma_d^2 = \sigma_s^2 + \sigma_i^2 . \tag{7.7}$$

If the *Rq* value to be displayed is to be equal to the *Rq* value of the surface, this requirement can be formulated in the form of equation (7.8)

$$\sigma_d < F\sigma_s \tag{7.8}$$

where *F* is the maximum allowed error (1.01 for 1 % error). From equation (7.8), the minimum standard deviation of the instrument σ_i can be calculated thus

$$\sigma_i < \sigma_s \sqrt{F^2 - 1} . \tag{7.9}$$

For example, if the surface roughness *Rq* is 100 nm and the maximum allowed deviation is 1 %, then the worst vertical standard deviation is approximately 14 nm. This estimation will now help in estimating the vertical resolution required. In the following paragraph, the relationship between the instrument noise σ_i and the vertical resolution is described.

If two single measurement points with depth values z_1 and z_2 with a standard deviation σ_i are provided the following calculation is possible. Each of the two depth values is affected by the instrument noise σ_i. Therefore, each depth value lies with a probability of 95 % within a $\pm 2\sigma_i$ range. A depth difference *D* is resolved if these intervals do not overlap. This is true if

$$D^2 > 8\sigma_i^2 . \tag{7.10}$$

So a depth difference *D* can be resolved if it is larger than approximately 2.8 times the standard deviation of the instrument. This gives the relationship between the standard deviation σ_i and the vertical resolution of the instrument (which can also be visualized in Fig. 7.15)

$$R_{vertical} = \sigma_i \sqrt{8} . \tag{7.11}$$

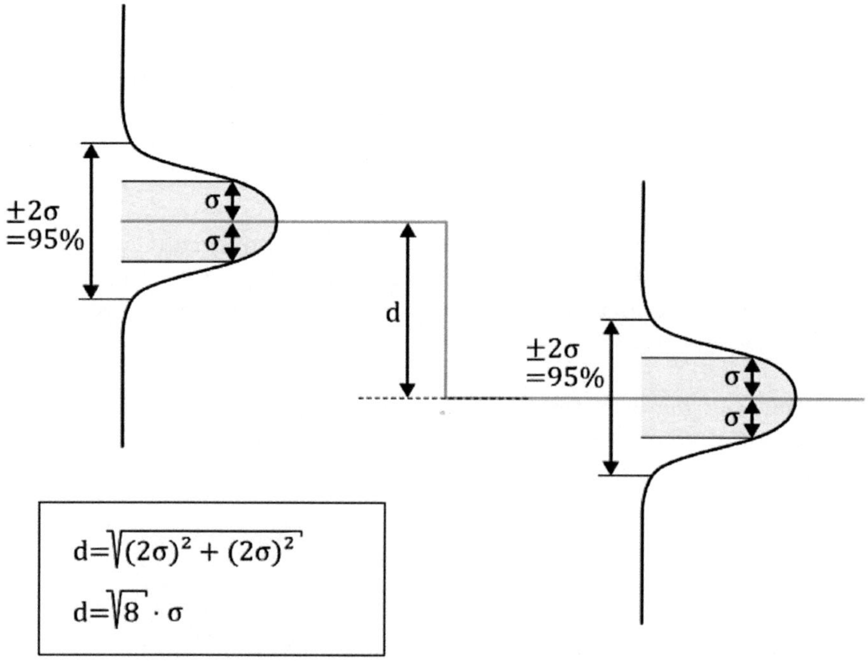

Fig. 7.15 Relationship between the vertical resolution *d* and the Gaussian instrument noise σ

Together with equation (7.9), this gives an approximate estimation of the lowest vertical resolution of the instrument for roughness measurement if the error should be less than 1 %

$$R_{vertical} = \frac{Rq}{2.5}, \tag{7.12}$$

and

$$R_{vertical} = \frac{Rz}{15}. \tag{7.13}$$

In equations (7.12) and (7.13), an approximate estimation of the required vertical resolution can be seen. Since other effects can influence the measurement, typically, a lower vertical resolution should be chosen than that provided in equations (7.12) and (7.13). This estimation has the following drawbacks:

- the accuracy of the instrument is not taken into account,
- the instrument noise is not only Gaussian distributed, and
- the surface roughness may not be Gaussian distributed, even if there are many surfaces that have a distribution that are approximately Gaussian.

Following a measurement, a quantification of the quality of the measurement result can be carried out with different metrics. If the result is not adequate, a readjustment of the available settings for the measurement should be carried out. One important metric for the quality of the measurement is the percentage of the area that is valid. Some instruments give a value for the quality of each 3D point. This can be used to:

- repeat the measurement with different settings if needed,
- use the measurement in the area where the quality is sufficient, and
- filter the dataset by using the quality information to get only the points that are adequate.

An example of the use of quality information is provided in Fig. 7.16 where bad measurement points are discarded due to quality information provided by the instrument.

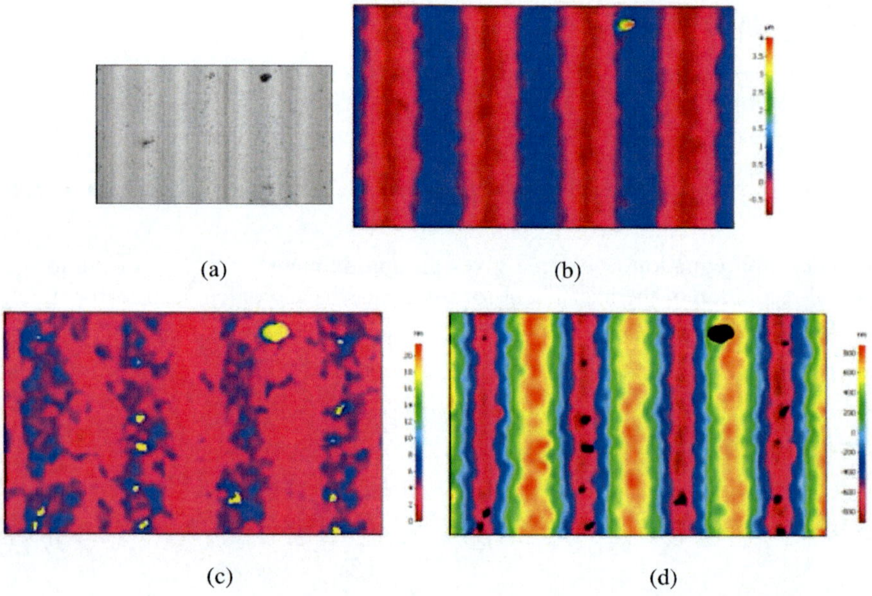

Fig. 7.16 (a) True colour image (converted to black and white for print) of a periodic roughness standard with a dust artefact. (b) Depth image where the dust artefact is shown in yellow and orange colour depth values. (c) Repeatability map where the dust artefact and some regions on the tool with little contrast have bad quality/repeatability values (yellow colour). (d) Depth map after points with bad repeatability have been filtered out

7.6 Limitations of the Technology

As with all methods for surface topography measurement, focus variation has some limitations. The following list of limitations should show the usability range of the technique. The reader should bear in mind that these limitations depend on the instrument used (different instruments will have different limitations).

7.6.1 Translucent Materials

As with other optical methods, translucent specimens will make a focus variation measurement more complex than just a top surface measurement. One solution for this problem is the use of replica materials.

7.6.2 Measurable Surfaces

Due to the fact that focus variation needs to measure a range of focus values, the images on the CCD sensor must contain enough contrast. This contrast amount is determined by the illumination, the sample surface and the optical system with the CCD sensor (see Sect. 7.3).

The contrast in the image of the CCD sensor is the result of two different effects. Firstly, it can be produced by small topography variations and secondly by material variations. This image generation is illustrated in Fig. 7.17.

Due to the fact that many samples do not have material contrast, a small topography variation (nanoscale roughness) must be present on the surface. In Fig. 7.18, a profile from a surface without and with nanoscale roughness can be seen. For focus variation nanoscale roughness of around Ra = 15 nm at λc = 2 µm must be present to allow topography measurements.

True colour images of surfaces with and without nanoscale roughness are shown in Fig. 7.19.

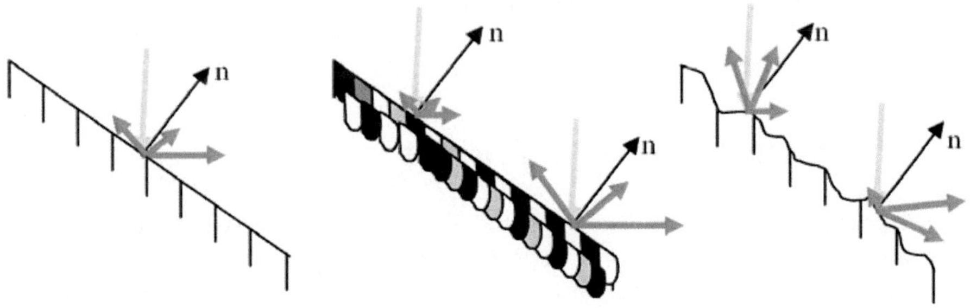

Fig. 7.17 Left: the contrast is very low since the surface has no material variations and no topography variations. Centre: contrast generated by material changes is shown. Right: the contrast generation by topography change can be seen

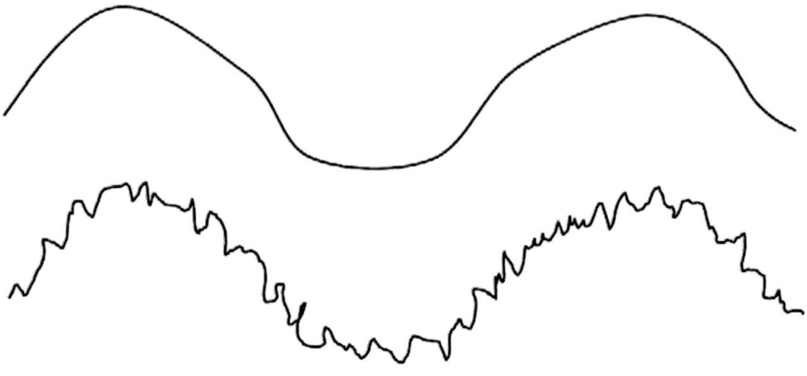

Fig. 7.18 Both profiles show a sinusoidal structure. The top profile shows a surface without superimposed nano-scale roughness. The bottom surface shows a surface with nano-scale roughness

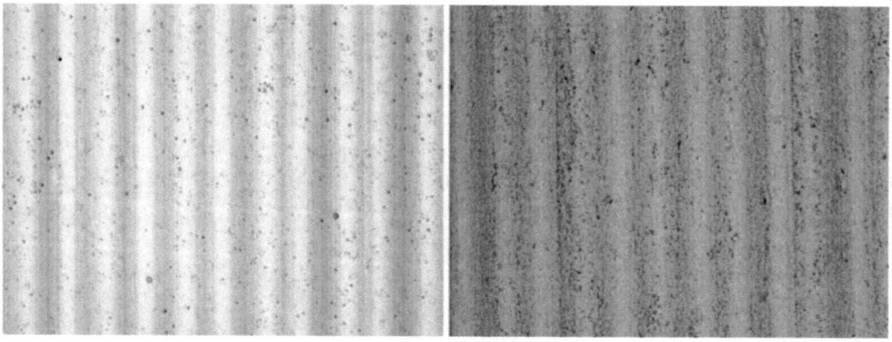

Fig. 7.19 Both images show a sinusoidal roughness standard with Ra = 100 nm and RSm = 10 μm at λc = 250 μm. Left: no significant nanoscale roughness. Right: a nanoscale roughness of Ra = 20 nm at λc = 2.5 μm

7.7 Extensions of the Basic Principles

The following section will describe some extensions to the basic focus variation principle. These extensions help in solving more complex applications such as those that may require stitching of measurements over large areas or extending the radiometric sensitivity.

7.7.1 Repeatability Information

One of the advantages of optical metrology is its flexibility. However, this flexibility can be a drawback if untrained users work with the instrument. Incorrect measurements due to incorrect settings can be the result. This situation can be improved if the system provides information about the quality of the measured data. One possibility to provide such information is to use Monte Carlo simulations in a similar fashion to that used for tactile coordinate measuring machines (Keck et al. 2004). Another method is described here.

The basic idea of the calculation of repeatability information is to fit a mathematical model to the focus points (see Sect. 7.2) (Danzl and Helmli 2008). This can be carried out by linear or nonlinear optimisation methods depending on the mathematical model used. After the model has been fitted, there are different ways to estimate the repeatability of the measurement. One possibility is to analyse the error function used for the optimisation in the region around its minimum. A measurement with a very distinct minimum has a smaller repeatability than a measurement where the error function is very flat in the region around the minimum. Another possibility is to estimate the repeatability based on methods from (non)-linear regression (Waldorp et al. 2006).

7.7.2 High Radiometric Data Acquisition

As described in the Sect. 7.2 focus variation only works if contrast is present in the image. To ensure that the measurement with focus variation works, the image on the CCD sensor has to resolve the required brightness change in the different areas. The brightness on the CCD sensor is different on flat or steep patches of the sample as shown in Fig. 7.20 (especially where specular reflections are dominant). If the slope difference on the sample is too high, the CCD sensor cannot resolve the brightness difference due to limited radiometric resolution. This brightness saturation does not allow the measurement of the focus information.

To overcome this problem, the following techniques can be used:

- high radiometric data acquisition,
- use of ring light, and
- use of diffuse reflection as opposed to specular reflection.

High radiometric data acquisition is a procedure to vary the exposure of the image of the CCD sensor and/or the amount of illumination to generate one image with higher dynamic range. The following sequence shows the procedure to generate one single high radiometric image that is acquired during the focus variation process.

1. Determine the required range of exposure for the CCD sensor.
2. Make several images within the determined exposure range.
3. Combine the individual images to one image with higher dynamic range.

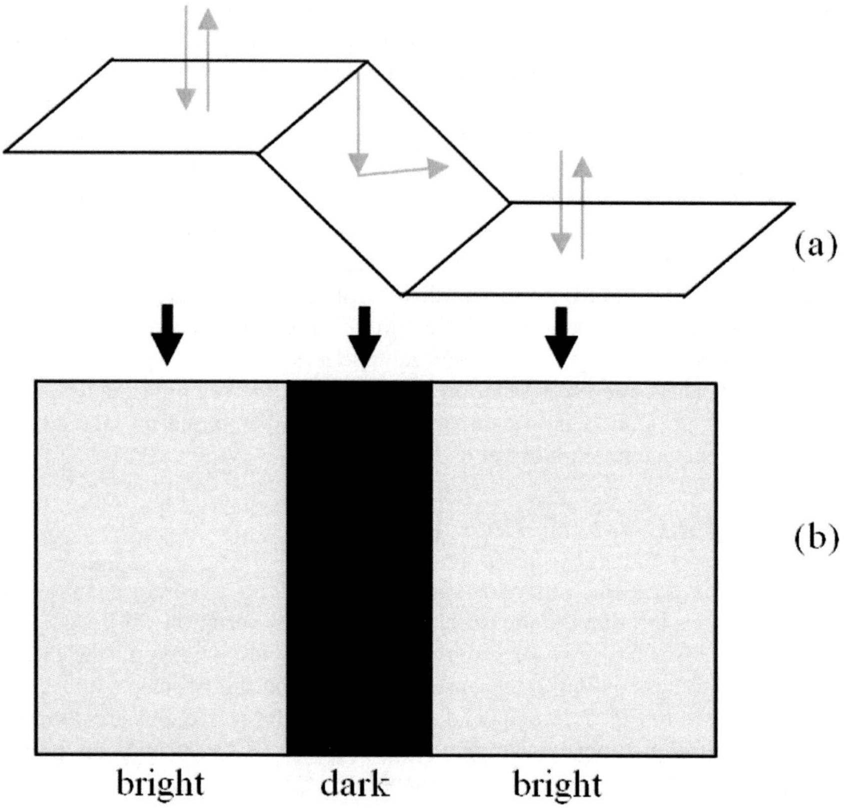

Fig. 7.20 The flat area of a sample results in higher brightness compared to steep areas; (a) shows the sample surface together with the illumination light ray and the reflected light rays; (b) shows the resulting image of the CCD sensor

7.7.3 2D Alignment

For many applications, a large area has to be measured. Very often the area is larger then the field of view of the chosen objective. To fulfil this requirement to measure a large area several individual measurements have to be performed and combined to one single dataset. To ensure a seamless result at the borders between the datasets, highly accurate displacement of the sample by the driving units has to be performed.

A way of reducing the cost of the measurement instrument is to use less accurate driving units in the x and y directions and to compensate the error using information in the overlap region between two adjacent single datasets. The following sequence can be used to generate a dataset with large field of view from a series of smaller individual datasets.

1. Acquire the individual datasets with overlap at coarse positions of the *xy* stage.
2. Refine the positions of each dataset using the information in the overlap region of the adjacent datasets.
3. Combine the individual datasets at the refined positions.

The refinement procedure needs to refine six parameters (three for translation and three for rotation). An advantage of the focus variation technology is the presence of colour information (on some instruments). This additional information increases the accuracy of the fine alignment of the datasets as shown in Fig. 7.21.

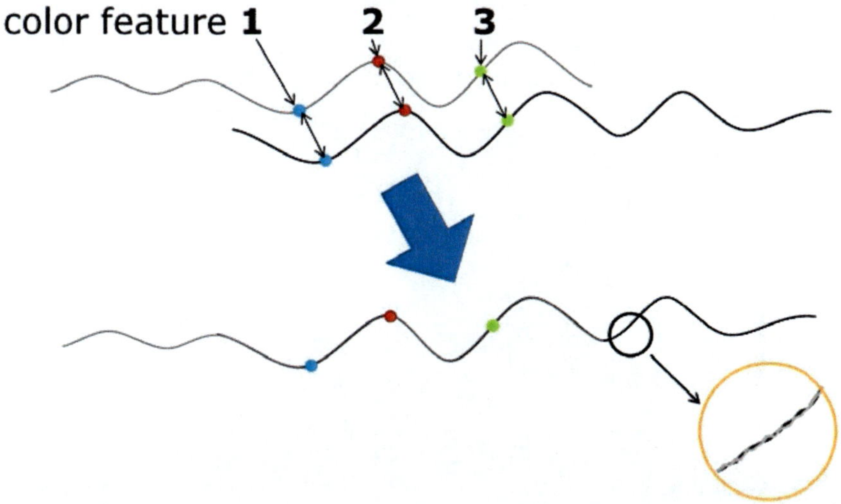

Fig. 7.21 Top: two profiles from adjacent datasets before alignment. The three points show corresponding colour features. Below: the two profiles after the alignment

7.7.4 3D Alignment

Almost all optical metrology instruments described in this book measure the sample from one direction. This is sufficient, if the surface texture is measured. For many form measurements (for example, drill bits, thread cutters), the sample has to be measured from different directions. One solution to perform such measurements is to rotate the sample using a rotation unit (Fig. 7.22). A combined dataset generated from the measured individual 3D datasets is then used for form and roughness measurements (see Fig. 7.23).

In order to ensure a highly accurate combined 3D dataset, the position of the individual 3D datasets must be controlled accurately. This can be achieved using a highly accurate rotation unit or by using software alignment within the overlap region of adjacent datasets (Fig. 7.24 and Fig. 7.25). The second method can be realized by the following procedure.

1. Measure the sample at different rotation angles.
2. Refine the positions of all individual datasets (three translation values and three rotation angles).
3. Combine the individual datasets at the refined positions.

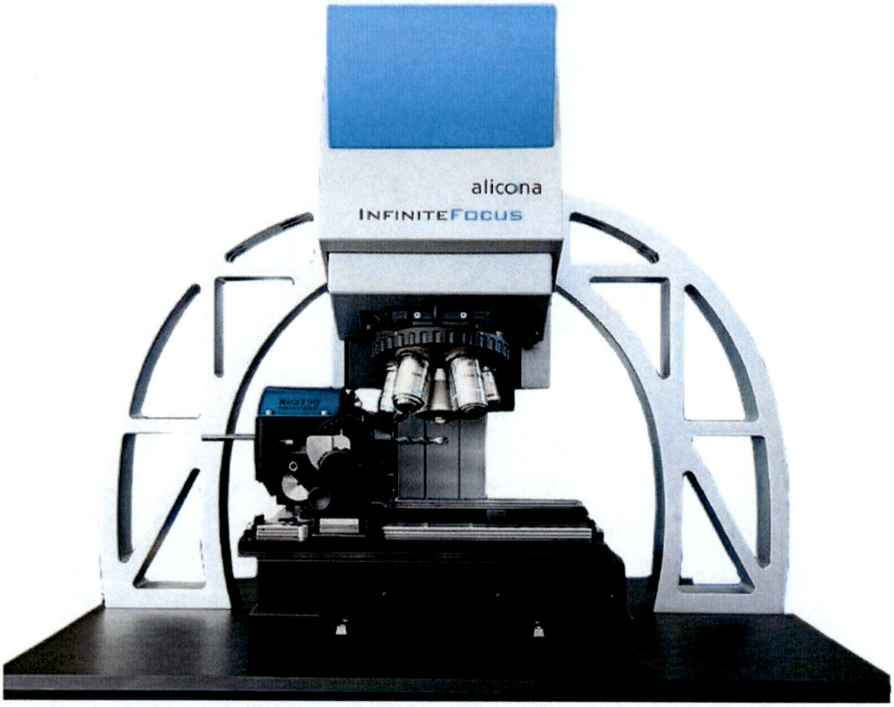

Fig. 7.22 A rotation unit mounted on the *xy* stage of a measuring instrument to generate 3D data

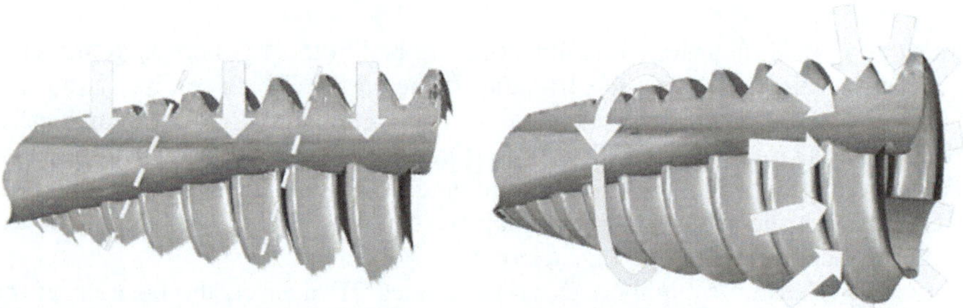

Fig. 7.23 Left: 3D dataset shows a measurement of a screw tap from one direction. Right: dataset shows the same sample measured from multiple directions

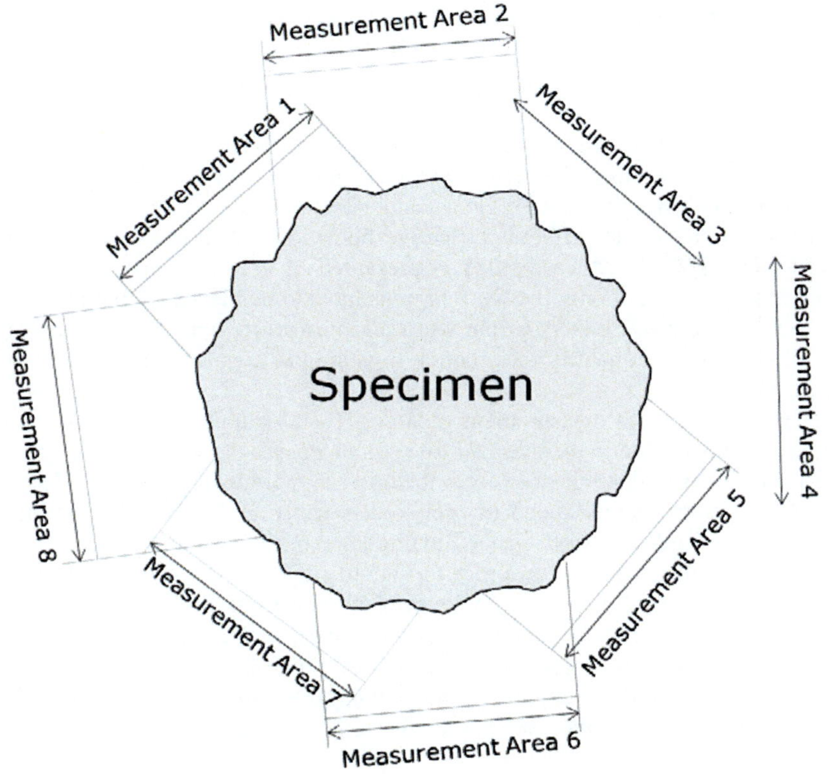

Fig. 7.24 The measurement positions at different angles with overlap for a circular sample

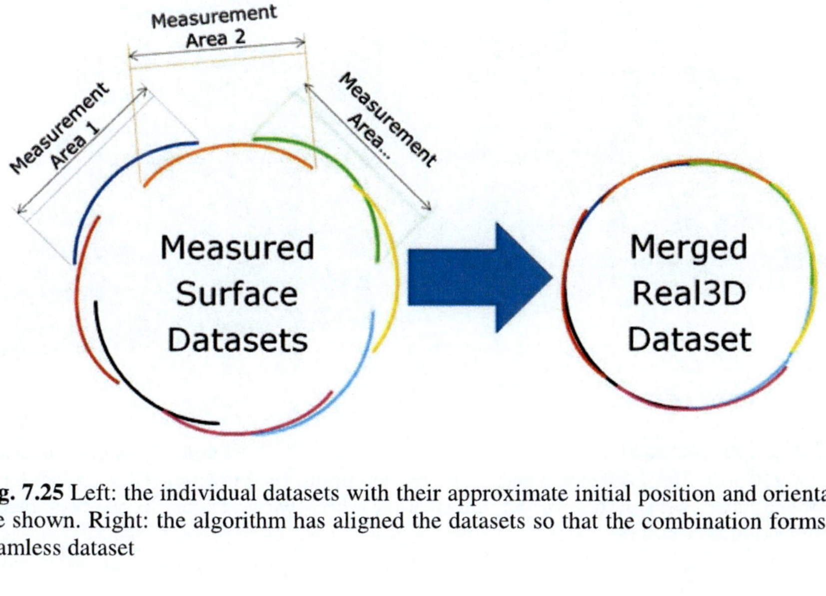

Fig. 7.25 Left: the individual datasets with their approximate initial position and orientation are shown. Right: the algorithm has aligned the datasets so that the combination forms one seamless dataset

7.8 Case Studies

7.8.1 Surface Texture Measurement of Worn Metal Parts

A common problem in the automotive industry is to identify wear on different components and to measure its influence on the function of the surface. Often this is a complicated task since worn regions cannot be easily identified, or are not easily measured due to the different reflective behaviour or their different topographical characteristics. A case study is presented of a wear measurement on a metal sample, where not only the worn regions have to be identified, but also various surface texture parameters in the worn and unworn regions are compared to each other in order to quantify the amount of wear, and to gain deeper insight into the wear process.

In Fig. 7.26, the 3D measurement of a cast metal sample obtained by focus variation is shown. Due to the overlaid true colour image, the worn regions (bright colour) can be easily distinguished from the unworn regions (dark colour). The 3D measurement with heights coded in pseudo-colours (Fig. 7.26b) does not allow this easy classification at first glance although worn regions can be identified to have fewer significant peaks. In order to quantify the amount and type of wear, various surface texture parameters have been obtained, for the worn and for the unworn regions. The region selection has been made on the basis of the true colour image. The resulting surface texture parameters in Tab. 7.2 show that the worn regions are significantly smoother (smaller *Sdq* value) and have less material in the peak region (smaller *Vmp* values) than the original parts.

(a)

Fig. 7.26 (a) 3D measurement of a metal sample with overlaid true colour image (converted to black and white for print) showing worn regions (bright) and unworn regions (dark). (b) 3D measurement with pseudo-colour information where worn regions have fewer peaks (yellow and orange colour)

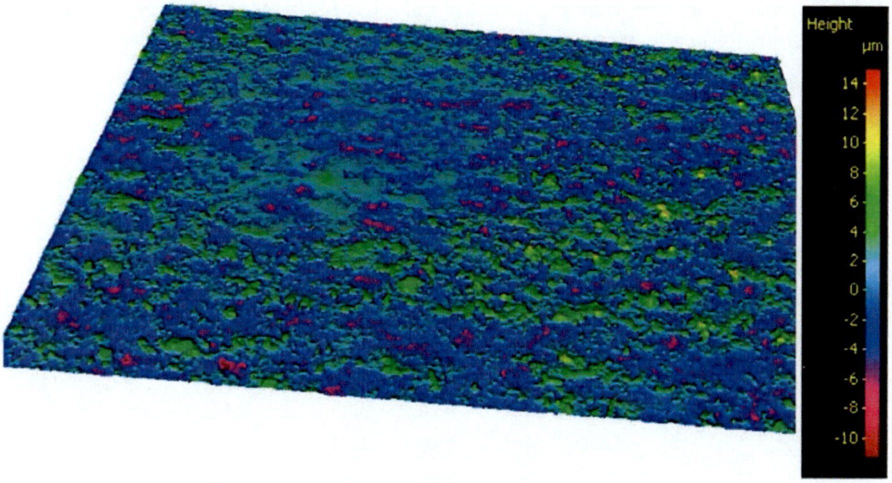

(b)

Fig. 7.26 (*continued*)

In order to check whether the mean height of the worn region is lower than those of the original parts (which means that the material has been removed), a surface profile (Fig. 7.27b) has been extracted along the profile path in Fig. 7.27a. The profile shows that the worn regions are approximately at the same height level as the other parts, but have significantly fewer peaks and less deep valleys. Again the true colour information allows a perfect assignment of profile parts to the worn and the unworn region.

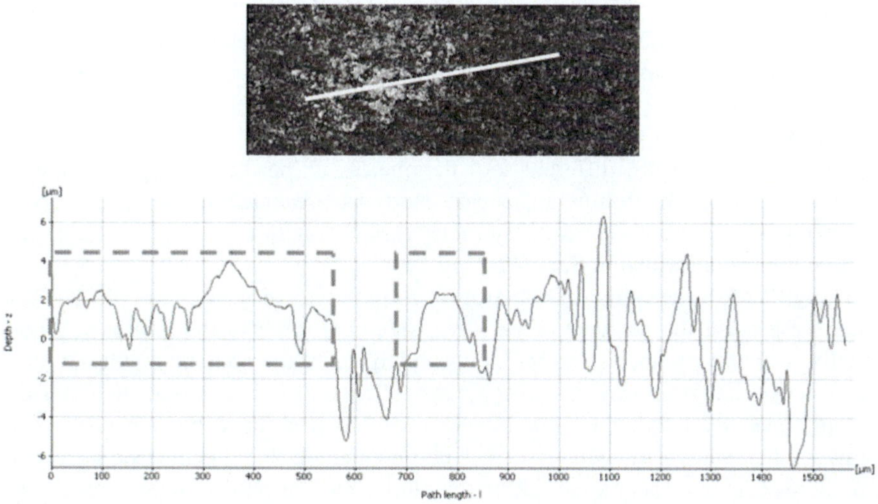

Fig. 7.27 Top: white profile path used to extract the surface profile at the bottom. Worn regions (white in the true colour image) are marked with red boxes in the height profile and show fewer significant peaks and valleys but are at a similar height level as the unworn parts

Table 7.2 Surface texture parameters for worn and unworn regions. Parameters significant for discrimination are marked bold. For example, the material volume in the peak region (*Vmp*) of the original parts is about twice those of the worn regions.

Name	unit	Wear	Original
Sa	μm	1.73	2.39
Sq	μm	2.16	2.99
Ssk		-0.66	0.29
Sku		3.28	2.93
Sdq		0.26	0.36
Sdr	%	3.12	6.10
Sk	μm	4.95	7.73
Spk	μm	1.26	3.33
Svk	μm	2.96	2.17
*Smr*1	%	5.91	11.92
*Smr*2	%	83.33	92.57
Vmp	ml/m²	0.07	0.16
Vmc	ml/m²	2.04	2.67
Vvc	ml/m²	2.19	3.89
Vvv	ml/m²	0.31	0.28

7.8.2 Form Measurement of Complex Tap Parameters

The 3D measurement of cutting tools, drill bits and milling cutters is an important aspect of quality control in the industrial process, since the form and wear of cutting edges have a significant influence on the quality of machined parts. The complete 3D measurement of such tools is only partly possible by tactile methods. This restriction is due to the complex geometry of such tools and the long measurement time. Optical methods are an ideal alternative for fast generation of dense 3D datasets to measure different tool parameters or perform comparisons with reference data. In the following, a case study is presented where the complex parameters of a tap are measured by focus variation with the aid of an additional rotation unit as described in Sect. 7.6.

In Fig. 7.28, a measured 3D dataset of a tap is shown in pseudo-colours. The colours represent the distance of each measurement point to the tool axis. Additionally, a cutting plane is shown along which a surface contour of the tap in a cross section has been obtained (Fig. 7.28b). Based on this contour, the outer diameter (2061 μm) and the core diameter (995.9 μm) have been measured as shown by the two circles. Moreover, the chipping angle (9.54°) has been measured (Fig. 7.28c). This measurement is carried out by measuring the angle between the centre of the inner circle, the cutting edge and a line that has been fitted to the region of the chipping surface.

Another measurement is the tap relief — the reduction of the radius from the front to the back part of a tap land. First, a contour has been extracted along a helical cutting surface as shown in yellow in Fig. 7.28d. Afterwards, the measurement points are converted to a profile where the radius of the measurement point is set in relation to the angle (Fig. 7.28e). Here only the measurement points of one revolution are shown. From this view, the relief can be obtained by measurement of the height difference (73 μm) at the marked positions. As for the other complex parameters, no special orientation of the measurement instrument is necessary as is the case for conventional systems (operating with back light). Instead, the parameters are directly measured from the 3D dataset.

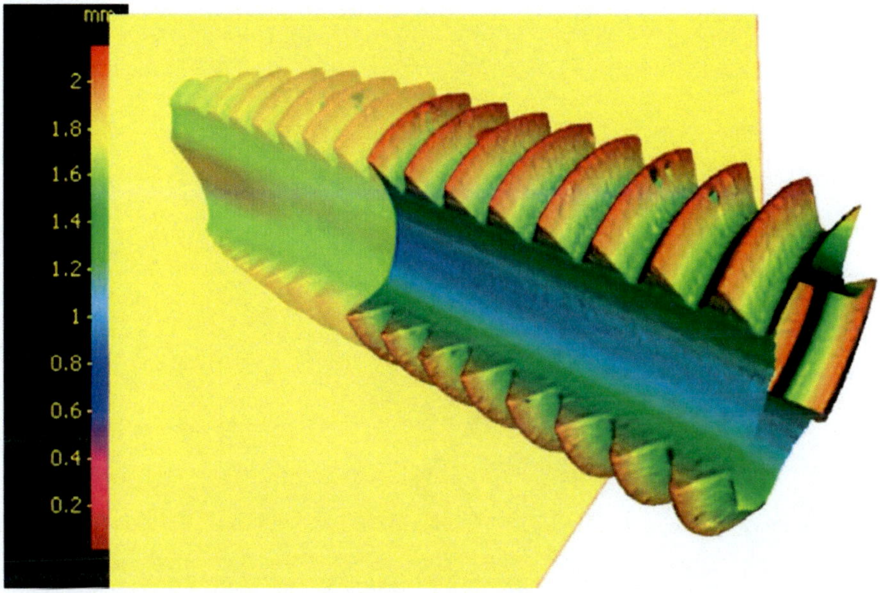

(a)

Fig. 7.28 (a) 3D dataset of a tap with cutting plane. (b) Cross section with measured outer and inner diameter. (c) Measurement of the chip angle. (d) 3D dataset with a helical cutting surface for relief measurement. (e) Measurement of the tap relief

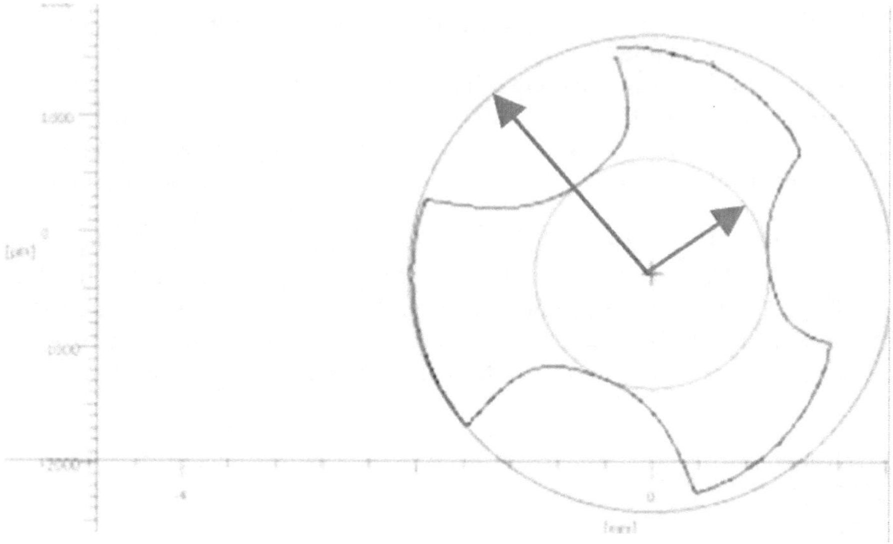

(b)

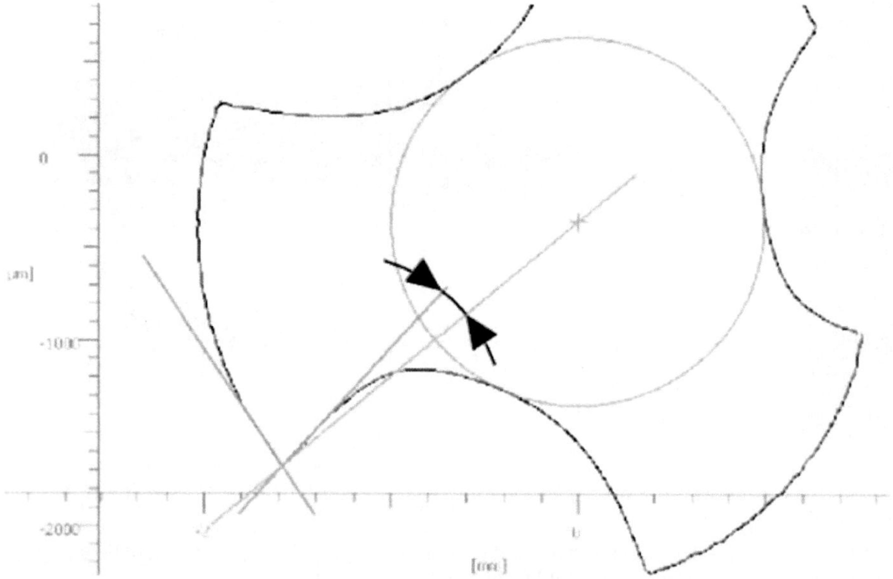

(c)

Fig. 7.28 (*continued*)

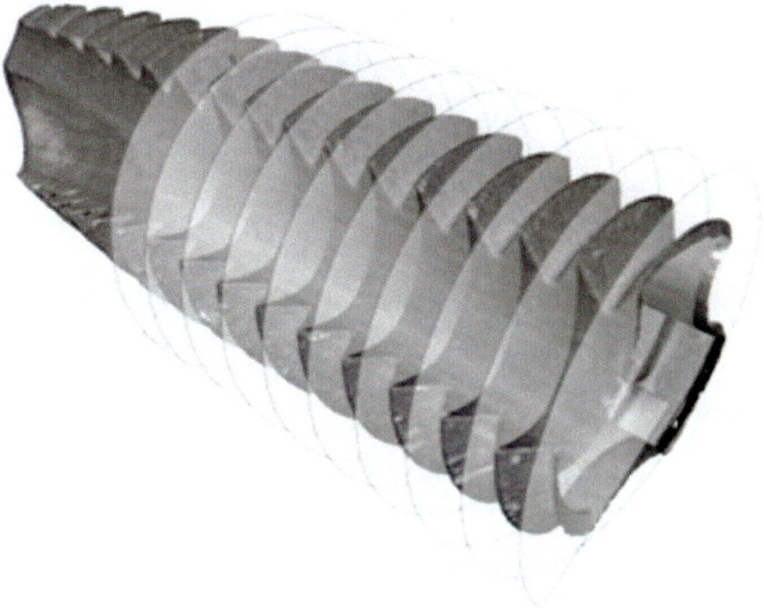

(d)

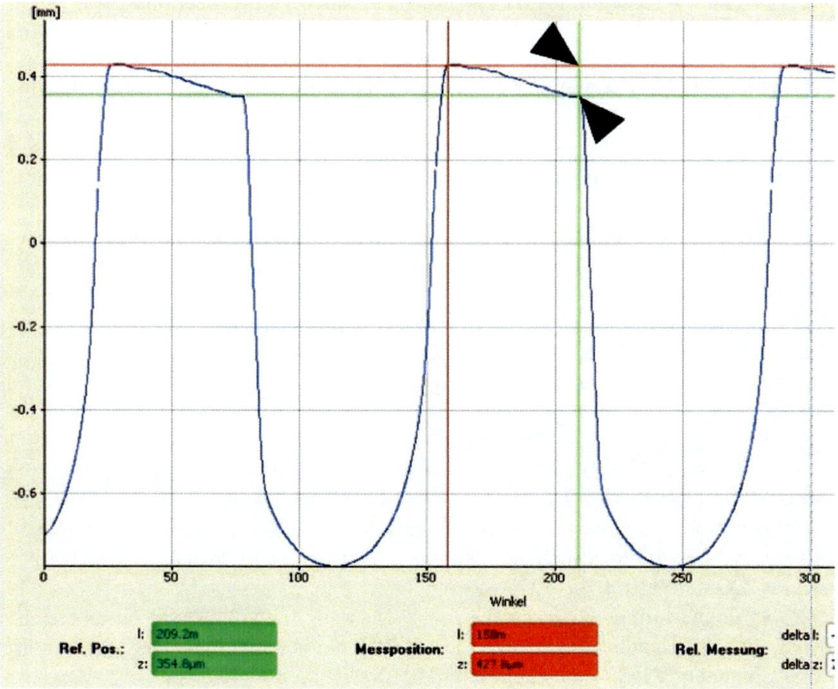

(e)

Fig. 7.28 (*continued*)

7.9 Conclusion

In this chapter, focus variation technology has been described together with the main components of an instrument, the limitations and possible extensions. To conclude, focus variation technology can be used for form and roughness measurement so long as the samples are not too smooth.

Acknowledgements

The author would like to thank Dr Reinhard Danzl and Ms Verena Harratzmüller (Alicona) for their great help with this chapter.

References

Danzl, R., Helmli, F.: Form measurement of engineering parts using an optical measurement system based on focus variation. In: Proc. 7th Int. euspen Conf., Bremen, pp. 270–273 (May 2007)

Danzl, R., Helmli, F., Scherer, S.: Automatic measurement of calibration standards with arrays of hemi-spherical calottes. In: 11th Int. Conf. Metrology and Properties of Engineering Surfaces, Huddersfield, pp. 41–46 (July 2007)

Danzl, R., Helmli, F.: Measurement of geometric primitives with uncertainty estimation. In: Proc. XII. International Colloquium on Surfaces, Chemnitz, pp. 328–335 (February 2008)

von Helmholtz, H.L.F.: Helmholtz's treatise on physiological optics. Optical Society of America (1924)

ISO 25178-6, Geometrical product specifications (GPS) - Surface texture: Areal - Part 6: Classification of methods for measuring surface texture. International Organization for Standardization (2010)

ISO/WD 25178-606, Geometrical product specification (GPS) – Surface texture: Areal – Nominal characteristics of non-contact (focus variation) instruments. International Organization for Standardization (2011)

Keck, C., Franke, M., Schwenke, H.: Werkstückeinflüsse in der Koordinatenmesstechnik. Technisches Messen 71, 81–92 (2004)

Leach, R.K.: Fundamental principles of engineering nanometrology. Elsevier, Amsterdam (2009)

Nayar, S.K.: Shape from focus. Department of Electrical and Computer Engineering, The Robotics Institute, Carnegie Mellon University, Pittsburgh, Pennsylvania 15213 (1989)

Niederöst, M., Niederöst, J., Scucka, J.: Automatic 3D reconstruction and visualization of microscopic objects from a monoscopic multifocus image sequence. International Archives of the Photogarmmetry, Remote Sensing and Spatial Information Sciences XXXIV 5/W10, 1682–1777 (2003)

Scherer, S.: Fokus-variation zur optischen 3-D-Messung im Mikro- und Nanobereich. In: Bauer, N. (ed.) Handbuch zur Industriellen Bildverarbeitung: Qualitätssicherung in der Praxis, Fraunhofer IRB Verlag, pp. 198–210 (2007)

Waldorp, L.J., Grasman, R.P.P.P., Huizenga, H.M.: Goodness-of-fit and confidence intervals of approximate models. J. Math. Psych. 50, 203–213 (2006)

8 Phase Shifting Interferometry

Peter de Groot

Zygo Corporation
Laurel Brook Road
Middlefield, Connecticut, USA

Abstract. Phase shifting interferometry is a well-established technique for areal surface characterisation that relies on digitisation of interference data acquired during a controlled phase shift, most often introduced by controlled mechanical oscillation of an interference objective. The technique provides full 3D images with typical height measurement repeatability of less than 1 nm independent of field size. Microscopes for interferometry employ a range of specialised interference objectives for roughness and microscopic form measurement.

8.1 Concept and Overview

From the earliest days of measuring microscopy, it was understood that a reference reflection superimposed on the object reflection would generate fringes of interference that relate to surface topography. Perhaps the simplest interference instrument is an ordinary, non-interferometric microscope observing an object with a cover slip positioned close to the object surface, to provide the necessary reference reflection. In ordinary white light illumination, the resulting Fizeau interference fringes are multi-coloured and have the strongest contrast when the cover slip is almost in physical contact with the object surface. The shapes of the fringes (straight, circular, irregular) are indicative of the surface shape. Microscopes specifically designed for interferometry employ a specialised objective with an internal beam-splitter and reference allowing convenient optical contact with a large physical working distance to the object. Such microscopes also often make use of a narrow-band filter or near-monochromatic light source to improve the fringe contrast over a wide range of measurement conditions.

Modern interference microscopes generally employ image detection using an electronic camera and computer analysis for convenience, measurement accuracy and automation. From 1980 to 1990, there were significant developments of these automated 3D measuring microscopes using the principles of phase shifting interferometry (PSI), first developed in the context of optical testing of lenses and mirrors. PSI acquires a sequence of images with a precisely controlled phase change between them, which when a few fringes are visible on the surface, manifests itself as a shift in fringe position between the images captured by a camera. The phase shifting is almost always generated by a mechanical motion of the interference objective, which allows for fast, non-contact metrology.

This chapter discusses the most common forms of interference microscopy for 3D surface measurement using an internal reference mirror. The common feature of PSI instruments is that, unlike differential or phase contrast methods, surface heights are directly proportional to interference phase.

8.2 Principles of Surface Measurement Interferometry

Interferometers take advantage of the wave properties of light to analyse surface characteristics, including in particular surface height variations. For evaluation of areal surface topography, interferometers separate source light so that it follows two independent paths, one of which includes a reference surface and the other the object surface. The separated light beams then recombine and are directed to a digital camera that measures the resultant light intensity over multiple image points simultaneously. The intensity of the recombined light exhibits high sensitivity to the differences in path lengths, effectively comparing the object surface with the reference surface with nanometre resolution.

The interference phenomenon may be understood by considering the classical Michelson interferometer shown in Fig. 8.1, here assumed equipped with a high-coherence light source such as a laser. Following the usual two-beam interference analysis (see for example Hariharan 1992), the interference signal observed at the square-law detector can be written

$$I(h,\zeta) = I_{DC} + I_{AC}\cos\left[K(h-\zeta)+\xi\right], \tag{8.1}$$

where I_{DC} and $I_{AC} < I_{DC}$ are fixed coefficients. The quantity

$$K = 4\pi/\lambda, \tag{8.2}$$

sometimes called the fringe frequency, corresponds to the rate at which the interference signal oscillates sinusoidally as a function of changes in sample surface height h or in position ζ of the reference mirror. In equation (8.2), λ is the source wavelength and ξ is a phase offset related to the reflection and transmission properties of the interferometer components. The inset graph in Fig. 8.1 shows the variation in detected intensity as a function of the difference $h-\zeta$. The periodic modulations are characteristic of interference phenomena, with a full cycle of modulation every half wavelength of motion of the reference mirror.

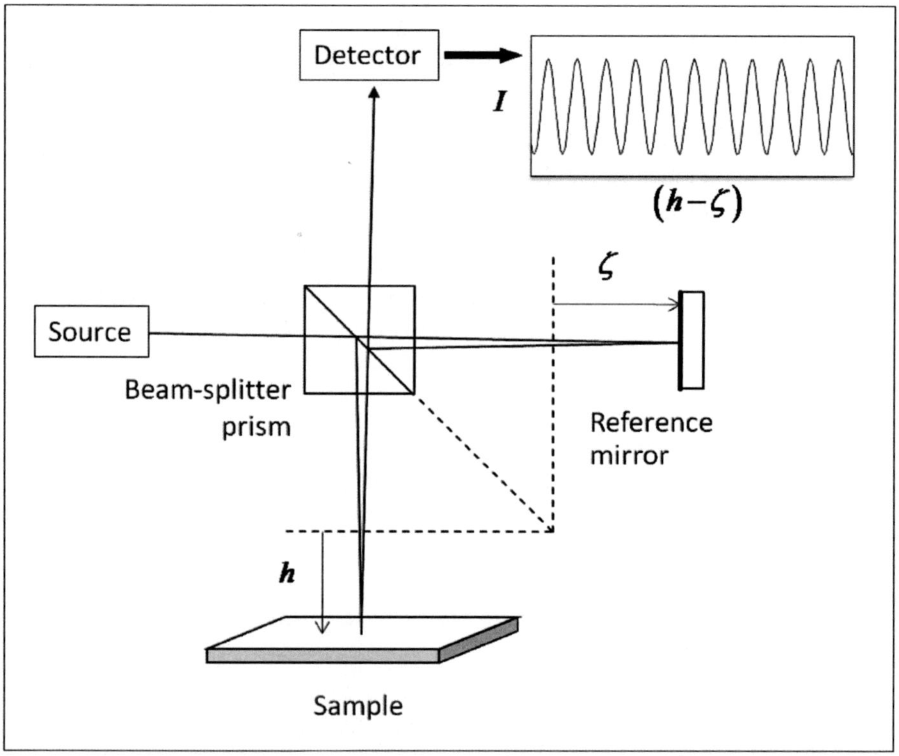

Fig. 8.1 Michelson interferometer

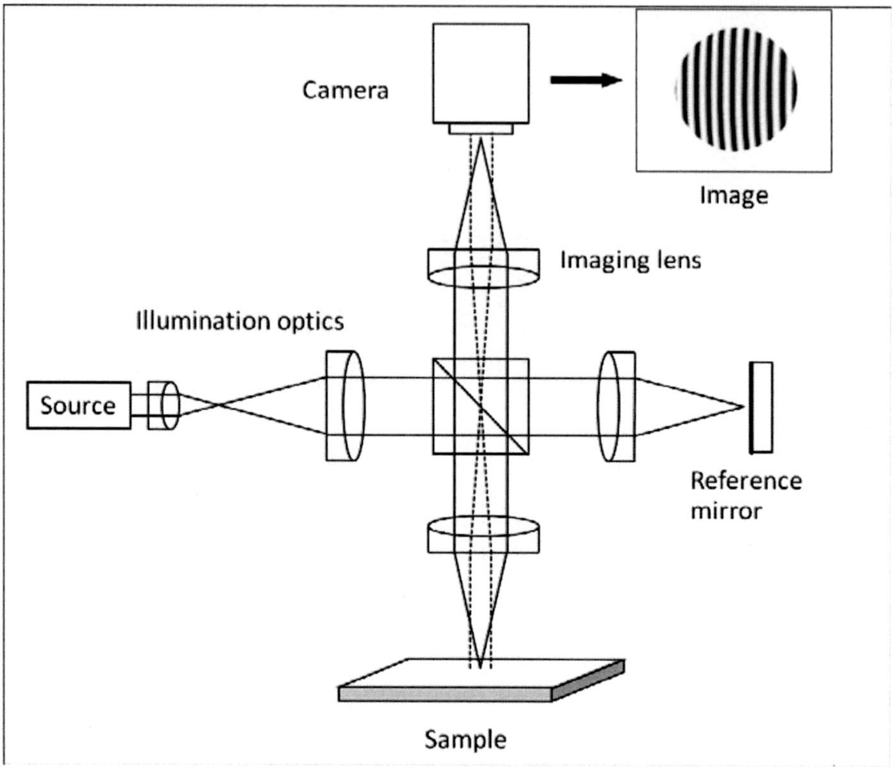

Fig. 8.2 Imaging interferometer for areal surface profiling

An upgrade to the optical system converts the Michelson interferometer into a tool for areal surface topography measurement. The addition of lenses and an electronic camera in Fig. 8.2 creates a digital image such that each camera pixel corresponds to a conjugate point on the object surface. In Fig. 8.2, a small amount of object tilt introduces a continuously varying surface height h in the horizontal direction, which shows up as light and dark bands or fringes of interference.

The interference fringe image of Fig. 8.2 can be directly interpreted, given that the fringes map out areas of equal surface height. Thus, the bright fringes correspond to areas where the argument of the cosine in equation (8.1) is an integer multiple of 2π, which from equation (8.2) means that the change in surface height between neighbouring fringes is $\lambda/2$. This method of surface characterisation by interferometry has its origin in visual interpretations of fringe patterns, and has modern incarnations in computerised methods based on analysis of high-density fringes (Takeda 1982).

8.3 Phase Shifting Method

The need for high accuracy surface measurement requires a more sophisticated approach than the visual interpretation of fringe patterns. Therefore, a digital processing of the electronic images refines the estimate of the phase

$$\theta = Kh + \xi \tag{8.3}$$

in the argument of the cosine in equation (8.1). Assuming that the offset ξ is constant across the field of view, it can be set to zero and the surface height map follows from the inversion

$$h = \theta / K . \tag{8.4}$$

Phase shifting techniques were first employed for optical testing in laser interferometers such as the Fizeau and Twyman-Green (Bruning 1974). In the 1980s, PSI was adapted to interference microscopy of fine surface features and roughness (Koliopoulos 1981, Wyant et al. 1984). An extensive review literature on PSI is now available (Creath 1988, Schreiber and Bruning 2007, Schmit et al. 2007).

PSI relies on controlled phase shifts usually imparted by the mechanical motion ζ of the reference mirror. Sometimes referred to as temporal PSI to distinguish the method from the spatial interpretation of individual fringe pattern images, PSI analyses the interference signals from each pixel individually using what amounts to heterodyne detection. For the case of a moving reference mirror, the interference signal in equation (8.1) becomes

$$I(\varphi) = I_{DC} + I_{AC} \cos(\theta + \phi), \tag{8.5}$$

where

$$\phi = -\zeta / 4\pi \tag{8.6}$$

is the phase shift in terms of the reference mirror displacement ζ.

In the most common PSI method, the reference mirror (or equivalent phase-shifting device) imparts a linear change in phase ϕ during data acquisition, leading to an intensity signal similar to that shown in the inset graph of Fig. 8.1 for every image pixel, but with a starting phase offset θ that depends on the local surface height. To analyse such a signal, one approach is to recognise that equation (8.5) expands to

$$I(\phi) = I_{DC} + I_{AC}\left[\cos(\theta)\cos(\phi) - \sin(\theta)\sin(\phi)\right]. \tag{8.7}$$

Fitting sine and cosine waves to the interference signal gives the $\sin(\theta)$ and $\cos(\theta)$ terms, respectively, in equation (8.7); providing quadrature signals that allow an accurate determination of θ. The following integrals illustrate one method to achieve quadrature signals by integration over a full cycle of phase ϕ

$$N = -\int_{-\pi}^{\pi} I(\phi)\sin(\phi)\,\mathrm{d}\phi$$
$$D = \int_{-\pi}^{\pi} I(\phi)\cos(\phi)\,\mathrm{d}\phi \quad . \qquad (8.8)$$

The phase θ then follows from

$$\tan(\theta) = N/D\,. \qquad (8.9)$$

This autocorrelation or synchronous detection technique is equivalent to Fourier analysis at a single frequency determined by the rate of movement of the reference mirror (Bruning 1974).

In practice, the signal $I(\varphi)$ is sampled discretely as a sequence of digitally captured camera frames. As an example, assume the acquisition of four sample intensities $I_{0,1,2,3}$ corresponding to four phase shifts $\varphi_{0,1,2,3}$ starting at $\phi_0 = -3\pi/4$ and equally spaced by $\Delta\phi = \pi/2$. Equation (8.8) becomes

$$N = I_0 + I_1 - I_2 + I_3$$
$$D = -I_0 + I_1 + I_2 - I_3 \quad . \qquad (8.10)$$

Table 8.1 Example PSI algorithms with uniform phase shift

P	$\Delta\phi$	$\tan(\theta)$
3	$\frac{\pi}{2}$	$\frac{I_0 - I_2}{-I_0 + 2I_1 - I_2}$
5	$\frac{\pi}{2}$	$\frac{2I_1 - 2I_3}{-I_0 + 2I_2 - I_4}$
7	$\frac{\pi}{2}$	$\frac{-I_0 + 7I_2 - 7I_4 + I_6}{-4I_1 + 8I_3 - 4I_5}$
13	$\frac{\pi}{4}$	$\frac{-3(I_0 - I_{12}) - 4(I_1 - I_{11}) + 12(I_3 - I_9) + 21(I_4 - I_8) + 16(I_5 - I_7)}{-4(I_1 + I_{11}) - 12(I_2 + I_3 + I_9 + I_{10}) + 16(I_5 + I_7) + 24I_6}$

PSI algorithms have anywhere from a minimum of three to as many as twenty sample intensities. Discrete sampling requires full normalization, the orthogonality of N, D and the suppression of the offset term I_{DC}. More refined algorithms include the use of Fourier theory (Freischlad and Koliopolous 1990, de Groot 1995a), and characteristic polynomials (Surrel 1996) to optimize performance. Design rules for PSI algorithms cover a wide range of mathematical methods. Table 8.1 provides example PSI algorithms employing a constant $\Delta\phi$ phase shift between P intensity samples. Review articles provide more comprehensive lists, together with the relative performances in the presence of error sources (Malacara *et al.* 1998, Schreiber and Bruning 2007).

The algorithms in Tab. 8.1 are designed for a linear phase shift. Alternative phase shift patterns have also been demonstrated. An example alternative is a phase shift which is itself sinusoidal, a technique that has the advantage of being less demanding on phase shift mechanisms than repeated linear ramps (Sasaki 1987, Dubois 2001). In this case, the phase shift becomes

$$\phi = u\cos(\alpha), \tag{8.11}$$

where it is now the sinusoidal phase α that is incremented in equal steps $\Delta\alpha$ between intensity samples. The resulting intensity pattern is a cosine of a cosine, which generates a signal comprised of multiple harmonics of the original phase modulation. Unlike linear phase shifting, in sinusoidal phase shifting the phase θ follows from an evaluation of the relative strengths of the odd and even harmonics. This can be carried out efficiently using algorithms that have the same basic form as that of equation (8.9), for example using the following eight-frame algorithm (de Groot 2009) designed for $\Delta\alpha = \pi/4$, a starting offset phase $\alpha_0 = \pi/8$, and phase shift amplitude $u = 2.93$

$$\tan(\theta) = \frac{-1.6647(I_1 - I_2 - I_5 + I_6)}{-(I_0 + I_3 + I_4 + I_7) + (I_1 + I_2 + I_5 + I_6)}. \tag{8.12}$$

Phase shifting methods do not necessarily require an accurate phase shift of any particular form. If the phase shifts are known or can be measured by a separate displacement sensor, the surface height follows from a generalised least-squares fit (Grievencamp 1984, Lai and Yatagai 1991). Alternatively, if there is sufficient information within each individual interference image to infer the phase shift between camera frames, data processing can accommodate a certain degree of unexpected random motion generated by mechanical vibrations (Deck 2009).

In addition to these alternative time-dependent phase shift methods, it has been proposed to use multiple detectors to acquire phase-shifted images simultaneously in an interference microscope using, for example, polarization encoding of the reference and measurement beams (Massig 1992).

8.4 Phase Unwrapping

The algorithms for PSI return a phase value within the 2π range of the arctangent function. As a consequence, the result of a PSI measurement includes an unintended ambiguity in surface height of an unknown integer value times the height change corresponding to one interference fringe. The basic task to remove this ambiguity is illustrated in Fig. 8.3. In Fig. 8.3, the vertical height scale is in units of wavelength. The wrapped data are modulo ½ the wavelength, equivalent to a 2π phase interval.

Methods of unwrapping or connecting images generated by interferometry form a discipline unto themselves (Ghiglia and Pritt, 1998). Experienced users of interferometers will readily confirm that one of the most common failure modes for these tools is a mistaken fringe order, sometimes referred to as a 2π error, often the result of surface texture or other surface properties that exceed the software's ability to perform a correct unwrapping. 2π errors manifest as spikes or false steps in the 3D images with a height equal to the height spacing corresponding to one interference fringe.

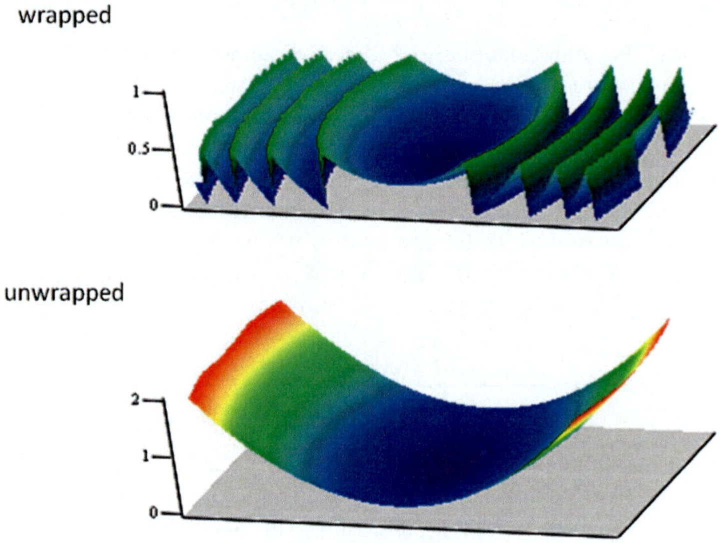

Fig. 8.3 Wrapped and unwrapped 3D images generated by PSI, the ordinates are in units of wavelength

8.5 Phase Shifting Error Analysis

PSI is a mature technique with an abundant literature regarding sources of measurement error, often targeting the task of identifying the best algorithm design methodology in the presence of detector nonlinearity, intensity noise and errors in the phase shift (Schwider et al. 1983, Creath 1988, Brophy 1990, de Groot 1995b, Schreiber and Bruning 2007). Fig. 8.4 compares the performance of the first three algorithms in Tab. 8.1 in the presence of phase shift calibration error (Fig.8.4, left) and 1 nm amplitude mechanical vibrational noise using a 100 Hz camera (Fig. 8.4, right).

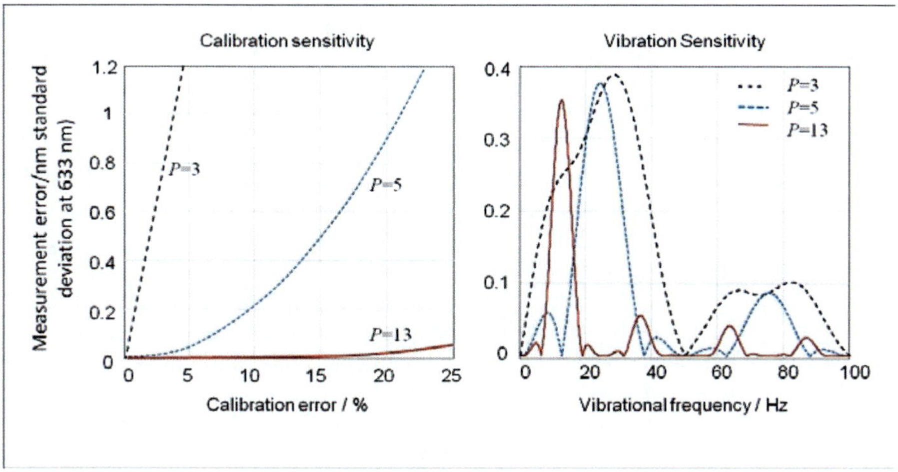

Fig. 8.4 PSI calibration sensitivity (left) vibrational sensitivity (right) for the first three algorithms in Table 8.1

8.6 Interferometer Design

The history of interferometric methods for characterisation of microscopic objects has been previously reviewed by Krug et al. (1964) and by Pluta (1993). The range of instrumentation for interference microscopy extends from the simple addition of a reference surface directly, to the object to the elaboration of complex, dedicated platforms employing variations of all known methods for division of wavefront and for division of amplitude adapted to high magnification instruments.

Modern interference microscopes have evolved to instrument platforms recognizable as refined versions of a conventional microscope, configured for digital data acquisition and low vibration, combined with one or more removable interference objectives that take the place of the conventional imaging microscope objective. This basic design provides flexibility and convenience, allowing the use of a turret of interference objectives of varying magnification and interferometer design depending on the needs of the application.

Fig. 8.5 illustrates the optical configuration of a microscope for interchangeable interference objectives, using a common interferometer design derived from the modified Michelson geometry of Fig. 8.2. Within the objective, a beam-splitter placed between the imaging optics and the sample establishes the reference and measurement paths. The beam-splitter is most often a prism. Variants of the basic design include those with fixed or adjustable dispersion compensation for use with samples having a transparent glass cover (Hedley 2001, Han 2006).

The system magnification is determined by the ratio of the tube lens focal length to the focal length of the objective. The objective magnification is defined in terms of a nominal unity-magnification tube lens focal length, which varies between 160 mm and 200 mm, depending on the manufacturer. Thus a 10× objective for a defined 200 mm tube lens has a focal length of 20 mm.

The working distance from the objective housing to the object is a function of the design of the objective, and relates to factors such as interferometer geometry, focal length, mechanical structure and lens design. A large working distance is generally preferred and can be a deciding factor in the choice of interferometer geometry.

At higher magnifications, usually for 20× and above, there is insufficient working distance to accommodate a large beam-splitter between the objective lens and the sample and the Michelson geometry is no longer practical. Fig. 8.6 shows a Mirau objective with a parallel beam-splitter and reference surface, both aligned with the optical axis of the imaging lenses (Mirau 1952). The reference is a small reflective disk, somewhat larger in diameter than the field of view of the objective, and typically comprised of a coating on an otherwise transparent supporting plate of identical thickness to the beam-splitter. The design inherently relies on a sufficiently large numerical aperture (see Chapter 2) to allow at least a portion of the illuminating and imaging rays to pass around the reflecting reference disk, often referred to as the central obscuration. The Mirau is by far the most common high-magnification interference objective in use today.

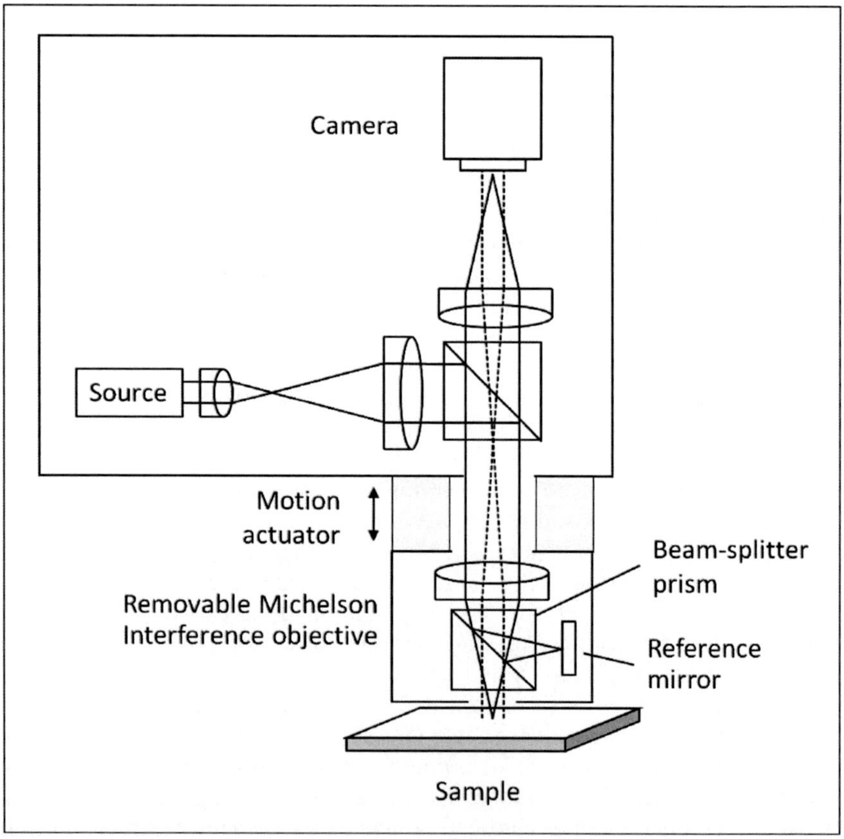

Fig. 8.5 Interferometer for areal surface measurement employing a Michelson-type interference objective

For very high magnifications of 100× and above, the presence of even the thin beam-splitter plate in the Mirau becomes awkward enough to require the movement of the beam-splitter to above the lenses, as shown in the objective of Fig. 8.7 (Linnik 1938). For the Linnik objective the imaging optics are included directly in the measurement and reference paths, similarly to Fig. 8.2. This makes the Linnik objective more challenging to design and to manufacture because the separate imaging objectives ideally must perfectly match in terms of aberrations and dispersion characteristics, which is much less an issue with the Michelson and Mirau designs. This complication is balanced by superior working distance and greater achievable numerical aperture at high magnification.

In practice, the Michelson is the preferred geometry for low magnifications of 10× and lower. The Mirau is common for 20× to 100×, while the Linnik is sometimes used at 50× and 100×.

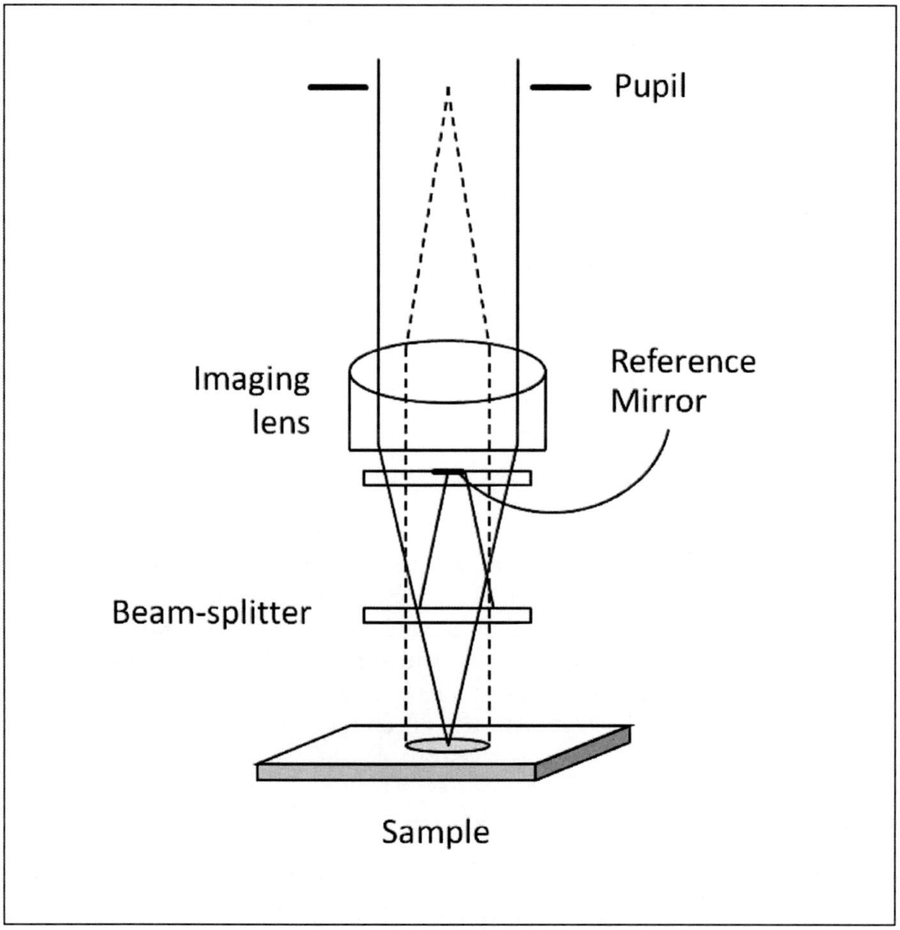

Fig. 8.6 Mirau interference objective

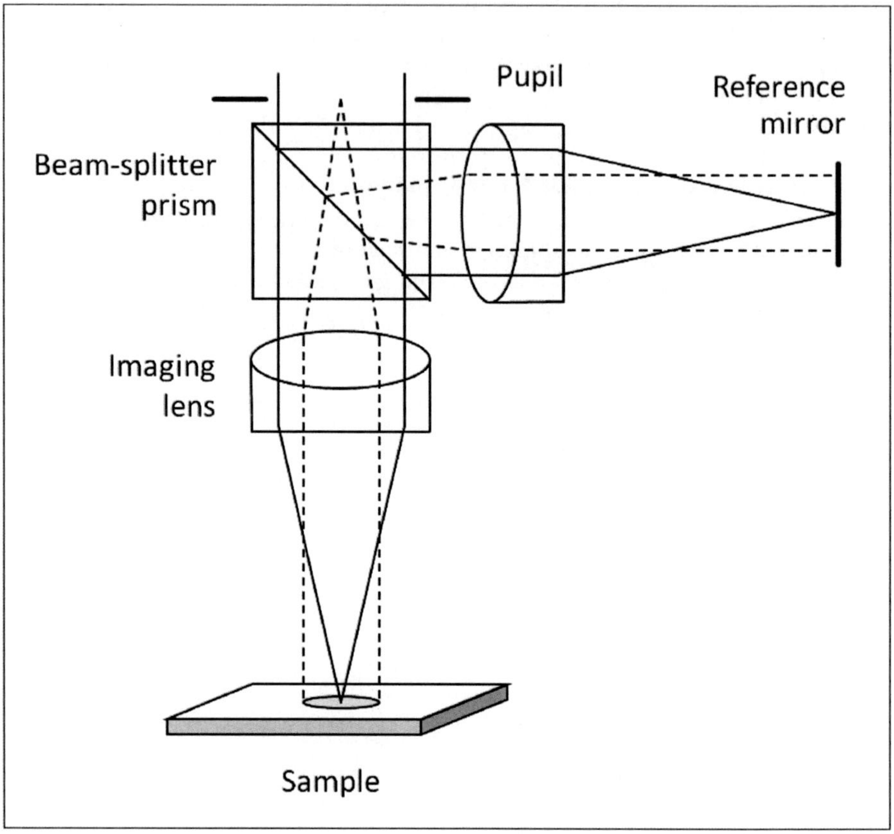

Fig. 8.7 Linnik interference objective

Not considered here, but at least of historical importance, are the many types of self-referencing interferometers, for which the object surface itself serves as both the reference and sample surfaces (Sommargren 1981). These instruments are relatively uncommon today but have the continuing interest of being insensitive to vibration, which is a chief limitation in interference microscopy for many production applications where the environment is uncontrolled.

8.7 Lateral Resolution

Areal surface characterisation requires, in addition to accurate measurements of surface height, a clear quantification of the instrument's ability to discern or resolve fine surface detail. Often referred to as lateral resolution, this performance specification for interference microscopes varies significantly with magnification and can have a strong effect on the perceived variations in surface height, particularly for roughness (see Chapter 2).

An informative method of describing the lateral resolving power of interference microscopes is by the instrument transfer function (ITF) (Takacs 1993, de Groot and Colonna de Lega 2006). The ITF describes how the instrument would respond to an object surface having a specific spatial frequency. The ITF is the 3D measurement analogy to the more conventional optical transfer function (OTF), the latter referring to an optical instrument's ability to generate 2D images.

The ITF tells us what the measured amplitude of a sinusoidal grating of a specified spatial frequency ν (in lines per millimetre) would be relative to the true amplitude of the sinusoid. In the limit of very small surface heights, the magnitude of the ITF for a PSI instrument is the same as the more familiar modulation transfer function (MTF). Using the result for the OTF, the ITF of a PSI instrument having an incoherent light source that fills the objective pupil is

$$\mathrm{ITF}(\nu) = \frac{2}{\pi}\left[\phi - \cos(\phi)\sin(\phi)\right] \tag{8.13}$$

where

$$\phi = \arccos\left(\frac{\lambda\nu}{2A_N}\right). \tag{8.14}$$

This ITF must be multiplied by the MTF of the camera, which has limited resolution related to pixel size. Fig. 8.8 shows the ITF for 5×, 20× and 100× microscope objectives (numerical aperture A_N equal to 0.13, 0.3 and 0.8, respectively).

Setting aside the contribution of the camera resolution, the ITF reaches zero for a spatial frequency given by $\phi = 0$ or

$$\nu_{0\%} = \frac{2A_N}{\lambda}, \tag{8.15}$$

which is the Sparrow criterion for optical resolution. The 50 % point is at

$$\nu_{50\%} = \frac{A_N}{1.22\lambda}, \tag{8.16}$$

which is approximately half the spatial frequency (twice the spatial period) of the Rayleigh criterion (Smith 1966, see also Chapter 2). Tab. 8.2 summarises the optical lateral resolution for common PSI objectives in terms of these two criteria.

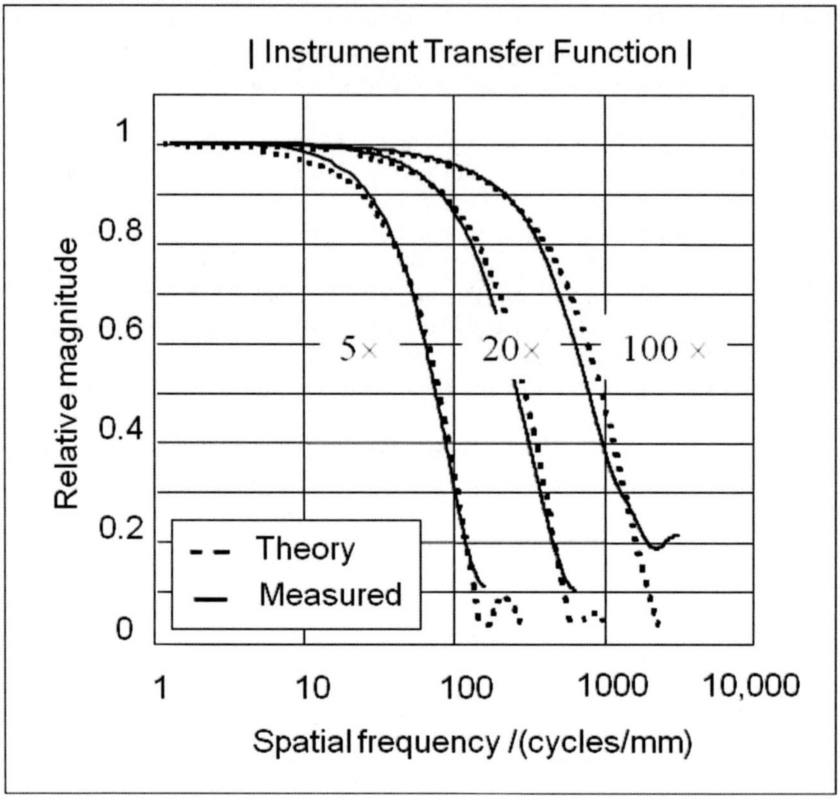

Fig. 8.8 Measured and theoretical magnitude of the ITF

Table 8.2 Optical lateral resolution of common interference objectives, assuming perfect optics

Magnification	A_N	Raleigh / μm	Sparrow / μm
1×	0.03	11.18	9.17
2.5×	0.08	4.47	3.67
5×	0.13	2.58	2.12
10×	0.30	1.12	0.92
20×	0.40	0.84	0.69
50×	0.55	0.61	0.50
100×	0.80	0.42	0.34

Camera resolution is usually matched to the optical resolution up to a magnification of approximately 10×, meaning that the pixel size in object space has the same lateral dimensions as listed in Tab. 8.2. Above a magnification of 20×, the camera resolution exceeds the optical resolution, becoming a negligible limit on the net ITF at 100×, as is apparent from Fig. 8.8.

However, having a particular numerical aperture does not mean that the objective will perform as tabulated in Table 8.2. Aberrations, focus errors or variations in the illumination can strongly influence the actual performance of the instrument. The actual ITF may be calculated experimentally by comparison of the spatial frequency content of a measured surface topography map divided by the theoretical spatial frequency content of the object surface. A convenient object for this comparison is a step height of no more than one-eighth the mean source wavelength in height (Takacs 1993, de Groot 2006).

8.8 Focus

Proper imaging, including focus, is essential for accurate surface topography measurement for all interferometric systems — this is particularly true for interference microscopes. Several important measurement factors depend on having the object in focus. The most obvious factor is a decline in lateral resolution with defocus, which results in the loss of high spatial frequencies of the ITF. Somewhat less expected is distortion of the 3D measurement. Fig. 8.9 illustrates the effect of

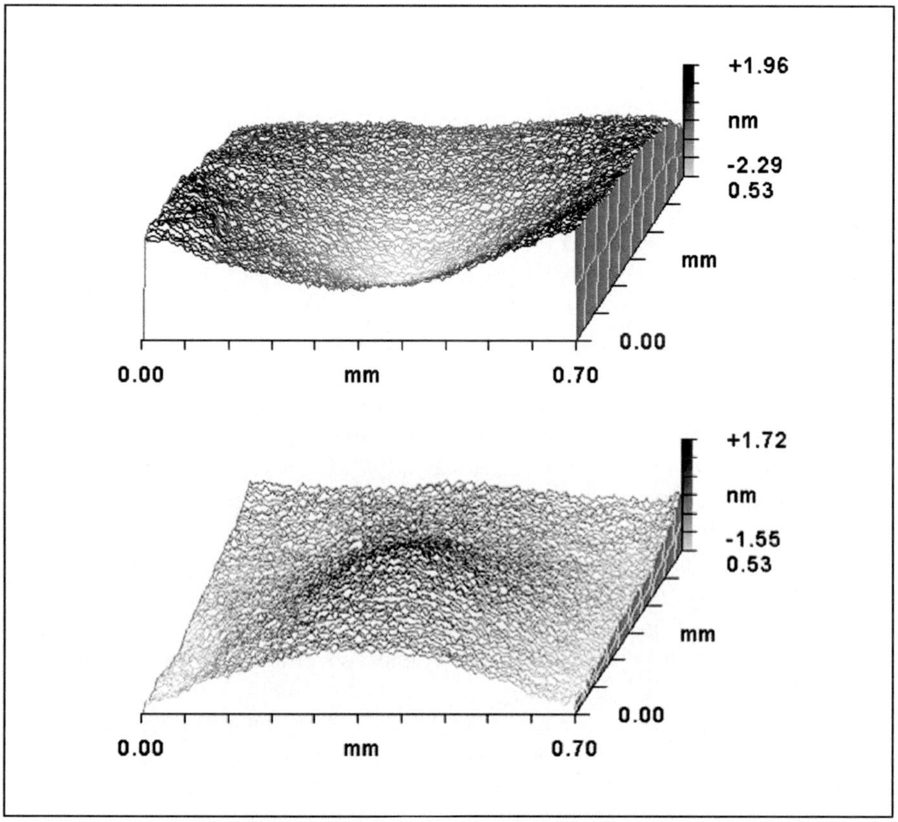

Fig. 8.9 Measured surface profiles of a SiC optical flat with ± 3 μm of defocus using a 10×, 0.25 numerical aperture objective

a focus error in a 10× objective, showing spherical distortion that is an artefact of the focus error (Chakmakjian et al. 1996). These effects become increasingly important at higher numerical aperture values typical of high-magnification objectives. As a guideline, the instrument should be adjusted to well within the depth of focus of the objective (see Chap. 2) as provided in Tab. 8.3. If available, the best focus technique is automated, particularly for production applications where manual intervention is unrealistic.

Table 8.3 Depth of focus for common interference objectives

Magnification	A_N	Focus / µm
1×	0.03	306
2.5×	0.08	49
5×	0.13	16
10×	0.30	3
20×	0.40	1.7
50×	0.55	0.9
100×	0.80	0.4

8.9 Light Sources

Although interferometers for optical testing of lenses and mirrors usually employ high coherence sources such as the HeNe 633 nm laser, with few exceptions interference microscopes use relatively lower coherence light. For many years filtered light from an incandescent source such as a halogen bulb was dominant for visible wavelength applications, not only for interferometry, but generally for microscopy. More recently, the most common light source has become a light-emitting diode (LED). Spectral filters of approximately 10 nm bandwidth are still common for instruments employing white-light LEDs. Single colour LEDs are also common, especially if the microscope is dedicated to PSI applications.

Whether incandescent or solid state, microscope illumination is almost universally of the Köhler geometry, wherein the light source is imaged into the pupil of the microscope objective. Optimal resolution and light efficiency together with minimum image artefacts generally favour completely filling the entrance pupil with source light, leading to an ITF as shown in Fig. 8.8.

The spatially extended illumination and typical spectral bandwidth of incandescent and LED sources limits the height range of PSI. In effect, the low coherence of the illumination leads to reduced interference signal strength or fringe contrast away from the best focus position. To maximise fringe contrast at best focus, the interference objective is adjusted so that best focus coincides with the zero difference in optical distance between the object and reference paths. Away from this

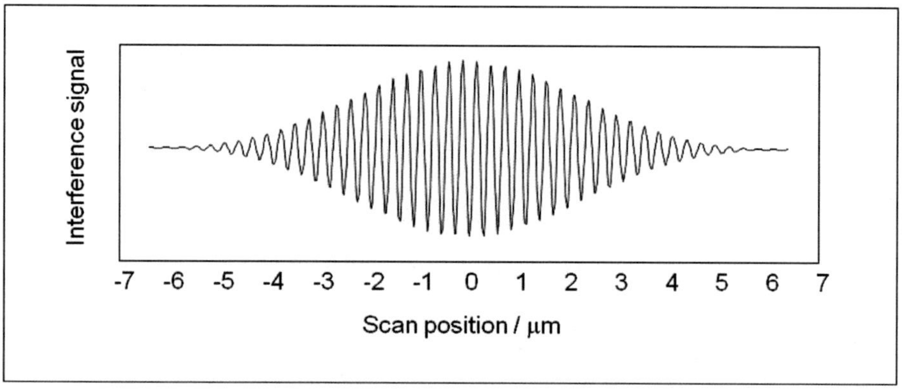

Fig. 8.10 Signal for an interference microscope near best focus

position, the fringe contrast declines as illustrated in Fig. 8.10 for a 10×, 0.3 numerical aperture objective and a 20 nm bandwidth, 550 nm light source. As a general rule, the height range for strong fringe contrast is greater for smaller spectral bandwidths (narrower light filters) and smaller numerical aperture (lower magnifications).

8.10 Calibration

The scaling of phase data to height data, in the simplest case of normal incidence illumination and viewing, follows equation (8.2), reproduced below as equation (8.17).

$$K = 4\pi/\lambda\,. \tag{8.17}$$

If the geometry conforms to this low numerical aperture case and the wavelength λ is known, then in principle the measurement is self-calibrated by virtue of the measurement principle.

In interference microscopy, however, the conditions for self-calibration are rarely met. As described in the previous section, the illumination is most often achieved by filling the objective pupil with incoherent light of broad spectral bandwidth. This results in a fringe frequency K that is more complicated to calculate as a consequence of the geometric obliquity factor (see Chap. 2) and the uncertain spectrum of the light source. Although in principle methods exist to calculate K, the most common method is to calibrate the tool by making measurements of step artefacts of calibrated height, often supplied by the instrument vendor.

An alternative to calibration using obliquity corrections or calibration artefacts is to perform an initial measurement using the coherence scanning method, as described in the Chap. 9. Fourier techniques may then be employed to relate the fringe frequency directly to the scan rate of the motion actuator shown in Fig.8.6.

It is important to underscore that no surface measuring instrument is fully calibrated simply by measuring its axial performance. Optical artefacts in the data due to the spatial frequency response of the instrument mean that for a true calibration a full understanding of the ITF is necessary. The topic of calibration is considered in greater detail in chapter 4.

8.11 Examples of PSI Measurement

As a first example of the use of PSI microscopy, Fig. 8.11 shows the interference image for a single camera frame when viewing a sinusoidal roughness artefact (PGN3 supplied by Mahr) having a height of 3 μm and lateral period of 120 μm, viewed with a 0.57 μm wavelength source. The image shows dense parallel fringes in the regions of high slope, and inflection points readily recognised as the peaks and valleys of the sinusoid. To measure the surface, the instrument that analysed this sample collected thirteen images and processed them according to the algorithm given in Tab. 8.1, resulting in the 3D map shown in Fig. 8.12.

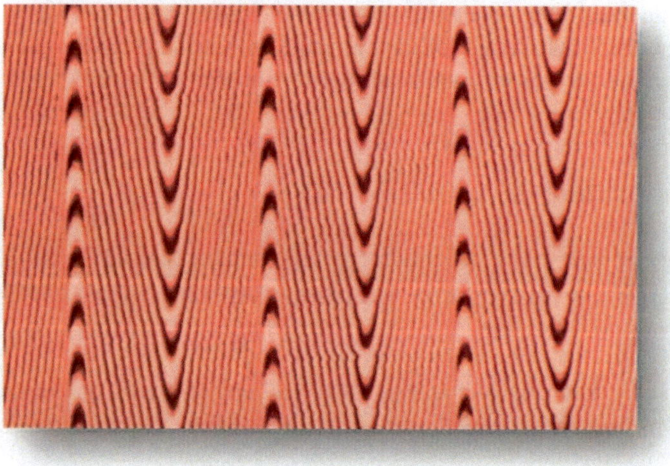

Fig. 8.11 Portion of the image of interference fringes for a sinusoidal roughness artefact

Fig. 8.12 3D image of the roughness artefact for which the fringe pattern appears in Fig. 8.11

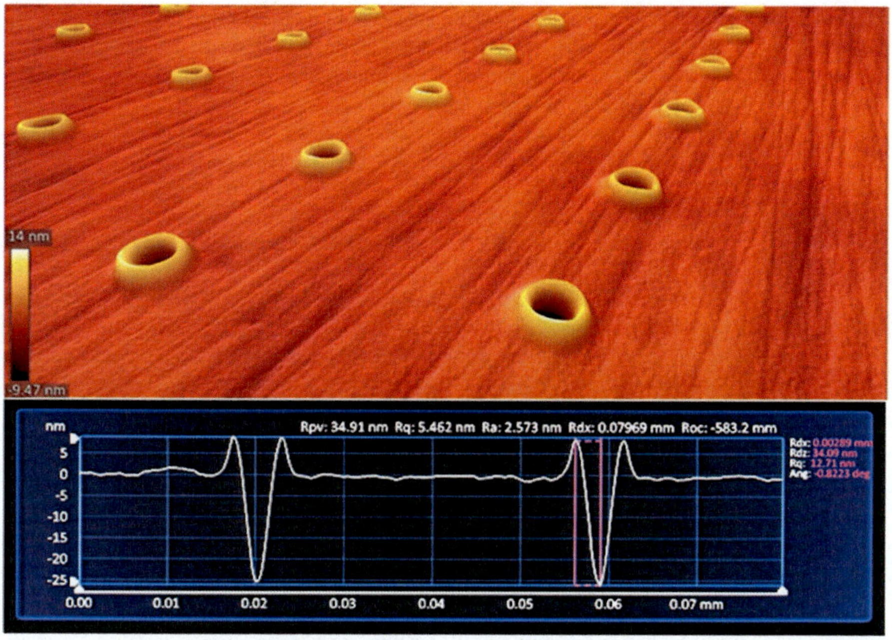

Fig. 8.13 3D PSI image of a laser textured disk (courtesy of Zemetrics)

A second example, shown in Fig. 8.13, is a 3D image of a magnetic disk substrate with laser texture bumps put on with a pulsed high-power YAG-laser. The image was obtained with PSI, using a 50× magnification Mirau-type interference objective and the 13-frame PSI algorithm listed in Tab. 8.1. The height sensitivity and data density illustrated by this example is typical of PSI.

References

Bruning, J.H., Herriott, D.R., Gallagher, J.E., Rosenfeld, D.P., White, A.D., Brangaccio, D.J.: Digital wavefront measuring interferometer for testing optical surfaces and lenses. Appl. Opt. 13, 2693–2703 (1974)

Brophy, C.: Effect of intensity error correlation on the computed phase of phase-shifting interferometry. J. Opt. Soc. Am. A 7, 537–541 (1990)

Chakmakjian, S., Biegen, J., de Groot, P.: Simultaneous focus and coherence scanning in interference microscopy. In: Technical Digest, International Workshop on Interferometry, pp. 171–173 (1996)

Creath, K.: Phase-measuring interferometry techniques. In: Wolf, E. (ed.) Progress in optics, vol. XXVI, pp. 349–393. Elsevier Science Publishers, Amsterdam (1988)

Deck, L.L.: Suppressing phase errors from vibration in phase-shifting interferometry. Appl. Opt. 48, 3948–3960 (2009)

de Groot, P.: Derivation of phase shift algorithms for interferometry using the concept of a data sampling window. Appl. Opt. 34, 4723–4730 (1995a)

de Groot, P.: Vibration in phase shifting interferometry. J. Opt. Soc. Am. A. 12, 354–365 (1995b) de Groot, P., Colonna de Lega, X.: Interpreting interferometric height measurements using the instrument transfer function. In: Proc. FRINGE 2005, pp. 30–37. Springer, Berlin (2006)

de Groot, P.: Design of error-compensating algorithms for sinusoidal phase shifting interferometry. Appl. Opt. 48, 6788–6796 (2009)

Dubois, A.: Phase-map measurements by interferometry with sinusoidal phase modulation and four integrating buckets. J. Opt. Soc. Am. A 18, 1972–1979 (2001)

Freischlad, K., Koliopoulos, C.L.: Fourier description of digital phase-measuring interferometry. J. Opt.Soc. Am. A 7, 542–551 (1990)

Greivenkamp, J.E.: Generalized data reduction for heterodyne interferometry. Opt. Eng. 23, 350–352 (1984)

Greivenkamp, J.E., Bruning, J.H.: Phase shifting interferometry. In: Malacara, D. (ed.) Optical shop testing, 2nd edn. John Wiley & Sons, Hoboken (1992)

Han, S.: Interferometric testing through transmissive media (TTM). In: Proc. SPIE, vol. 6293, pp. 629301–629305 (2006)

Hariharan, P.: Basics of interferometry. Academic Press, London (1992)

Hedley, J., Harris, A., Burdess, J., McNie, M.: The Development of a workstation for optical testing and modification of IMEMS on a wafer. In: Proc. SPIE, vol. 4408, pp. 402–408 (2001)

Koliopoulos, C.: Interferometric optical phase measurement techniques. University of Arizona, PhD Thesis (1981)

Krug, W., Rienitz, J., Schulz, G.: Contributions to interference microscopy. Hilger & Watts Ltd (1964)

Lai, G., Yatagai, T.: Generalized phase-shifting interferometry. J. Opt. Soc. Am. A 8, 822–827 (1991)

Linnik, V.P.: Ein apparat fur mikroskopisch-interferometrische untersuchung reflektierender objekte (mikrointerferometer). Akademiya Nauk S.S.S.R. Doklady (1933)

Malacara, D., Servin, M., Malacara, Z.: Interferogram analysis for optical testing. Marcel Dekker, New York (1998)

Massig, J.H.: Interferometric profilometer sensor. US Patent 5,166,751 (1992)

Mirau, A.H.: Interferometer. US Patent 2,612,074 (1952)

Pluta, M.: Advanced light microscopy, vol. 3. Elsevier, Amsterdam (1993)

Sasaki, O., Okazaki, H., Sakai, M.: Sinusoidal phase modulating interferometer using the integrating-bucket method. Appl. Opt. 26, 1089–1093 (1987)

Schmit, J., Creath, K., Wyant, J.C., Malacara, D.: Optical shop testing. In: Malacara, D. (ed.) Surface profilers, multiple wavelength and white light interferometry, 3rd edn., ch. 15, pp. 667–755. John Wiley & Sons, Hoboken (2007)

Schreiber, H., Bruning, J.H.: Phase shifting interferometry. In: Malacara, D. (ed.) Optical Shop Testing, 3rd edn. John Wiley & Sons, Hoboken (2007)

Schwider, J., Burow, R., Elssner, K.-E., Grzanna, J., Spolaczyk, R., Merkel, K.: Digital wave-front measuring interferometry: some systematic error sources. Appl. Opt. 22, 3421–3432 (1983)

Smith, W.J.: Modern optical engineering, p. 139. McGraw-Hill, New York (1966)

Sommargren, G.E.: Optical heterodyne profilometry. Appl. Opt. 20, 610–618 (1981)

Surrel, Y.: Design of algorithms for phase measurements by the use of phase stepping. Appl. Opt. 35, 51–60 (1996)

Takacs, S.P., Li, M., Furenlid, K., Church, E.: A step-height standard for surface profiler calibration. In: Proc. SPIE, vol. 235, pp. 65–74 (1993)

Takeda, M., Hideki, I., Kobayashi, S.: Fourier-ransform method of fringe-pattern analysis for computer based topography and tnterferometry. J. Opt. Soc. Am. 72, 156–160 (1982)

Wyant, J.C., Koliopoulos, C.L., Bushan, B., George, O.E.: An optical profilometer for surface characterization of magnetic media. ASLE Trans. 27, 101 (1984)

9 Coherence Scanning Interferometry

Peter de Groot

Zygo Corporation
Laurel Brook Road
Middlefield, Connecticut, USA

Abstract. Height-dependent variations in fringe visibility related to optical coherence in an interference microscope provide a powerful, non-contact sensing mechanism for 3D measurement and surface characterisation. Coherence scanning interferometry extends interferometric techniques to surfaces that are complex in terms of roughness, steps, discontinuities, and structure such as transparent films. Additional benefits include the equivalent of an autofocus at every point in the field of view and suppression of spurious interference from scattered light.

9.1 Concept and Overview

Chap. 8 describes the structure and principles of interference microscopy in the context of PSI - a technique for high-resolution measurement by evaluation of interference phase. This chapter considers coherence scanning interferometry (CSI), which in addition to, or as a substitute for the detection of phase, evaluates changes in interference signal strength related to optical coherence. In the simplest conceptualisation, surface heights are inferred by noting where the interference effect is strongest. Thus one defining feature of CSI instruments, with respect to PSI instruments, is that by design, interference fringes are only strongly observed over a narrow surface height range.

Fig. 9.1 shows how the appearance of the interference fringes varies as an interference objective is scanned vertically in the figure along its optical axis. A visual interpretation leads to the conclusion that the outside edge of the sample must be lower than the centre, as is evident from the scan position ζ being lower for high contrast fringes at the edge as compared to those at the centre. Although modern instruments employ a variety of optical configurations and data processing methods to extract surface data, a height dependent variation in fringe contrast is a common behaviour of all CSI instruments.

In practice, CSI instruments are automated systems with electronic data acquisition that provide a signal for each image pixel as a function of scan position. The resulting signals appear as in Fig. 9.2, which shows the signal itself and an overall modulation envelope with a peak position that at least conceptually provides a non-contact optical measurement of surface height.

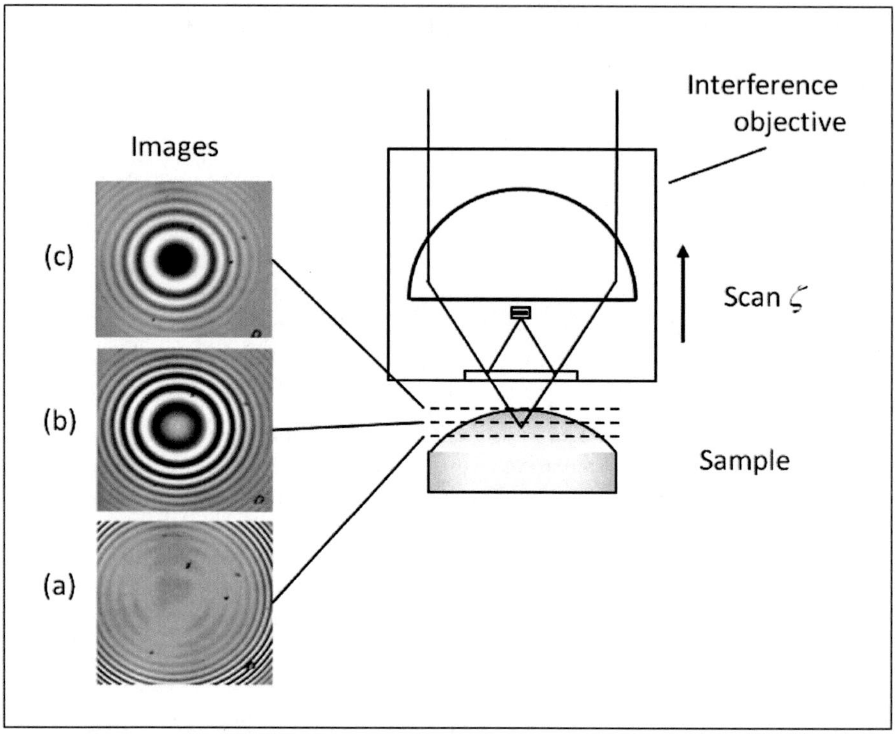

Fig. 9.1 Images of interference fringes on a curved surface with low coherence illumination

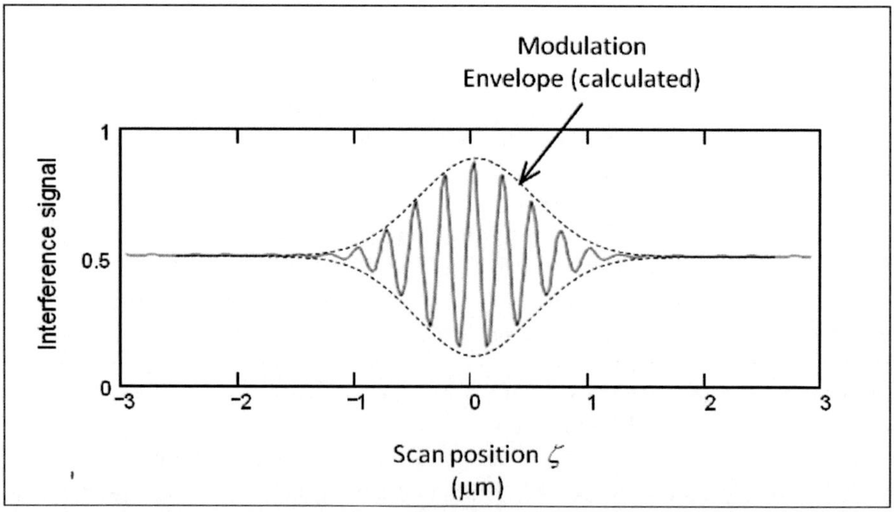

Fig. 9.2 CSI signal for a single pixel showing the modulation envelope

The modern history of CSI begins with the work of Davidson et al. (1987), who applied the idea to smooth surface, semiconductor applications, in part with the goal of improving the lateral resolution on fine features. A significant step was the realization that a CSI system could provide useful measurements of surfaces with sufficient roughness to generate completely random speckle (Dresel et al. 1992, Häusler and Neumann 1993, Caber 1993). The ability to measure technical surfaces inaccessible to conventional interferometry launched commercial products for both rough surface testing and high-precision measurement without the need to spatially unwrap fringes. Additional benefits of CSI include the equivalent of an autofocus at every point in the field of view and suppression of spurious interference from scattered light. Today the majority of interference microscopes for evaluation of areal surface topography operate according to the CSI principle. There is an extensive literature available on CSI, including review articles (Schmit et al. 2007), good practice guides (Leach et al. 2008, Petzing et al. 2010) and a draft ISO standard (ISO/CD 25178-604).

9.2 Terminology

Few technologies have as many names for the same basic functional principles as CSI. Tab. 9.1 below lists just a subset of terms that have appeared in the technical, patent or commercial literature that all relate to a variant of CSI on a microscope. These terms serve to tag specific data processing methods and differentiate commercial product offerings.

Table 9.1 Summary of recognized terms for CSI

Acronym	Term
CPM	Coherence probe microscope
CSM	Coherence scanning microscope
CR	Coherence radar
CCI	Coherence correlation interferometry
HSI	Height scanning interferometer
MCM	Mirau correlation microscope
RSP	Rough surface profiler
RST	Rough surface tester
SWLI	Scanning white light interferometry
TD-OCT	Time-domain optical coherence tomography (full field)
VSI	Vertical scanning interferometry
EVSI	Enhanced VSI
HDVSI	High-definition VSI
WLI	White light interferometry
WLSI	White light scanning interferometry
WLPSI	White light phase shifting interferometry

9.3 Typical Configurations of CSI

Figure 9.3 illustrates a generic CSI instrument. The object has height features h that vary over the object surface. A mechanical scanner provides a smooth, continuous scan of the interference objective in the z direction. During the scan, a computer records intensity data I for each image point or pixel in successive camera frames.

Light sources for CSI are incoherent with a broadband spectrum (white light), spatial extent, or both. A classic example is an incandescent lamp such as a tungsten halogen bulb, while currently the most common source is a white-light LED. The Köhler illumination optics, as shown in Fig. 9.3, images the light source into the pupil of an interference objective. The aperture stop controls the numerical aperture of the illumination, while the field stop controls the illuminated area of the object surface. Most often, the illumination fills the pupil to minimise spatial coherence and maximise lateral resolution. For dynamically moving objects, the light source may be flashed stroboscopically to freeze the object motion (Nakano 1995, Novak and Schurig 2004).

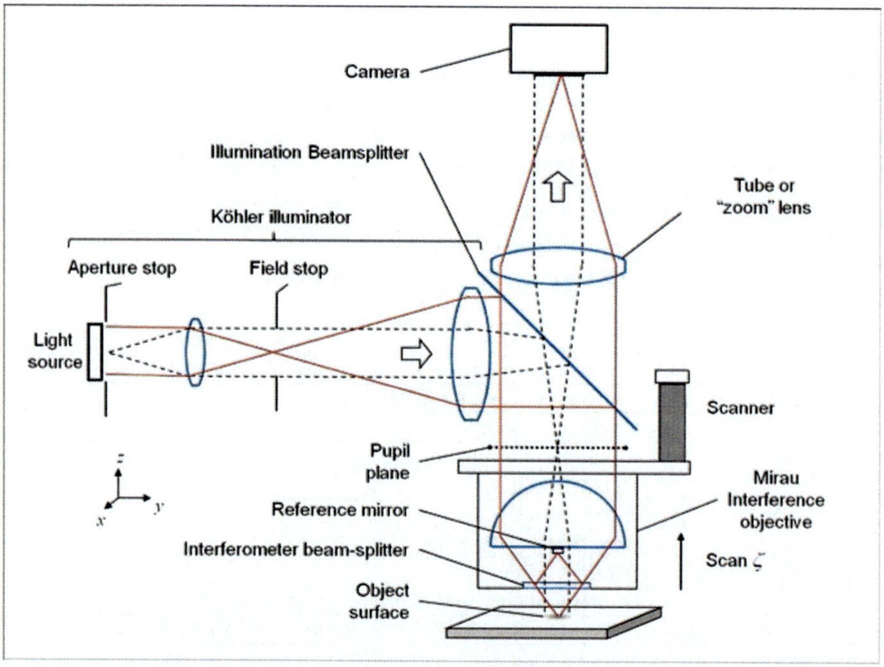

Fig. 9.3 Geometry of an interference microscope suitable for CSI

CSI instruments, just as PSI microscopes, are most often configured in a manner similar to a conventional microscope with the normal objective replaced by a two-beam interference objective; although a few systems have a Twyman-Green geometry (Dresel et al. 1992, de Groot et al. 2002a). Interference objectives are most commonly of the Michelson, Mirau, or Linnik type, as described for PSI in Chap. 8.

The interference objective should be compatible with a low coherence light source. The two interferometer paths should be balanced for dispersion of refractive index with wavelength, and the position of best focus should be coincident with the position of zero optical path difference. For some applications, a non-flat reference surface may be used that better matches the sample shape (Biegen 1990).

The scanner shown in Fig. 9.3 moves the interference objective or the objective turret; in other cases an actuator moves the object moves. Although less common, it is feasible to move the reference mirror, beam-splitter or some combination of optical elements within the objective in place of translating the entire interference objective (Colonna de Lega 2004). Generally, the scan motion is along the optical axis of the objective perpendicular to the sample surface, i.e., in the direction, shown in Fig. 9.3. The scan length is typically between 10 μm and 200 μm for piezo-electric scanners, and several millimetres for motorized scanners.

9.4 Signal Formation

In CSI using a microscope, the height-dependent fringe contrast is most often a consequence of both the spatial extent of the incoherent illumination in the pupil and a broadband optical spectrum. Therefore, the signal modelling should include both of these effects. This section provides the mathematical description of signal generation in CSI together with some examples of limit cases.

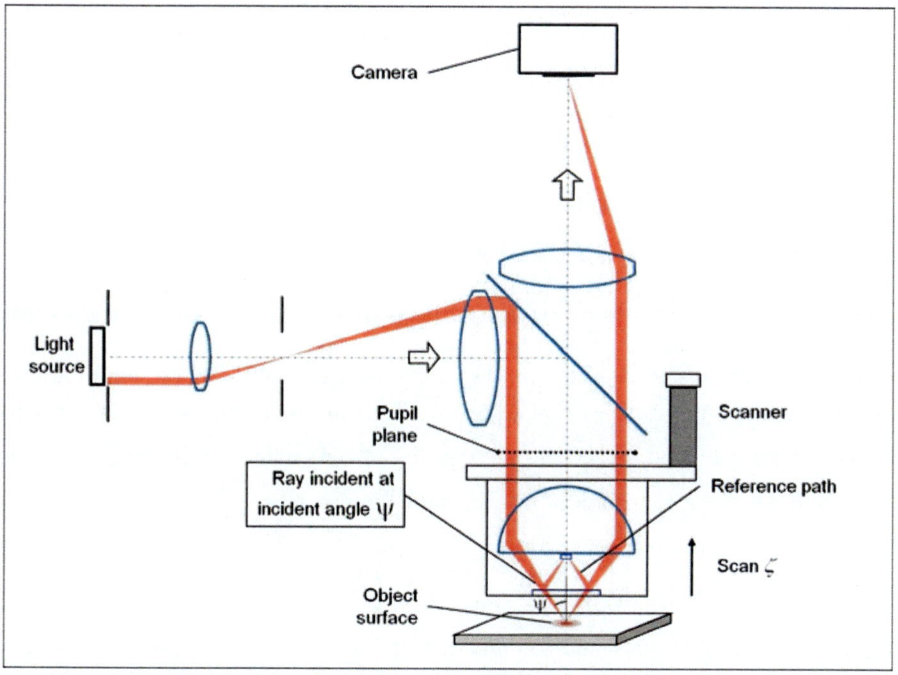

Fig. 9.4 Path of a single ray bundle through a CSI instrument

A straightforward model of the signal generation assumes a randomly polarized, spatially incoherent illumination and a smooth surface that does not scatter or diffract the incident light (Davidson et al. 1987, Sheppard and Larkin 1995, Abdulhalim 2001). The total signal is the sum of all the incoherent interference contributions of the ray bundles passing through the pupil plane of the objective and reflecting from the object and reference surfaces.

The incoherent superposition involves a summation of interference signals over a range of fringe frequencies K expressed in radians per micrometre. These frequencies refer to the rate at which interference fringes pass by as a function of the objective scan position ζ. The fringe frequency depends on the source wavelength λ and, assuming that the objective is scanned with respect to the sample, on the angle of incidence ψ according to

$$K(\beta,k)=2k\beta \tag{9.1}$$

where the directional cosine is

$$\beta=\cos(\psi) \tag{9.2}$$

and

$$k=2\pi/\lambda \tag{9.3}$$

is the angular wavenumber for a spectral contribution. The factor of two in equation (9.1) is the result of the reflection measurement, which doubles the optical path length change for a corresponding change in surface height.

The ray bundle at an incident angle Ψ is shown in Fig. 9.4 corresponding to a specific image point and position in the pupil. Assuming perfect contrast, the DC-normalised contribution to the interference signal is

$$g(\beta,k,\zeta)=1+\cos\left[-K(\beta,k)\zeta+\phi(\beta,k)\right]. \tag{9.4}$$

Here the phase value relates to a surface height h according to

$$\phi(\beta,k)=K(\beta,k)h+\varpi(\beta,k), \tag{9.5}$$

where the offset ϖ relates to the optical properties of the CSI system and of the surface being measured.

The incoherent superposition concept consists mathematically of summing the interference contributions $g(\beta,k,\zeta)$ over all angular wavenumbers k, weighted by the illumination and detection spectrum $V(k)$, and over all directional cosines β, weighted by the distribution of light in the pupil $U(\beta)$

$$I(\zeta)=\int_0^\infty\int_0^1 g(\beta,k,\zeta)U(\beta)V(k)\,\beta\,d\beta\,dk. \tag{9.6}$$

It can be shown (de Groot and Colonna de Lega 2004) that the integration over all points in the pupil leads to a net signal that can be written more simply as an inverse Fourier transform

$$I(\zeta) = \int_{-\infty}^{\infty} q(K) \exp(-iK\zeta) \, \mathrm{d}K \,, \tag{9.7}$$

where the non-zero, positive part of the frequency spectrum is

$$q(K) = \rho(K) \exp(i\,Kh) \tag{9.8}$$

and the coefficients $\rho(K)$ are computed from the characteristics of the surface and the system for each couple (β, k) corresponding to a specific frequency K.

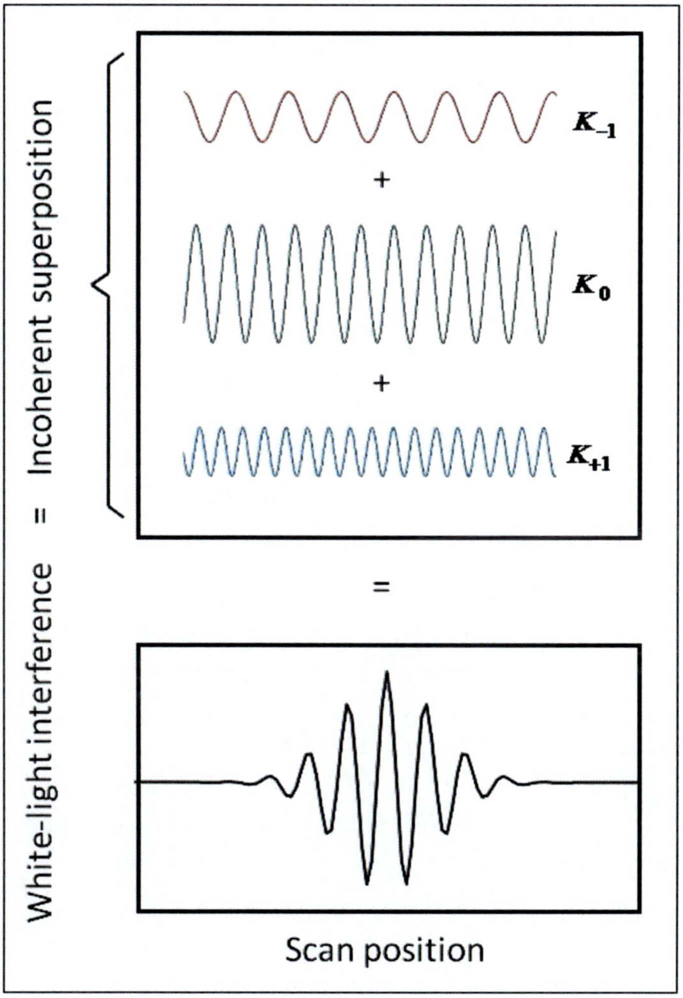

Fig. 9.5 Signal formation by incoherent superposition

Writing the intensity as a sum of interference patterns of various fringe frequencies K as in equation (9.7) allows for a useful qualitative understanding of the signal generation process. Referring to Fig. 9.5, the summation over a range of frequencies results in a peak signal strength where all of the individual contributions are mutually in phase. This position is sometimes referred to as the stationary phase point. Generally, we can represent the CSI signal as either the intensity scan $I(\zeta)$ or the Fourier frequency spectrum $q(K)$, with the components of the Fourier spectrum directly calculable from the incoherent superposition model.

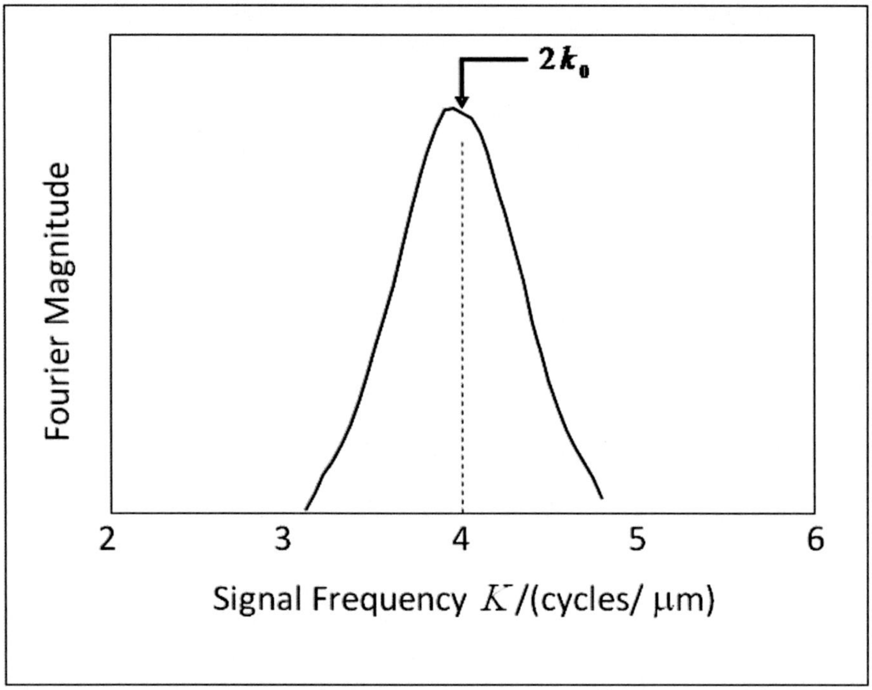

Fig. 9.6 Fourier magnitude for small numerical aperture. The range of frequencies is the result of the spectral bandwidth of the white light source

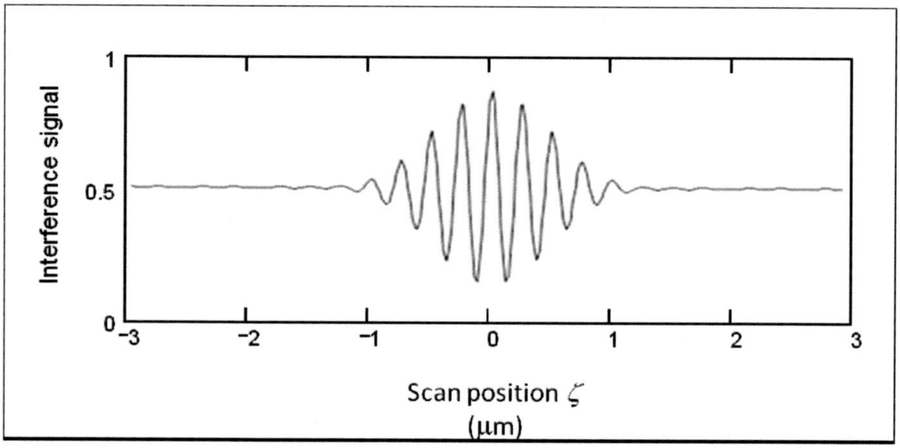

Fig. 9.7 White light interference signal corresponding to the Fourier spectrum in Fig. 9.6

Certain simplifying limit cases are of practical and conceptual value. A familiar configuration is a low numerical aperture with broadband or "white"-light. In this case, the magnitudes of the Fourier coefficients are directly proportional to the source spectral distribution V

$$\left| q(K>0) \right| \propto V(k) \tag{9.9}$$

where at normal incidence the frequency K is simply twice the angular wavenumber k . Fig. 9.6 shows an example where $A_N = 0.2$, the spectral bandwidth of the light source is 100 nm, and the centre wavelength is $\lambda_0 = 500$ nm. The corresponding interference signal shown in Fig. 9.7 has interference fringes with high visibility near the zero scan position, but low visibility elsewhere in the scan. Note that there is one fringe for every 250 nm, equivalent to four interference cycles per micrometre, as expected from Fig. 9.6.

A somewhat less familiar but no less relevant case is narrow bandwidth at high numerical aperture. The frequency spectrum in this case simplifies to

$$\left| q(K>0) \right| \propto \beta\, U(\beta) \ . \tag{9.10}$$

Now the coherence properties are determined primarily by the light distribution in the pupil plane, rather than by the optical spectrum. As an example, the frequency spectrum shown in Fig. 9.6 for a 20 nm bandwidth centered at a wavelength $\lambda_0 =$ 500 nm is broadened towards the lower frequencies by a wide, $A_N = 0.6$ pupil (approximately 50× magnification). Although not a white light interferometer, a high

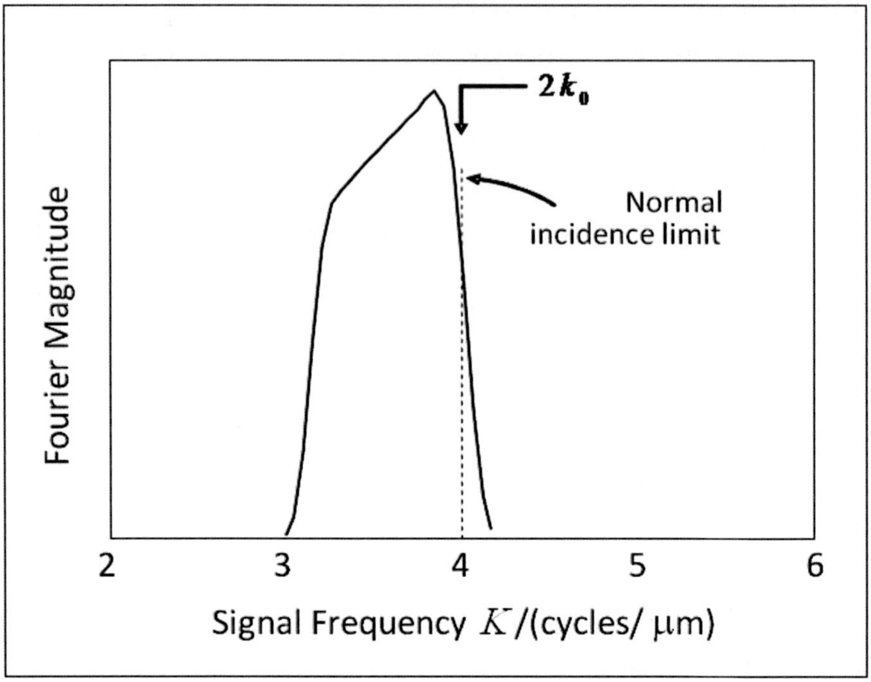

Fig. 9.8 Fourier magnitude for a spectrally narrow band interference microscope with high numerical aperture

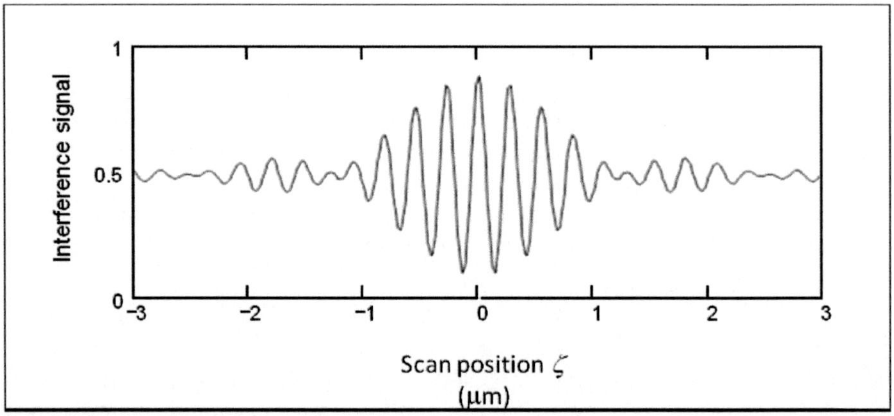

Fig. 9.9 Signal corresponding to the frequency spectrum of Fig. 9.8, for a narrow bandwidth light source and 50× magnification

numerical aperture, monochromatic system nonetheless creates similar signals (see Fig. 9.7) and functions in much the same way as a white light instrument.

In practice, CSI microscopes operate in a compromise regime, wherein both spatial coherence and spectral bandwidth contribute to the signal shape. Typically, the low numerical aperture, white-light condition is approximately satisfied when using objectives of a magnification of 5× or lower; while for a magnification of 50× or higher focus effects contribute strongly to signal shape.

9.5 Signal Processing

An advantageous feature of the CSI technique with respect to PSI is the ability to measure surfaces without the risk of interference fringe order, or 2π, errors. Among other benefits, this allows for the measurement of unpolished surfaces that generate poorly defined interference fringes. Signal processing for CSI is invariably optimised to take best advantage of this phenomenon.

One data processing approach is to characterize the CSI signal according to the qualitative characteristics of a rapidly oscillating interference fringe and an overall variation in signal strength, which localizes the interference effect to a specific value of $h-\zeta$. In some specific idealised cases, the interference signal (equation (9.7)) is mathematically separable into a constant offset I_{DC} and a co-sinusoidal carrier signal at a frequency K_0 modulated by a slowly varying I_{AC} modulation envelope (see Fig.9.2) approximately as

$$I(\zeta)=I_{AC}(\zeta)+I_{AC}(\zeta-h_{x,y})\cos\left[K_0(\zeta-h)\right]. \tag{9.11}$$

For the white light fringes, the peak of the envelope is at the position of zero group velocity optical path difference (OPD). This envelope peak position usually differs from the scan position where the argument of the cosine in Eq.(9.11) is equal to zero by an amount sometimes referred to as the phase gap, as a consequence of dispersion in the optical system and other effects.

It is worth emphasising that this concept of a slowly modulated carrier signal is only an approximation corresponding to idealised conditions. In real CSI instrumentation, there are scan-dependent phase distortions in addition to amplitude modulations. However, equation (9.11) has proven itself to be a useful starting point for many signal processing strategies. A common idea is to determine the scan position for which the signal strength or fringe visibility is maximum, thereby sectioning smooth surface images according to surface height (Balasubramanian 1982). Envelope detection usually involves a form of digital filter, using demodulation techniques based on communications theory (Caber 1992) or other methods to remove the carrier and synthesise a signal exhibiting only the envelope (Haneishi 1984, Kino and Chim 1990).

An alternative to demodulation and peak searching that offers greater noise resistance is to estimate the overall signal position using the centroid of the square of the signal derivative (Larkin 1996, Ai and Novak 1997). An example calculation of the centroid position for a specific pixel is

$$H = \frac{\sum_z (I_z - I_{z-1})^2 \zeta_z}{\sum_z (I_z - I_{z-1})^2}, \tag{9.12}$$

where z is the data frame index.

Envelope detection is an effective method for surface topography measurement on both smooth and rough surfaces, but it suffers from sensitivity to error sources such as optical aberrations, diffraction, vibration and noise, to a much greater degree than conventional interference phase techniques such as PSI (see Chapter 9). However, since the CSI signal also contains interference fringes, the ultimate surface topography repeatability for CSI is the same as that of PSI. Most modern CSI microscopes offer an enhanced analysis mode that couples envelope detection with the much finer resolution of interference phase estimation to improve the precision of CSI on surfaces that are smooth (for example $<\lambda/10$ RMS surface roughness). In this mode of operation, the envelope detection serves only to resolve fringe order, according to ideas that date back to the earliest use of white light interferometry (Michelson 1893).

To resolve fringe order using the coherence information, a first analysis of CSI data using envelope detection or other coherence-based analysis leads to a first surface height topography map $H(x, y)$. A second analysis, either of the same CSI data or of a separate data acquisition, provides an interference phase map θ for an interference frequency K_0 corresponding to the effective mean wavelength. The phase θ carries with it an inherent fringe order ambiguity of an integer multiple of 2π. The combined surface topography map is then

$$h(x, y) = \frac{\theta(x, y)}{K_0} + \frac{2\pi}{K_0} \text{round}\left(\frac{A(x, y) - \langle A \rangle}{2\pi}\right), \tag{9.13}$$

where

$$A(x, y) = \theta(x, y) - K_0 H(x, y) \tag{9.14}$$

is the phase gap in units of phase at the frequency K_0, between the envelope detection and phase analysis techniques. In Eq.(9.14), the round function returns the nearest integer to its argument, and the parentheses $\langle\ \rangle$ represent a field average (de Groot and Deck 1995).

Errors in the evaluation of equation (9.13) or its equivalent lead to evident erroneous measurement heights in CSI topography maps known as fringe-order or 2π errors. These errors are most common with surfaces having different optical properties across the field of view, film structures, sharp edges or high surface roughness. A considerable amount of effort has been dedicated to a proper treatment of the fringe-order and it remains to be a central point of software development for CSI (Harasaki 2000, de Groot et al. 2002b).

An approach to simultaneous envelope detection and phase analysis is to correlate the intensity data $I(\zeta)$ at each pixel with a complex kernel $f(\zeta')$ designed to be sensitive to the CSI signal

$$S(\zeta)=\int_{0}^{\infty} f(\zeta')I(\zeta+\zeta')\,\mathrm{d}\zeta' . \tag{9.15}$$

In discrete form, where the camera frames are indexed by z, equation (9.15) becomes

$$S_z = \sum_{z'=-N'/2}^{N'/2} f_{z'} I_{z+z'} . \tag{9.16}$$

For this example the symmetric complex kernel has a set of $N'+1$ real coefficients $c_{z'}$ and $N'+1$ imaginary coefficients $s_{z'}$ indexed by the index z' used in the summation

$$f_{z'} = c_{z'} + i s_{z'} . \tag{9.17}$$

The CSI signal envelope and phase as a function of scan position follow from the magnitude $|S_z|$ and phase argument $\arg(S_z)$ evaluated at the peak (or centroid) of the envelope.

Suitable correlation kernels are available in PSI algorithms (Larkin 1996, Sandoz et al. 1997). For example, a well-known PSI algorithm (see Table 9.1) valid when the sampling corresponds to five data points per fringe, reads

$$\begin{aligned} c &= (0,\ \ 2,\ \ 0,\ \ -2,\ \ 0) \\ s &= (-1,\ \ 0,\ \ 2,\ \ 0,\ \ -1) \end{aligned} . \tag{9.18}$$

The envelope is then

$$|S_z| = \sqrt{(2I_{z-1} - 2I_{z+1})^2 + (-I_{z-2} + 2I_z - I_{z+2})^2} . \tag{9.19}$$

After evaluation of the envelope shape to determine the optimal z_0 location in the scan, the phase evaluated with respect to this position is

$$\phi = \arctan\left(\frac{2I_{z_0-1} + 2I_{z_0+1}}{-I_{z_0-2} + 2I_{z_0} - I_{z_0+2}}\right) . \tag{9.20}$$

Variations of this basic idea include wavelet techniques (Sandoz 1997), correlation with a kernel that represents as nearly as possible the shape of the CSI signal itself (Lee-Bennett 2004), and least-squares fitting of a kernel to the CSI signal (de Groot 2008).

As an alternative to envelope detection, the CSI signal can be analysed by directly examining its frequency content (de Groot and Deck 1995). A digital Fast-Fourier

Transform of the experimental intensity $I(\zeta)$ recovers the frequency spectrum $q(K)$ of equation (9.8). A line fit (Fig. 9.10) to the argument of $q(K)$ as a function over a range of frequencies K yields a first estimate of the surface height

$$H = slope \qquad (9.21)$$

as well as the phase gap defined in equation (9.14)

$$A = intercept \qquad (9.22)$$

where *slope* and *intercept* follow from the line fit. The phase is then

$$\theta = A + K_0 H \qquad (9.23)$$

and equation (9.13) yields the final high resolution surface height h. The nominal carrier frequency K_0 is automatically calibrated with respect to the scan increment, obviating the need for any previous knowledge of the optical spectrum shape or aperture effects.

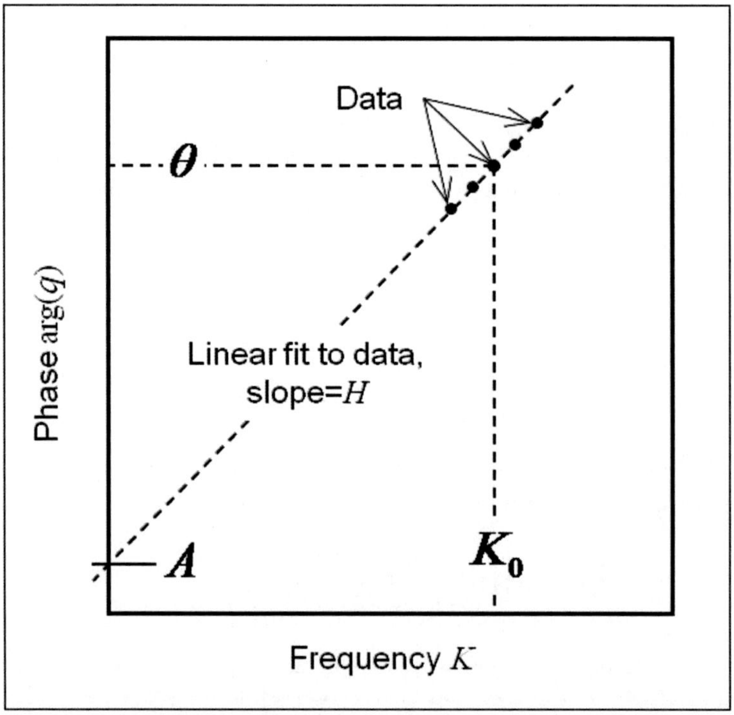

Fig. 9.10 Frequency-domain analysis of CSI signals

9.6 Foundation Metrics and Height Calibration for CSI

Conceptually, a CSI instrument behaves like a highly parallel optical contact probe. The software records the location of the CSI signal at each pixel with respect to a scan position, resulting in a surface topography map, the scale and linearity of which are directly tied to knowledge of the scan motion. Therefore, in CSI, the mechanical scanner, rather than the wavelength of light determines the basic unit of measure.

The metrology remains tied to the scan mechanism even when incorporating interference phase to improve the accuracy of CSI measurements. This is a consequence of requiring that the envelope detection and phase estimation measurements agree in scale to combine them into a final result (equation (9.13)), which is complicated by such effects as the numerical aperture, which influences the effective wavelength (see Section 2.4). Consequently, a frequency domain analysis, post-processing of the CSI signals or an appropriate setup procedure calibrates the phase measurement with respect to the scan increment.

Given that CSI measurement relies on knowledge of the scan motion, some instruments equip the scanner with electronic sensors such as capacitance sensors, inductive sensors, displacement interferometers or optical encoders, which are used in feedback systems to improve scan linearity.

Overall scale calibration can be accomplished with a calibration artefact, such as a step height or an artefact with rectangular grooves. Alternatively, a displacement interferometer with a well-established laser wavelength directly connects the measurement of scan displacement to a wavelength standard (de Groot et al. 2002a, Schmit et al. 2003).

9.7 Dissimilar Materials

The optical properties of the materials that make up the object surface are integral to the CSI measurement process. In particular, assuming that the material has an index of refraction with at most a linear dispersion in wavenumber, a phase change on reflection (PCOR) can be identified (Dubois 2004) that influences the measured interference fringe phase, and a rate of change of PCOR or dispersion, which influences the position of the modulation envelope.

Tab. 9.2 quantifies the change in apparent surface height resulting from PCOR (phase shift) and the rate of change of PCOR (envelope shift), for a 570 nm light source with a 100 nm bandwidth, at low numerical aperture. Note that there is no simple correlation between the envelope shift and the phase shift. If the object surface material is uniform, there is generally no identifiable error in the final surface topography map, assuming that an overall DC offset is irrelevant. For surfaces composed of a mixture of materials having different optical properties, these errors may in principle be corrected using a prior knowledge of the PCOR and rate of change of PCOR, and the modelling outlined in Sect. 9.5.

Table 9.2 Change in apparent height resulting from materials

Material	Envelope shift / nm	Phase shift / nm
Bare glass	0	0
Silicon	-3	0
Aluminum	-9	-13
Chrome	-15	-13
Platinum	-11	-18
Copper	-1	-31
Cobalt	-11	-18

9.8 Vibrational Sensitivity

CSI measurements acquire data over time, which means that other time-dependent phenomena, such as mechanical vibrations, tend to be convolved into the data. The environment can, therefore, be an important contributor to measurement error.

Fig. 9.11 shows the sensitivity of CSI to a 10 nm amplitude sinusoidal mechanical vibration as a function of vibrational frequency. For these simulated data, the 560 nm light source has a 120 nm bandwidth. The vibrational frequency is normalised to the interference fringe scan rate (for example, 10 Hz vibration at a scan rate such that five interference fringes pass per second would be a normalised frequency of two). The very high sensitivity of envelope detection to vibration, approximately 10× greater than for phase measurement, mandates that the CSI instrument be situated in an environment isolated from sources of vibration, particularly at a normalised frequency of two.

In some cases of practical interest, high measurement errors related to stable, single-frequency vibrations can be reduced by altering the scan rate to avoid the peak sensitivities evident in Fig. 9.11. Software post processing can also improve data quality by interpretation of the CSI signals to detect and compensate vibration (Deck 2007, Munteanu and Schmit 2009).

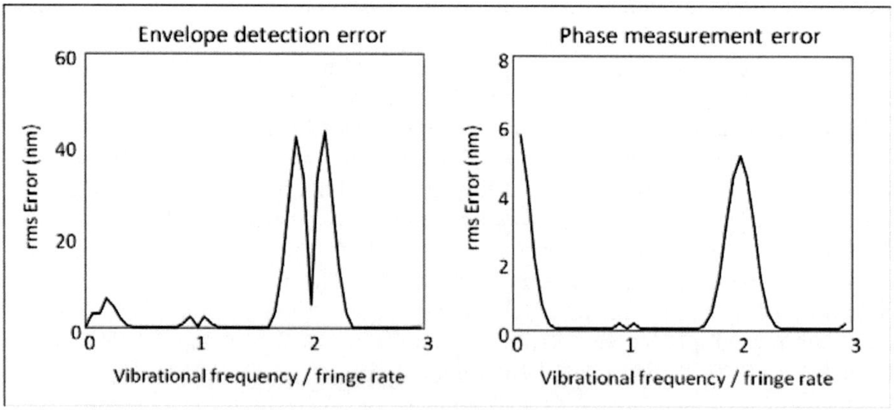

Fig. 9.11 Sensitivity of CSI to vibration

9.9 Transparent Films

One of the unique benefits of CSI recognized early in its development is the ability to separate multiple reflections from semi-transparent film structures on surfaces (Lee and Strand 1990).

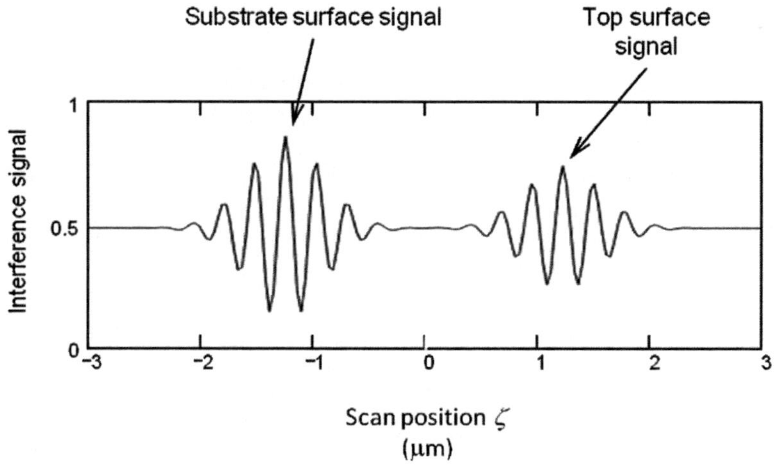

Fig. 9.12 CSI signal in the presence of a single layer transparent film a few micrometres in thickness

From Fig. 9.12, it is apparent that for a sufficiently thick single-layer film, there are two clearly identifiable modulation envelopes corresponding to surface reflections from the film boundaries. Therefore, an approach to generating surface topography maps over films is to identify the right-most signal as the top surface signal, as shown in Fig. 9.12. Further, if the refracting properties of the film are known, the substrate or other secondary surfaces below the top surface can be mapped for height by analysis of the signals that follow the top-surface signal, yielding additional information such as 3D film thickness maps (Bosseboeuf and Petigrand 2003). Alternatively, if the physical thickness of a film layer is known, the refracting properties of the film material may be determined by analysis of the CSI signal.

When analysing film thickness maps, the location of the substrate or secondary-surface signal is influenced by two competing effects: the axial group velocity optical path length, which is longer in a film than in air, and the position of best focus, which is shorter in a film than in air. These two competing effects strongly influence the shape and position of CSI signals resulting from reflections within a film.

If the film is too thin (less than 1 μm thick), the separate signals shown in Fig. 9.12 coalesce and it is difficult to clearly separate them. Depending on the instrument configuration, there is a lower limit to an analysis based entirely on signal separation.

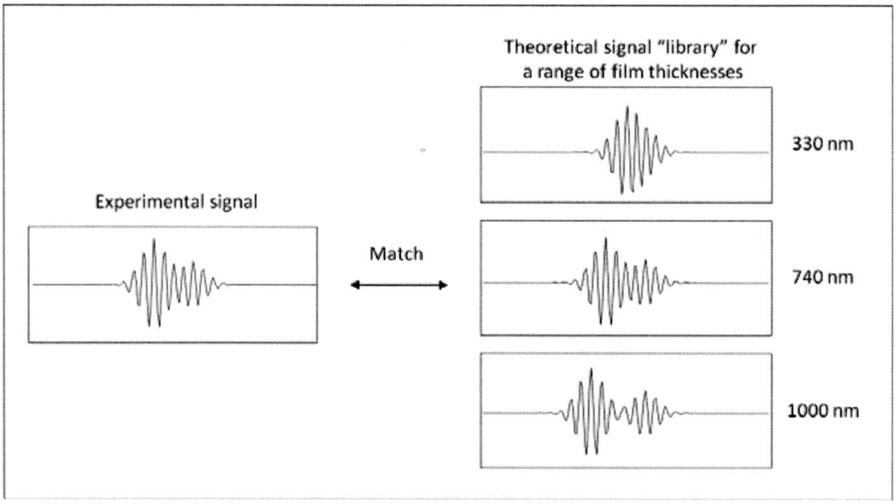

Fig. 9.13 Model-based CSI for film thickness analysis

An approach to analysis of films thinner than 1 μm is to compare predicted signals based on modelling of the system comprised of the interferometer and the object surface structure, using some basic assumptions regarding the number of film layers and their optical properties. The modelling includes a free parameter, such as film thickness, resulting in a range of possible model signals to which the experimental data are compared. Fig. 9.13 illustrates the concept for a silicon dioxide on silicon grating with a 0.8 numerical aperture system operating at $\lambda = 570$ nm with a 100 nm bandwidth. Kim and Kim (1999) demonstrated a model-based approach using Fourier phase values from equation (9.8) to evaluate goodness of match between the experimental and theoretical model signals. The concept has been extended to a variety of both time- and frequency-domain search methods and has found application in the semiconductor and flat panel display industries (Colonna de Lega and de Groot 2005).

9.10 Examples

Two examples provide some insight into the capabilities of CSI. Fig. 9.14 shows a computer screen capture of the measurement of a read-write data storage hard disk drive. The screen has four windows, showing clockwise from the lower right the interference fringes for a single camera frame, the surface heights map for the air-bearing surfaces, the slider surface with the leading-edge taper removed, and a 3D image of the completed analysis.

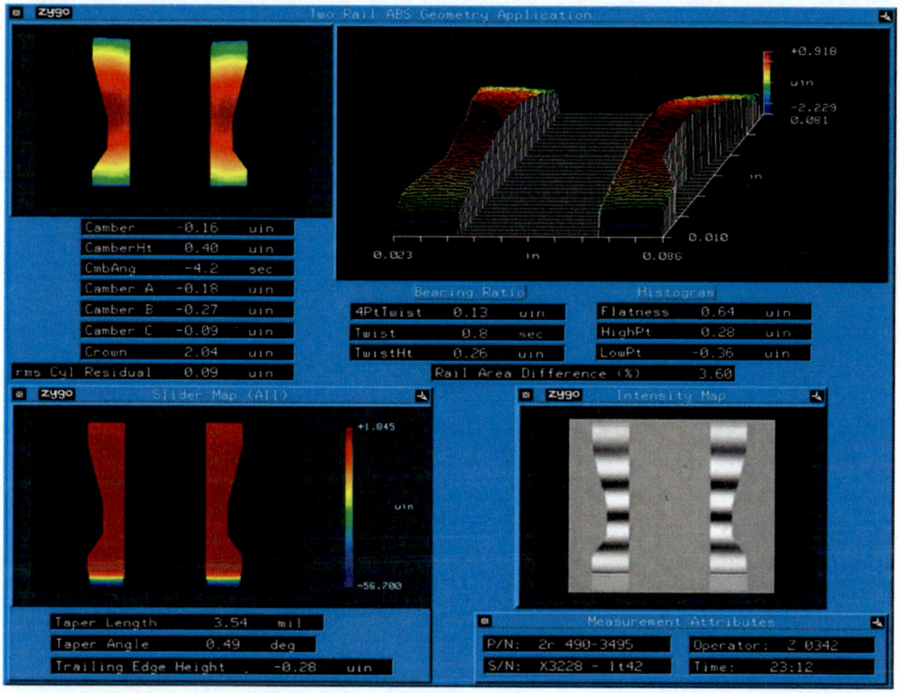

Fig. 9.14 Example CSI data acquisition and analysis of a data storage read-write slider

Fig. 9.15 shows the rough surface of a solder bump used for connections on a semiconductor device. Here the four windows, clockwise from the lower right, show a synthesised image of the bump, a cross sectional profile, the colour-coded surface heights map, and a 3D image.

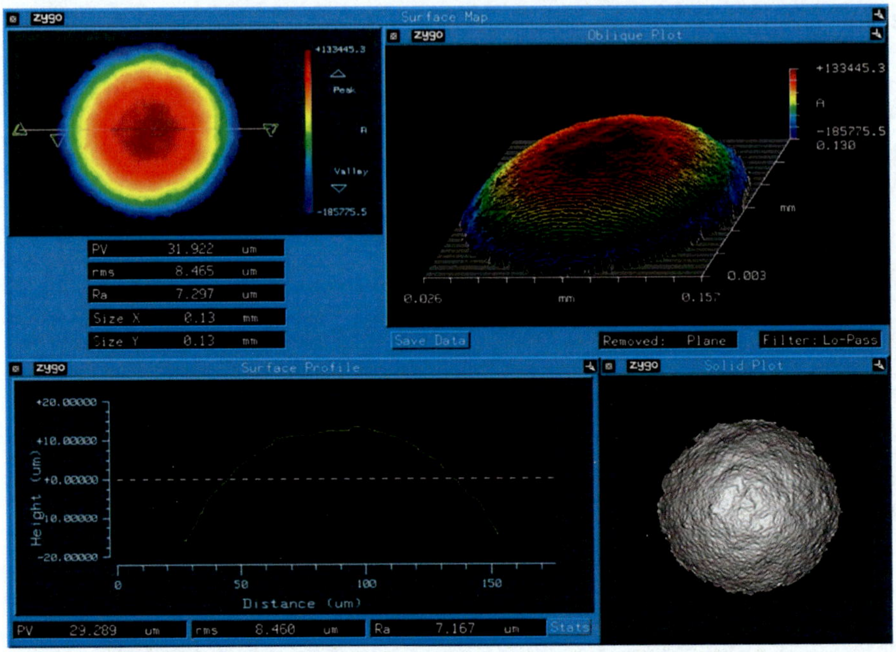

Fig. 9.15 Example CSI data acquisition and analysis of a solder bump

9.11 Conclusion

CSI has evolved to become the dominant technique in interference microscopy and development continues. When compared to phase shifting methods, CSI has a superior tolerance for variations in surface texture and greater capability for measuring surface features and structures such as step heights. An additional benefit of CSI is that data for every image pixel are gathered at the accurate point of best focus for that pixel. CSI maintains the basic interferometry advantage of subnanometre vertical resolution regardless of the numerical aperture or field of view of the microscope. The technique continues to develop in capability, performance and flexibility, including advanced methods for transparent film and other surface structure analyses.

References

Abdulhalim, I.: Spectroscopic interference microscopy technique for measurement of layer parameters. Meas Sci. Technol. 12, 1996–(2001)

Ai, C., Novak, E.: Centroid approach for estimation modulation peak in broad-bandwidth Interferometry. U.S. Patent 5,633,715 (1997)

Balasubramanian, N.: Optical system for surface topography measurement. US Patent 4,340,306 (1982)

Biegen, J.F.: Interferometric surface profiler for spherical surfaces. US Patent 4,948,253 (1990)

Bosseboeuf, A., Petigrand, S.: Application of microscopic interferometry techniques in the MEMS field. In: Proc. SPIE, vol. 5145, pp. 1–16 (2003)

Caber, P.J.: Interferometric profiler for rough surfaces. Appl. Opt. 32, 3438–3441 (1993)

Colonna de Lega, X.: Surface profiling using a reference-scanning Mirau interference microscope. In: Proc. SPIE, vol. 5532, p. 106 (2004)

Colonna de Lega, X., de Groot, P.: Optical topography measurement of patterned wafers. In: Proc Characterization and Metrology for ULSI Technology, American Institute of Physics, pp. 432–436 (2005)

Davidson, M., Kaufman, K., Mazor, I., Cohen, F.: An application of interference microscopy to integrated circuit inspection and metrology. In: Proc. SPIE, vol. 775, pp. 233–240 (1987)

Deck, L.: High precision interferometer for measuring mid-spatial frequency departure in free form optics. In: Proc. SPIE, vol. TD04, TD040M-1 (2007)

Dresel, T., Haeusler, G., Venzke, H.: Three-dimensional sensing of rough surfaces by coherence radar. Appl. Opt. 31, 919–925 (1992)

Dubois, A.: Effects of phase change on reflection in phase-measuring interference microscopy. Appl. Opt. 43, 1503–1507 (2004)

de Groot, P., Deck, L.: Surface profiling by analysis of white-light interferograms in the spatial frequency domain. J. Mod. Opt. 42, 389–401 (1995)

de Groot, P., Colonna de Lega, X., Grigg, D.: Step height measurements using a combination of a laser displacement gage and a broadband interferometric surface profiler. In: Proc. SPIE, vol. 4778, pp. 127–130 (2002a)

de Groot, P., Colonna de Lega, X., Kramer, J., Turzhitsky, M.: Determination of fringe order in white light interference microscopy. Appl. Opt. 41, 4571–4578 (2002b)

de Groot, P., Colonna de Lega, X.: Signal modeling for low coherence height-scanning interference microscopy. Appl. Opt. 43, 4821–4830 (2004)

de Groot, P.: Method and system for analyzing low-coherence interferometry signals for information about thin film structures. US Patent 7,321,431 (2008)

Häusler, G., Neumann, J.: Coherence radar - an accurate 3-D sensor for rough surfaces. In: Proc SPIE, vol. 1822, pp. 200–205 (1993)

Haneishi, H.: Signal processing for film thickness measurements by white light interferometry. Graduate thesis, Department of Communications and Systems Engineering, University of Electro-communications, Chofu, Tokyo (1984)

Harasaki, A., Wyant, J.C.: Fringe modulation skewing effect in white-light vertical scanning interferometry. Appl. Opt. 39, 2101–2106 (2000)

ISO/CD 25178-604, Geometrical product specification (GPS) – Surface texture: Areal – Part 604: Nominal characteristics of non-contact (coherence scanning interferometry) instruments. International Organization for Standardization (2011)

Kim, S.-W., Kim, G.-H.: Thickness-profile measurement of transparent thin-film layers by white-light scanning interferometry. Appl. Opt. 38, 5968–5973 (1999)

Kino, G.S., Chim, S.S.C.: Mirau correlation microscope. Appl. Opt. 29, 3775–3783 (1990)

Larkin, K.G.: Efficient nonlinear algorithm for envelope detection in white light interferometry. J. Opt. Soc. Am. A4, 832–843 (1996)

Leach, R., Brown, L., Jiang, X., Blunt, R., Conroy, M.: Guide to the measurement of smooth surface topography using coherence scanning interferometry. Measurement Good Practice Guide No. 108, National Physical Laboratory (2008)

Lee-Bennett, I.: Advances in non-contacting surface metrology. In: Proc. Optical Fabrication and Testing (OSA), paper OTuC1 (2004)

Lee, B.S., Strand, T.C.: Profilometry with a coherence scanning microscope. Appl. Opt. 29, 3784–3788 (1990)

Michelson, A.A.: Comparison of the international metre with the wavelength of the light of cadmium. Astronomy and Astro-Physics 12, 556–560 (1893)

Munteanu, F., Schmit, J.: Iterative least square phase-measuring method that tolerates extended finite bandwidth illumination. Appl. Opt. 48, 1158–1167 (2009)

Nakano, K., Yoshida, H., Hane, K., Okuma, S., Eguchi, T.: Fringe scanning interferometric imaging of small vibration using pulsed laser diode. Trans. SICE 31, 454–460 (1995)

Novak, E., Schurig, M.: Dynamic MEMS measuring interferometric microscope. In: Proc. SPIE, vol. 5180, pp. 228–235 (2004)

Petzing, J., Coupland, J.M., Leach, R.K.: The measurement of rough surface topography using coherence scanning interferometry. Measurement Good Practice Guide No. 116, National Physical Laboratory (2010)

Sandoz, P., Devillers, R., Plata, A.: Unambiguous profilometry by fringe-order identification in white-light phase-shifting interferometry. J. Mod. Opt. 44, 519–534 (1997)

Sandoz, P.: Wavelet transform as a processing tool in white-light interferometry. Opt. Lett. 22, 1065–1067 (1997)

Schmit, J., Krell, M., Novak, E.: Calibration of high-speed optical profiler. In: Proc. SPIE, vol. 5180, pp. 355–364 (2003)

Schmit, J., Creath, K., Wyant, J.C.: Optical shop testing. In: Malacara, D. (ed.) Surface profilers, multiple wavelength and white light interferometry, 3rd edn. ch.15, pp. 667–755. John Wiley & Sons, Hoboken (2007)

Sheppard, C.J.R., Larkin, K.G.: Effect of numerical aperture on interference fringe spacing. Appl. Opt. 34, 4731–4733 (1995)

10 Digital Holographic Microscopy

Tristan Colomb[1] and Jonas Kühn[2]

[1] Lyncée Tec SA, PSE-A, 1015 Ecublens, Switzerland
[2] Ecole Polytechnique Fédérale de Lausanne, Applied Photonics Laboratory (LOA), 1015 Ecublens, Switzerland

Abstract. Digital holographic microscopy (DHM) is an interferometric technique that measures 3D topography from a single image, acquired in a few microseconds. With these features, DHM performs real-time measurements, up to twenty frames per second in live mode, and is only limited by the camera acquisition rate in post processing mode. For periodic displacements, a stroboscopic module enables 3D displacement measurement up to 25 MHz and short laser pulses of 7.5 ns. The lateral resolution is limited by the numerical aperture of the microscope objective; similar to standard optical microscopes. As DHM does not have a mechanical scan, the vertical calibration is given by the wavelength. DHM is also insensitive to vibration, enabling a vertical resolution of 0.1 nm and a repeatability of 0.001 nm. DHM is, therefore, ideal in certification metrology and for fast topographic imaging in many applications, for example quality control of surface texture, and height and displacement of MEMS and MOEMS. This chapter presents the basic principles of DHM, instrumentation, use and good practice, limitations and extensions of the basic principles, i.e. dual-wavelength DHM to increase the measurement range, infinite focus and DHM reflectometry to measure the topography, thickness and refractive indexes of semi-transparent multi-layered structures, and applications.

10.1 Introduction

Digital holographic microscopy (DHM) is a relative newcomer in the high-resolution microscopy field. DHM is based on the principle of holography that was discovered at the end of the 1940s by Dennis Gabor, who received a Nobel Prize for this work (Gabor, 1948). Indeed, the present capabilities in digital image processing and acquisition in the form of the charge-coupled device (CCD), allow DHM to take advantage of advanced image acquisition techniques. The numerical processing capability of present computers contributes also to give DHM several advantages compared with other optical methods in terms of measurement rates, robustness and ease of use. DHM has two different modes allowing the analysis of samples in reflection or transmission, which is particularly interesting when investigating transparent samples such as micro-optics or, outside the scope of this book, biological samples. The transmission mode has no equivalent in the field of interference microscopy. DHM has diffraction limited transverse resolution and sub-nanometre vertical resolution, as with optical interferometry and more particularly with interference microscopy techniques (see Chap. 8 and Chap. 9), which also use a reference wave to measure the phase of optical waves. But there are major differences between DHM and interference microscopy.

First, in DHM 3D surface topography is obtained in a single acquisition without a mechanical or spectral scan. This is in contrast to PSI (see Chap. 8), which needs several image acquisitions for different positions of the reference wave, or CSI (see Chap. 9), which requires a vertical scan of the sample or the objective. The fact that a scan is not required is a big advantage regarding robustness against vibration, since the acquisition is performed within the time frame of the shutter aperture (below a millisecond) and at the refresh rate of the CCD camera. From the single hologram capture, standard personal computer power and optimized software enable more than twenty frames per second (fps) reconstructions of a 1024×1024 pixel hologram. This unique temporal feature makes DHM attractive for quality control on the production floor and for dynamical analysis using fast cameras. Furthermore, the single acquisition feature coupled with pulsed laser sources simplifies stroboscopic imaging and allows displacement measurement to be carried out on samples excited up to 25 MHz.

Second, the vertical measurement calibration depends uniquely on the wavelength of the source, a physical parameter that can be easily certified and stabilized in contrast to mechanical displacements, for example piezo-electric actuators used in standard commercial PSI instruments that could have long-term drift.

Third, multiple light sources of different wavelengths can be combined in DHM to give several micrometres of vertical measurement range, while sub-nanometre resolution is preserved (Kühn et al, 2007).

Last, DHM is able to reconstruct the entire complex wavefront (amplitude and phase). From this information, it has been recently demonstrated that, under certain conditions, reflectometry allows the measurement of topography, thicknesses and the refractive indexes of transparent multi-layered micro-structures (Colomb et al, 2010).

10.2 Basic Theory

This section presents the basic theory of digital holographic microscopy, from the acquisition to the reconstruction. The acquisition architecture or the reconstruction procedures could be slightly different in the literature. For simplicity, only one variant of the acquisition architectures acquisition will be presented and the different numerical procedures reconstruction will not be discussed here. But more details can be found in the recent review (Kim, 2010).

10.2.1 Acquisition

The DHM architectures, depicted in Figure 10.1, are Mach-Zehnder configurations to analyze samples in (a) reflection mode or (b) transmission mode (Michelson configurations are also possible for reflection mode). The beam emitted by a light source (different types are described in Sect. 10.3.1) is expanded by a beam expander BE. Then, a first beam-splitter separates the object wave $\boldsymbol{O}$ and the reference wave $\boldsymbol{R}$ in to two different arms. The object wave is reflected by, or transmitted through, the object being measured, and magnified by a microscope objective (MO) to produce an image behind or in front of the camera (CCD or CMOS). Finally, the object wave is combined with the reference wave using a second beam-splitter to interfere on to the camera plane. The resulting hologram is written

$$I_H = \boldsymbol{R}^2 + \boldsymbol{O}^2 + \boldsymbol{R}^*\boldsymbol{O} + \boldsymbol{R}\boldsymbol{O}^* \tag{10.1}$$

where * denotes the complex conjugate. The first two terms in equation (10.1) are the zero-order terms, and the third and fourth terms are the real and virtual images, respectively.

The condenser lens C optimizes the beam size and focus position to illuminate the object with a collimated beam in reflection (focused on the back focal plane of objective, MO) and to cover the entire objective pupil in transmission. Lens RL is used to match the reference and object wave curvatures. In place of RL, the same objective can be used as in a Linnik configuration (Mann et al, 2005) (see Sect. 8.6). The optical path retarder OPR is used to match the optical path lengths in the reference and object arms when a low coherence length source is used.

10.2.2 Reconstruction

The reconstruction procedure has two different steps: the reconstruction of the complex wave in the hologram plane and its numerical propagation to the plane where the object is focused.

First, the real or virtual image has to be extracted from equation (10.1). When the reference and object waves are collinear ($\theta = 0$, in-line digital holography), a phase-shifting procedure involving several image acquisitions as in PSI is necessary. In contrast, in off-axis digital holography, the adjustment of the angle θ allows the separation of the different orders of diffraction in the frequency domain. The frequencies of the zero-order term (Kreis and Jüptner, 1997; Liu et al, 2002; Pavillon et al, 2009) and twin-image term (Cuche et al, 2000; Weng et al, 2010) can be easily filtered to provide the real image

$$I_H^f = \boldsymbol{R}^*\boldsymbol{O} = \mathrm{FFT}^{-1}\{\mathrm{FFT}\{I_H\}W\}, \tag{10.2}$$

where FFT is the fast Fourier transform, FFT^{-1} is the inverse fast Fourier transform and W is the filtering window in the frequency domain.

Second, the digitally filtered hologram is illuminated with a digital reference wave $\boldsymbol{R}_D$, which is a numerical copy of the optical wave, and the resultant wavefront

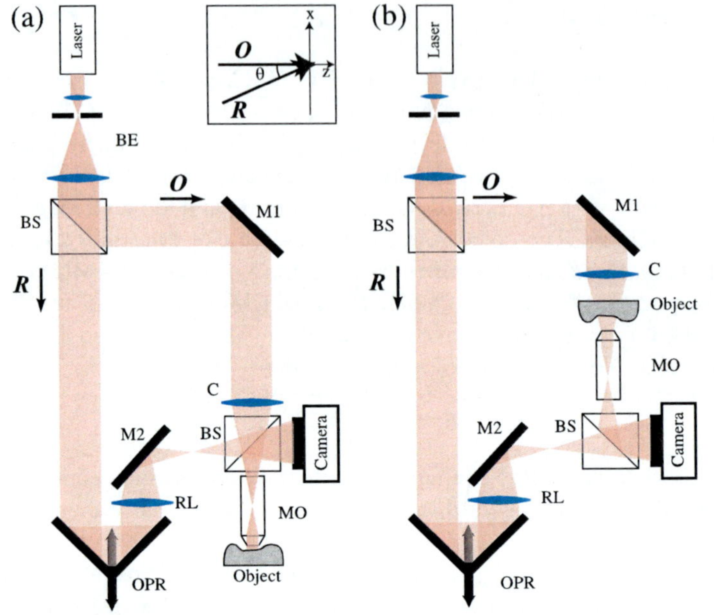

Fig. 10.1 Basic DHM architectures in (a) reflection and (b) transmission configurations. BE, beam expander; BS, beam-splitters; M1,M2 mirrors; OPR, optical path retarder; C, condenser lens; RL, lens in the reference arm; camera, digital camera (CCD, CMOS); MO, microscope objective; ***R***, the reference wave and ***O*** the object wave. Inset, details of the off-axis geometry defined by the angle θ between ***R*** and ***O***.

$\boldsymbol{R}_D\boldsymbol{R}^*\boldsymbol{O}$ is propagated numerically in the Fresnel approximation using one of the following formulations: single Fourier transform (Cuche et al, 1999; Ferraro et al, 2004), angular spectrum (de Nicola et al, 2005; Mann et al, 2005) or convolution (Schnars and Juptner, 2002; Colomb et al, 2006c). The usual Fresnel propagation equation is generalized by the introduction of numerical lenses $\Gamma_H(x,y)$ and $\Gamma_I(\xi,\eta)$ placed, respectively, in the hologram and image planes as shown in Figure 10.2. This gives the reconstructed wavefront (Colomb et al, 2006c)

$$\Psi(\xi,\eta) = \Gamma_I \cdot A \cdot e^{\frac{i\pi}{\lambda d}(\xi^2+\eta^2)} \cdot \int\int \Gamma_H I_H^f e^{\frac{i\pi}{\lambda d}\left[(x-\xi)^2+(y-\eta)^2\right]} dxdy, \tag{10.3}$$

where $A = \frac{e^{i2\pi d/\lambda}}{i\lambda d}$ and d is the reconstruction distance. These numerical lenses allow the simulation of the reference wave ($\Gamma_H = \mathbf{R}_D$ and $\Gamma_I = 1$ corresponds to the usual propagation equation (Cuche et al, 1999)), but can be generalized for higher order aberration compensation. There are two ways to define these numerical lenses, either by using a physical reference hologram recorded on a flat surface in reflection and in air in transmission (Ferraro et al, 2003; Colomb et al, 2006b), or by fitting a polynomial function (standard, Zernike) in the assumed flat areas in the field of view of the object (Colomb et al, 2006a,c; Miccio et al, 2007). A combination of the two methods is also possible. DHM can be configured by computing Γ_H as the inverse phase of a reference hologram I_{Ref} acquired with the reference object wave

$$\Gamma_H = e^{-i\arg[I^f_{Ref}]}, \tag{10.4}$$

where arg[x] is the phase of x. This procedure compensates for all phase aberration terms, in particular the tilt aberration introduced by the off-axis geometry in the hologram plane and the different curvature between the reference and object waves (Colomb et al, 2006c). Once DHM is configured with Γ_H, Γ_I is used principally to compensate numerically for object tilt as shown in Figure 10.2, as opposed to using a tilt stage for mechanical correction. But, it can also be used to compensate numerically for defocus aberration without the need for mechanical adjustment of the instrument as for PSI for example (Sect. 8.8).

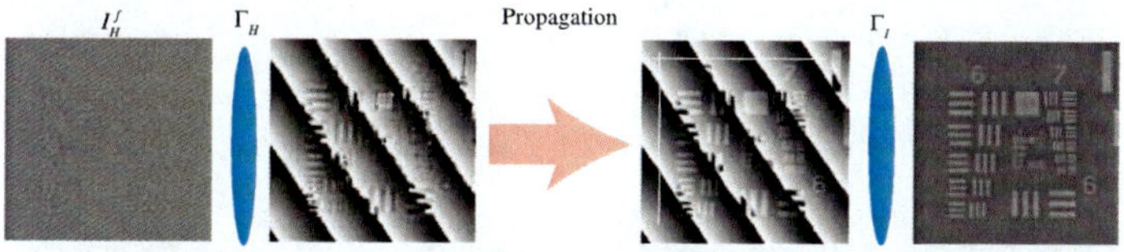

Fig. 10.2 Reconstruction principle. The images represent the reconstructed phase of the wavefront. The filtered hologram I^f_H is multiplied by the numerical lens Γ_H and then propagated to the image plane. When aberrations are introduced by the object tilt, the second numerical lens Γ_I compensates for them with a polynomial fit of the values along white profiles defined in assumed flat areas.

The reconstruction distance d in equation (10.3), defines the position of the reconstructed plane along the propagation direction as presented in Fig. 10.3. Therefore, for a single hologram acquisition, different planes of the object can be focused, numerically increasing the depth of field (Ferraro et al, 2005). Finally, the reconstructed phase contrast image φ_O is converted in to a topographic measurement by the following relations for reflection (h_r) and transmission (h_t) modes

$$h_r(\xi,\eta) = \frac{\varphi_O(\xi,\eta)}{4\pi n_m}\lambda, \tag{10.5}$$

$$h_t(\xi,\eta) = \frac{\varphi_O(\xi,\eta)}{2\pi(n_s - n_m)}\lambda, \tag{10.6}$$

where λ is the source wavelength, n_s is the refractive index of the sample and n_m is the refractive index of the surrounded medium, usually air, but n_m could be oil or water when immersion objectives are used. Fig. 10.4 presents some examples of pseudo-colour 3D perspectives obtained from topographic measurements.

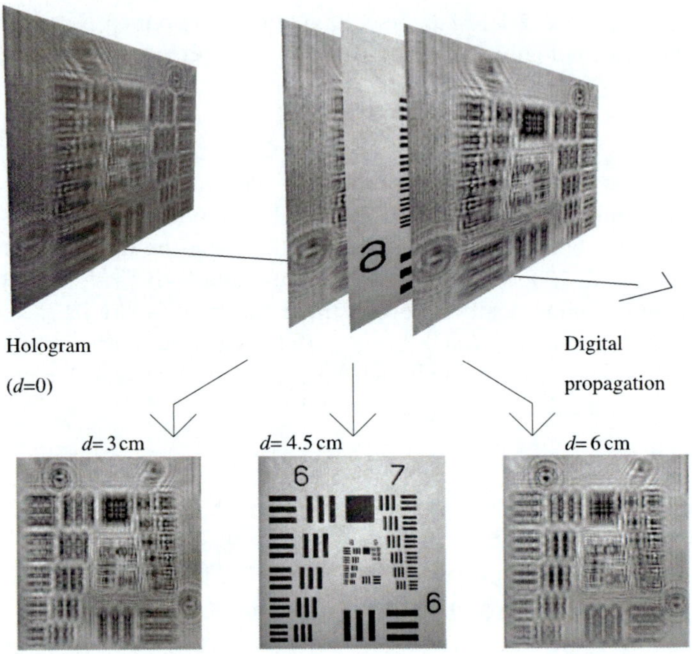

Fig. 10.3 Digital focusing along the propagation direction.

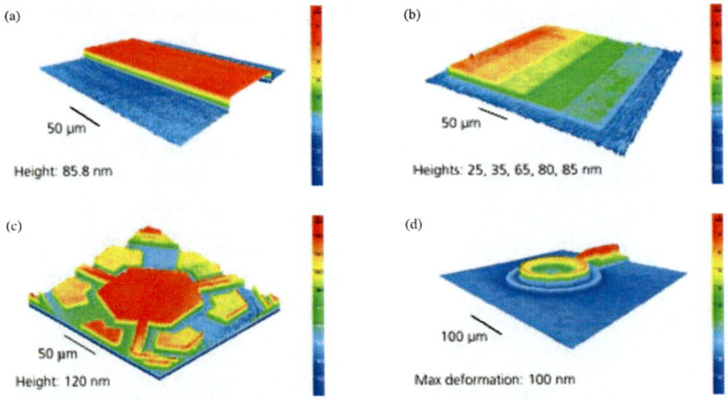

Fig. 10.4 Pseudo-color 3D perspective obtained with DHM. (a) is a certified step of 85 nm; (b) different step heights test target; (c) micro-mirror and (d) deformed micro-pump.

10.3 Instrumentation

Despite much research in to DHM, there is only one commercial manufacturer of DHM instruments, examples of which are presented in Fig. 10.5. The DHM instruments and some of its components' characteristics depend on measurement accuracy and applications. The next sections describe the different variants of these components as a function of the needs.

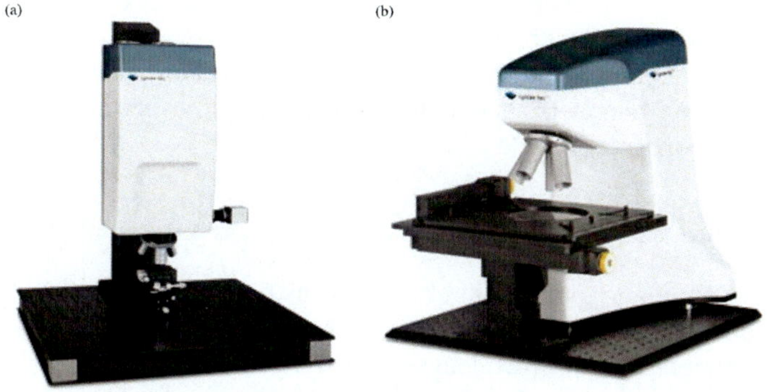

Fig. 10.5 Commerical DHM instruments manufactured by Lyncée Tec (www.lynceetec.com). (a) reflection DHM R1000 model and (b) transmission DHM T1000 model.

10.3.1 Light Source

The primary characteristic of the light in DHM is the wavelength. A smaller wavelength allows for a better vertical resolution (equations (10.5) and (10.6)), and is also advantageous for lateral resolution. On the contrary, larger wavelengths are useful for measuring large sample heights due to the 2π phase ambiguity, for example the infrared wavelength (10.6 μm) used by Allaria et al (2003). Spectral characteristics of the sample, such as absorption, can determine the wavelength choice. For example, infrared wavelengths allow transmission measurement through silicon (Delacrétaz et al, 2009). Finally, dual-wavelength DHM (see Sect. 10.6.1) is an effective alternative when a large wavelength is required, preserving the single-wavelength axial resolution (Parshall and Kim, 2006; Kühn et al, 2007; Khmaladze et al, 2008) and the use of standard visible range digital cameras. In this configuration, the wavelength difference is also an important parameter to define the synthetic wavelength.

The second important characteristic of the light source is the coherence length. The DHM off-axis geometry, in contrast to PSI, imposes the constraint that the coherence length is sufficiently large to have interference over the entire field of view of the camera. This explains why, for simplicity and practical reasons, the first DHM development researchers used monochromatic long coherence sources (Cuche et al, 1999; Takaki and Ohzu, 1999). Later, reduced coherence length sources were investigated, for example VCSELs (Mico et al, 2006), femtosecond lasers used with diffractive lenses (Martìnez-Cuenca et al, 2009) or Ti:sapphire lasers (Balciunas et al, 2009) amongst others, in order to minimize the influence of parasitic interference and to decrease their associated noise. Nevertheless, using a larger coherence length also has some practical advantages: less sensitive z-positioning of the sample and a larger out-of-plane measurement.

The emission characteristics of the source could also be important. Indeed, for stroboscopic techniques (see Sect. 10.6.2) a pulsed source is required to neglect in-plane or out-of-plane displacement of the sample during the hologram acquisition: the shorter the pulse, the higher the investigated displacement speed can be as shown in microscopic interferometry by Petitgrand et al (2001).

Finally, the light source choice is a compromise between all the previous parameters and other parameters such as source power, cost, overall dimensions and availability on the market.

10.3.2 Digital Camera

The detector is an important element of a DHM instrument. The choice of detector is driven by different parameters, some of which are dependent on the source.

The first parameter is the wavelength sensitivity. Indeed, in the visible range, CCD or CMOS cameras are commonly used, but for infrared, for example, other types of detectors are necessary, for example pyroelectric detectors (Allaria et al, 2003) or InGaAs/CMOS hybrid cameras (Delacrétaz et al, 2009).

Second, structural parameters such as pixel size, sampling interval and the sensitive area, that influence the reconstruction image (Jin et al, 2008), have to be taken in to consideration with the imaging performance, in particular shot noise or structural noise (Charrière et al, 2006b, 2007).

Finally, camera shutter and acquisition rate are important criteria depending on the application. Indeed, for most applications 15 fps to 30 fps are sufficient, which suggests a CCD detector. For faster applications (300 fps to several thousands frames per second), when stroboscopic techniques are not possible due to non-periodic deformation or irreversible processes (rupture, evaporation), CMOS cameras are required (Coppola et al., 2009).

10.3.3 Microscope Objective

DHM operates with any standard objective, including long distance or immersion (oil, water) objectives. The choice of objective is a compromise between field of view, magnification and numerical aperture. As with standard optical microscopy, the DHM lateral resolution is diffraction limited and depends on the objective numerical aperture, and the source wavelength (see Sect. 2.5). The lateral resolution can be as low as 300 nm with oil immersion objectives (with a numerical aperture of 1.4).

For samples with high slope angles, such as micro-lenses (Fig. 10.6(a-d)), a high numerical aperture microscope objective is needed, limiting the field of view. In this case, a stitching procedure can be applied as shown in Fig. 10.6(e).

10.3.4 Optical Path Retarder

When reduced coherence length sources are used, the optical path length difference between the reference and object arms has to be shorter than the coherence length,

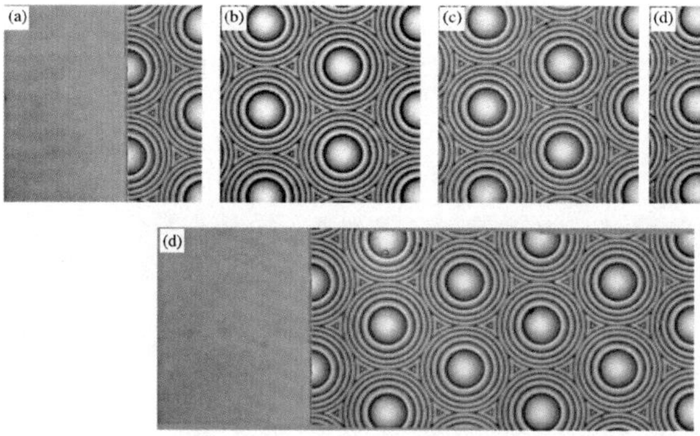

Fig. 10.6 Micro-lens wafer stitching. (a-d) phase reconstruction of the successive acquired holograms and (e) the stitching reconstruction.

which is achieved by adjustment of the optical path retarder OPR in Fig. 10.1. The optical path retarder allows accurate compensation for a modification of the optical path of the object wave, for example when the glass thickness is different from one microscope objective to another, use of an immersion liquid, or sample thickness in transmission DHM. In comparison this compensation is usually performed by the introduction of a glass plate in the reference arm of a Linnik microscope objective for standard PSI or CSI instruments.

10.4 Instrument Use and Good Practice

10.4.1 Digital Focusing

One unique feature of DHM is its ability to digitally focus for a single hologram acquisition. Indeed, in contrast to other optical techniques, the DHM object image formed by the objective is not focused on to the CCD plane. Rather, the focus is adjusted numerically by propagating the field with the reconstruction distance d (equation (10.3), Fig. 10.3).

The ability to digitally focus is particularly interesting when the depth of focus of the objective is smaller than the object height. First, in transmission, it allows depth localization, for example bubbles inside the thickness of a glass watch window (Fig. 10.7) or applications to particle image velocimetry (Ooms et al, 2005).

Second, the digital reconstruction distance allows perfect parfocality to be maintained among the various microscope objectives installed on the DHM.

Third, a small imprecision on the z-position of the sample can be compensated by the digital focusing without the need for a new data acquisition. Similarly, surfaces outside the depth of focus of the microscope objective can be focused digitally from the same hologram.

Finally, this unique capability can increase numerically the depth of focus of the microscope objective to provide infinite focus measurement as presented in Sect. 10.6.4.

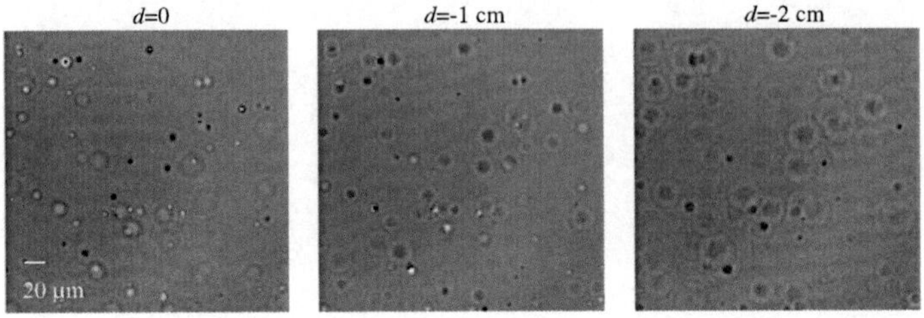

Fig. 10.7 Bubble defects in a watch glass window measured with transmission DHM and 10× microscope objective. By changing the reconstruction distance, different bubbles are focused from the same acquired hologram.

10.4.2 DHM Parameters

DHM instruments are used in combination with reconstruction software that enables the setting of numerical reconstruction parameters (reconstruction distance, numerical lenses) and parameters of the components of the microscope (the camera, the source, the optical path retarder and the mechanical stage).

The microscope objective conditions the optical path retarder position (glass thickness), the reconstruction distance (parfocality), the camera parameters (shutter, brightness, gain), the numerical lens parameters and mechanical stage displacement for the stitching procedure. The combination of the parameters associated with each microscope objective in a different dataset configuration, allows the microscope objective to be changed and a reconstruction to be quickly and easily achieved, similar to a standard optical microscope.

10.4.3 Automatic Working Distance in Reflection DHM

To calibrate the configuration of a microscope objective, the z-position of the sample and the reconstruction distance are adjusted together to achieve the best image quality. Then, if a reduced coherence length is used, the optical path retarder position is optimized. In other words, the fringe contrast of the hologram is maximized for the best reconstruction distance to sample position combination. A simple procedure for automatic working distance control consists simply in optimizing the interference contrast by moving a motorized z-axis stage. This is a very interesting procedure when automatic measurement is necessary, for example for on-line production control. Furthermore, the focus position can be retrieved even on an ideal surface such as a mirror, for which focus algorithms fail.

10.4.4 Sample Preparation and Immersion Liquids

Sample preparation is generally not necessary with DHM. In certain cases, use of immersion liquids offers important advantages by improving the lateral resolution (as with optical microscopes).

In the reflection mode, the surrounding medium refractive index n_m (equation (10.5)) can be changed with an immersion liquid to increase the apparent height of the sample and, therefore, the axial resolution.

In the transmission mode, a surrounding medium refractive index n_m close to the sample refractive index n_s (equation (10.6)) reduces the apparent height, allowing the measurement of high aspect ratio samples (Kühn et al, 2006). With $n_m = n_s$, the immersion liquid can be used to suppress one-sided structures (Fig. 10.8). Finally, by using two different known refractive index immersion liquids, the sample refractive index can be accurately measured (Rappaz et al, 2005).

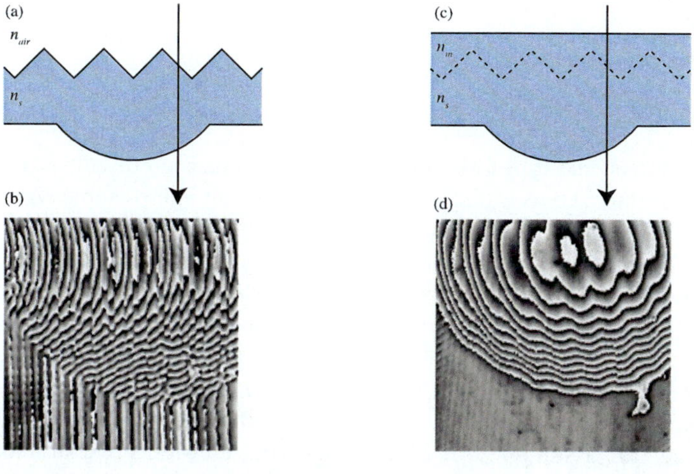

Fig. 10.8 Use of an immersion liquid to suppress one-sided micro-optics. (a,b) conventional transmission measurement principle and phase reconstruction, (c,d) use of the immersion principle and phase reconstruction.

10.5 Limitations of DHM

10.5.1 Parasitic Interferences and Statistical Noise

The use of the off-axis geometry in DHM for real-time imaging implies a larger coherence length source than the in-line geometry. Therefore, even with a reduced coherence length source, DHM cannot completely avoid parasitic interference effects and statistical noise such as shot noise. Spatial filtering can be used to suppress parasitic interference effects having spatial frequencies far removed from the spatial frequencies in the spectrum of the object surface (Cuche et al, 2000). To reduce shot noise, one solution is to use the averaging effect of multiple holograms (Baumbach et al, 2006) but the real-time advantage is clearly lost.

10.5.2 Height Measurement Range

As the reconstructed phase is defined modulo 2π, the maximum height measurement without ambiguity is half the wavelength in the reflection mode and one wavelength in the transmission mode. When the object slopes are sufficiently small to have a continuous signal, phase unwrapping procedures can overcome this ambiguity and enable the reconstruction of the topography. In contrast, for abrupt steps greater than half a wavelength (reflection), the phase unwrapping procedures can fail.

Nevertheless, the measurement range can be improved in transmission mode by using an immersion liquid (Sect. 10.4.4), and in reflection mode by using two wavelengths (Sect. 10.6.1) in order to create a synthetic wavelength of several micrometres. The use of a synthetic wavelength allows for the same accuracy as with a single wavelength.

10.5.3 Sample Limitation

The limitations on the types of sample that can be measured with DHM can be of two types, either the recorded wavefront provides no interpretable signal, or the assumptions to compute the topography from the data are not satisfied.

The first limitation type corresponds to scattering objects for which the reflected or transmitted wavefront phase is too disturbed to bring a topographic measurement, in particular rough samples such as paper, non-polished mechanical parts and plastic parts.

The second type are samples for which the measured data cannot be interpretable as topography from the usual linear relation with the phase (equation (10.5)), for instance when the assumption of a single material semi-reflective structure is not satisfied. Examples of such surfaces include composite surfaces and multiple semi-transparent underlying layers with thicknesses smaller than the coherence length of the source. Indeed, for semi-transparent layers, the recorded object wavefront is the sum of the multiple contributions of the reflected waves between the different layers. Therefore, the reconstructed phase depends on the multiple reflections and is no longer simply proportional to the topography of the surface.

As described in Sect. 10.6.3, the basic principles of DHM can be extended to take into account these multiple reflections and perform DHM reflectometry in order to determine layer thicknesses, refractive index and topography.

10.6 Extensions of the Basic DHM Principles

In this section, some extensions of the basic principles of DHM are presented. Some of these extensions solve the DHM limitations presented in the previous section.

10.6.1 Multi-wavelength DHM

10.6.1.1 Extended Measurement Range

As discussed in Sect. 10.5.2, the measurement range in reflection DHM is limited to half the wavelength for single-wavelength acquisition. An extension of the basic procedure consists of using a multi-wavelength acquisition to increase the height range from several micrometres to millimetres. The description is restricted to dual-wavelength for simplicity, but the concept can be generalized to multiple wavelength acquisitions.

Simultaneous dual-wavelength (Kühn et al, 2007) or successive acquisition (Wagner et al, 2000; Gass et al, 2003; Parshall and Kim, 2006) of two holograms with two different wavelengths allows the synthetic wavelength phase reconstruction

$$\Phi = \varphi_1 - \varphi_2 = 2\pi\frac{x}{\lambda_1} - 2\pi\frac{x}{\lambda_2} = 2\pi x\left(\frac{\lambda_2 - \lambda_1}{\lambda_1\lambda_2}\right) = 2\pi\frac{x}{\Lambda} \tag{10.7}$$

where $x = 2h_r$ is the optical path length (twice the height of the topography in reflection, for an homogeneous sample in air), φ_i is the reconstructed phase for the wavelength λ_i and Λ is the synthetic wavelength defined as

$$\Lambda = \frac{\lambda_1\lambda_2}{\lambda_2 - \lambda_1}. \tag{10.8}$$

This synthetic wavelength is much larger than the original pair of wavelengths; the smaller the difference $(\lambda_1 - \lambda_2)$, the larger the synthetic wavelength. The corresponding synthetic phase obtained with equation (10.7) enables the resolution of much larger structures by removing the phase ambiguity in the range of the synthetic wavelength Λ, thus increasing the vertical measurement range. An example of dual-wavelength acquisition in presented in Fig. 10.9. A multiplexed hologram containing the interference coming from the two wavelengths, is spatially filtered to reconstruct in parallel the two different phase images Fig. 10.9(a,b). The difference between these phase reconstructions gives the synthetic phase Fig. 10.9(c) and finally the topography. As with a single wavelength, simultaneous dual-wavelength acquisition allows real-time imaging up to 15 fps.

It should be noted that the uncertainty on the dual-wavelength topographic measurement, estimated by the standard deviation of the measurement in an assumed flat area, is a consequence of the error related to single-wavelength measurement uncertainties, but is amplified compared with single-wavelength topographic measurement. Indeed, considering two Gaussian noise distributions and quasi-identical single-wavelength phase standard deviations σ_i, the phase difference standard deviation becomes $\sigma_\Phi = \sqrt{2}\sigma_i$ (Colomb et al, 2009). The relationship between the topographic measurement standard deviation computed from the synthetic wavelength (σ_{OPL_s}) or from single wavelength (σ_{OPLi}) is

$$\sigma_{OPL} = \frac{\Lambda}{2\pi}\sigma_\Phi = \frac{\sqrt{2}\Lambda}{2\pi}\sigma_i = \sqrt{2}\frac{\Lambda}{\lambda_i}\sigma_{OPLi}, \tag{10.9}$$

and demonstrates this amplification.

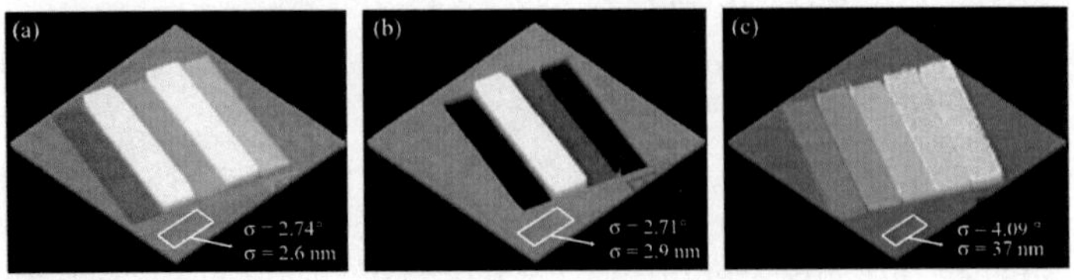

Fig. 10.9 Reconstructed phase images for (a) $\lambda_1 = 680$ nm and (b) $\lambda_2 = 760$ nm and (c) the 3D representation deduced from the synthetic phase ($\Lambda = 6.4$ μm with a standard deviation of 37 nm due to the error propagation.

10.6.1.2 Mapping

As seen in Fig. 10.9(c), the dual-wavelength method removes the phase ambiguity over an extended range, although with a corresponding increase in the topographic error (equation (10.9)). Nevertheless, single-wavelength accuracy can be retrieved by using the synthetic phase only to solve the phase ambiguity for one wavelength, then adding the correct integer multiple of 2π to the corresponding single-phase map φ_i (Fig. 10.10(a)) (Schnars and Juptner, 2002; Gass et al, 2003; Parshall and Kim, 2006; Kühn et al, 2008). Clearly this technique only works when the synthetic topographic error is smaller than half the single wavelength. The graph in Fig. 10.10(b) shows that the dual-wavelength unwrapped topography conserves the single-wavelength accuracy over the range of the synthetic wavelength.

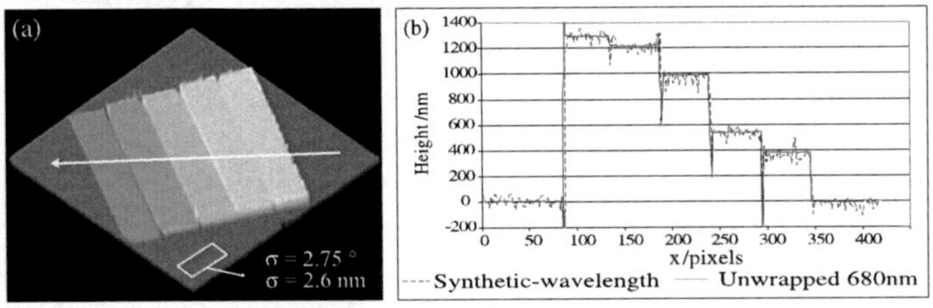

Fig. 10.10 Dual-wavelength unwrapping recovers single-wavelength accuracy as shown in the areal representation (a) and on the graph (b) along white arrow profile.

10.6.2 Stroboscopic Measurement

For very high frequency repetitive displacements and when the digital camera acquisition rate is not fast enough, the stroboscopic method can be used as an extension of DHM or phase-shifting interferometry in general (Conway et al, 2007; Leval et al, 2005; Petitgrand and Bosseboeuf, 2004; Cuche et al, 2009).

As DHM only needs a single hologram acquisition for topography and displacement determination, the stroboscopic method can easily be used to enable the measurement of displacements at up to 25 MHz (Cuche et al, 2009). The principle of stroboscopic measurement is illustrated in Fig. 10.11. The periodic movement of an object is driven by an external signal of period T. During the exposure time of the digital camera, a pulsed laser source illuminates the object periodically at frequency $1/T$, integrating the signal on to the CCD when the object is at the same position t_i ($i = 1, N$). Different holograms are recorded for N positions to provide the entire movement period. To neglect the object displacement during the pulse exposure, the pulse width τ is chosen as short as possible (for example, 7.5 ns in. (Cuche et al, 2009)).

It should be noted that the stroboscopic technique is able to measure not only periodic displacements produced by a periodic excitation, but also non-periodic displacements that can be reproduced periodically, as analyzing the process of transition or damping in the starting or stopping of the excitation signal.

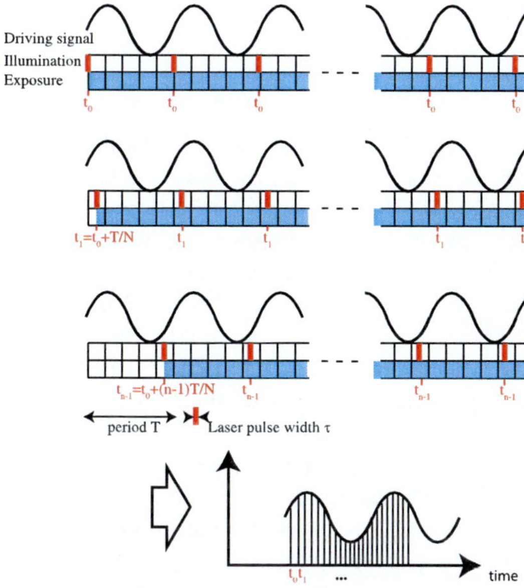

Fig. 10.11 Stroboscopic DHM principle.

10.6.3 DHM Reflectometry

As described in Sect. 10.5.3, for semi-transparent layered structures, the multiple reflections and transmission from the underlying interfaces modify the effective reflected wavefront, and the topography cannot be computed from equation (10.5).

Nevertheless, a recent extension of DHM uses the entire complex wavefront reconstruction (phase and amplitude) to perform reflectometry measurements. Usually, reflectometry methods analyze the reflected amplitudes for different wavelengths and use fitting procedures to deduce layer thicknesses and refractive indexes.

It has been demonstrated (Colomb et al, 2010) that DHM is able to measure the topography, determine layer thicknesses and/or refractive indexes for multi-layered micro-structures (up to three interfaces). The main limitation of this method is that it is a non-absolute measurement (the object has to contain a reference area). Furthermore, an approximate knowledge of the object is necessary to define the theoretical function used to fit the experimental data.

DHM reflectometry fills the measurement gap between spectroscopic ellipsometry, having excellent axial resolution, and, for example confocal microscopy or CSI, which are limited to approximately a micrometre axial resolution (see Sect. 5.3 and Sect. 9.9). DHM reflectometry is particularly interesting for measuring the structured surfaces of MEMS, integrated optics, integrated circuits or optoelectronic devices. Furthermore, even if ellipsometry can achieve sub-nanometre resolution, the real-time measurement capability of DHM makes the method advantageous for quality control on production lines for example.

10.6.4 Infinite Focus

An extension of the numerical focusing capability of DHM is the numerical increase of the depth of focus, as already demonstrated elsewhere for amplitude images obtained in reflection mode (Ferraro et al, 2005). But the complex wavefront in reflection or transmission mode can be reconstructed in focus, to provide the topography in focus as with CSI (Chap. 9) or imaging confocal (Chap. 11), from a single hologram as demonstrated by Colomb et al (2010). The principle consists of using a first evaluation of the topography of the sample from a single distance reconstruction. This shape is then converted into a matrix of reconstruction distances according to geometrical optics. Finally, a scan of the numerical reconstruction distance allows the computation of all the wavefront pixels in focus. For example, an immersed retroreflector is measured in transmission and reconstructed with single reconstruction distances (Fig. 10.12(a-c)) or in infinite focus in Fig. 10.12(d). The profile lines depicted in Fig. 10.12(e) show that the infinite focus reconstruction allows the measurement of the correct retroreflector topography.

10.6.5 Applications of DHM

The extended and basic principles of DHM allow a range of different metrological and surface control applications to be realized. Some of these applications are described in this section.

10.6.5.1 Topography and Defect Detection

Topography, geometry and defect detection are systematic needs in microtechnology when controlling the quality of a component. Defects, such as scratches, cracks, material inhomogeneities and others are often very difficult if not impossible to measure with standard techniques such as scanning stylus instruments or conventional

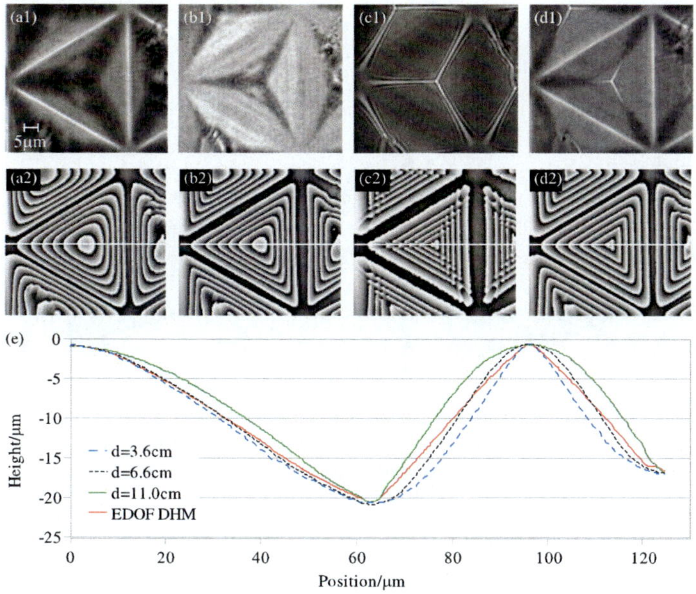

Fig. 10.12 Amplitude (1) and phase (2) reconstructions of a high aspect ratio retroreflector immersed in distilled water measured in transmission with a 60× MO, NA= 1.3 for different reconstruction distances, (a) 3.6 cm (b) 6.6 cm and (c) 11 cm and (d) with the extended depth of focus method. (e) Height profile computed from the phase measured along the white lines defined in a2, b2, c2 and d2.

optical microscopes. For a stylus instrument, a two-dimensional scan dramatically increases the image acquisition time, can damage the sample and cannot always be used on ductile samples. In the case of a conventional microscope, the contrast of the defects on the two-dimensional image is often very small and rarely quantitative. DHM allows fast and quantitative quality control (Fig. 10.13(a,b)).

The numerical lenses presented in Sect. 10.2.2 can be generalized to allow compensation of any shaped form and allow a separate calculation of shape information, and of the surface quality of the samples (see for example the suppression of a ball point pen tip curvature in Fig. 10.13(c,d)). The short acquisition time (a few microseconds) makes DHM systems insensitive to external vibrations and the acquisition rate up to 20 fps ideal for quality control on the production floor. The measurements can be carried out in reflection or transmission modes depending on the specimen's optical characteristics (reflective or transparent).

10.6.5.2 Roughness

A specific instrument that uses DHM technology that is dedicated to measuring surface roughness is shown in Fig. 10.14. This instrument integrates a 8× 0.5 numerical aperture microscope objective and two low-coherence length laser diodes,

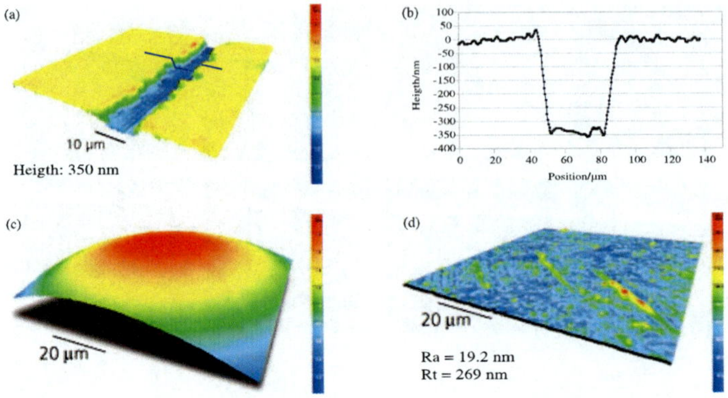

Fig. 10.13 Defect detection example, (a) scratch, reconstructed areal perspective and (b) quantitative profile along the blue line in (a), (c) topography of a ball point pen tip from which the curvature is suppressed (d) to study surface defects and roughness.

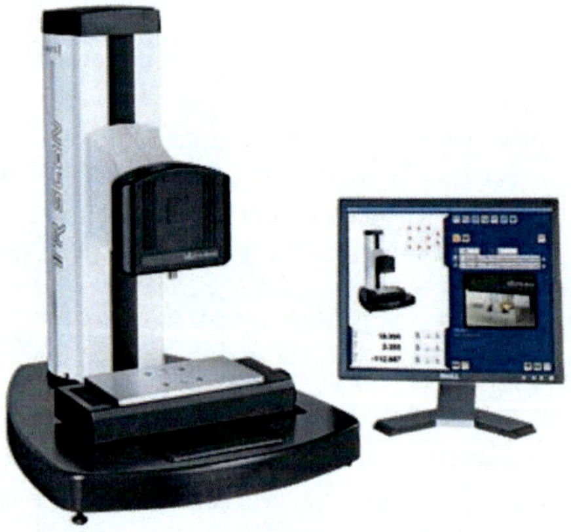

Fig. 10.14 A dedicated roughness measuring instrument that uses DHM technology.

at wavelength $\lambda_1 = 763$ nm and $\lambda_2 = 665$ nm, that can be operated to allow dual-wavelength measurements in order to extend the dynamic range to some micrometres (Sect. 10.6.1) (Kühn et al, 2010). This instrument was validated for roughness measurements by comparing the values obtained on calibrated material measures in the range of 20 nm *Ra* to 900 nm *Ra* with a stylus instrument and a chromatic confocal instrument (see Chap. 5).

In order to carry out the above comparison, the existing ISO requirements on how to extract profiles data are used for the DHM measurement. Multiple profiles (here up to 300) are extracted on a length of $6\lambda c$ ($\lambda c = 0.8$ mm for arithmetic roughness parameter *Ra* above 200 nm or $\lambda c = 0.3$ mm for smaller roughness values). The slowly-varying shape of the sample (waviness) defined as a spatial frequency equal to or smaller than $1/5 = \lambda c/L_{profile}$ is subtracted from the central portion of each profile of length $L_{profile} = 5\lambda c$. As the maximum field of view is approximately 650 $\mu m \times$ 650 μm with 1024 $\times$ 1024 pixels, long-profile data are obtained by the stitching of different transverse scan reconstructed phase images. The overall measurement principle, consisting of stitching (Fig. 10.6), phase offset adjustment and tilt correction (Fig. 10.2) is depicted in Fig. 10.15 with an aerospace test-target as an example. An example of roughness profile extraction on a 460 nm *Ra* sample is presented in Fig. 10.16.

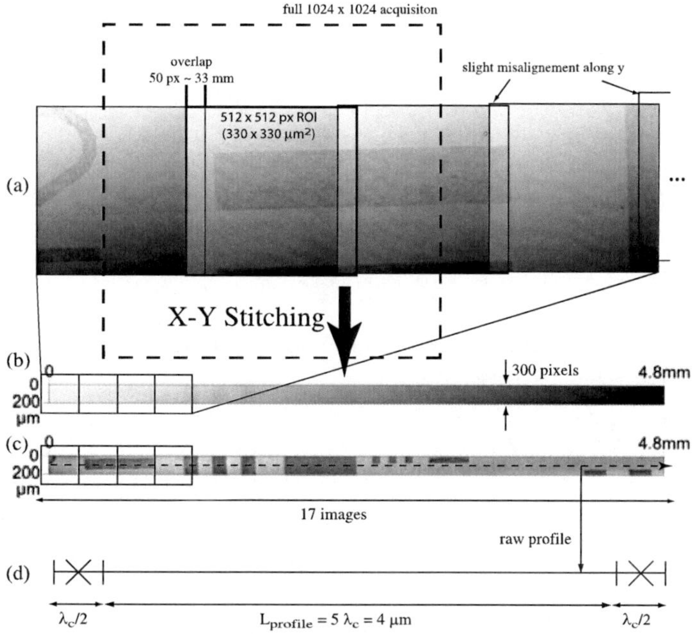

Fig. 10.15 Principle of the method for roughness measurement, demonstrated on an aerospace test target. (a) Example of four raw-phase images (512×512 pixels region of interest) sequentially acquired along the x direction, with field of view and overlap values indicated; (b) assembled 4.8 mm long "mosaic" result through stitching process from (a), with phase offsets adjusted in between each phase images; (c) same as (b), but with tilt corrected for the whole stitched area; (d) one-dimensional profile extracted along the dashed arrow in (c), illustrating the roughness portion according to ISO requirements.

A comparison between a stylus instrument, a chromatic confocal instrument and DHM is presented in Tab. 10.1 and demonstrates that DHM compares well with these roughness measurement techniques. Measurements correspond well with *Ra* values measured on a set of PTB certified material measures manufactured by Halle GmbH. Single-wavelength DHM performs well due to the interferometric accuracy. Then switching to the dual-wavelength mode for roughness above 100 nm *Ra* enables the measurement of surface roughness up to a micrometre with a peak accuracy around 800 nm *Ra*. However, the DHM drawback seems to lie in the 'transition

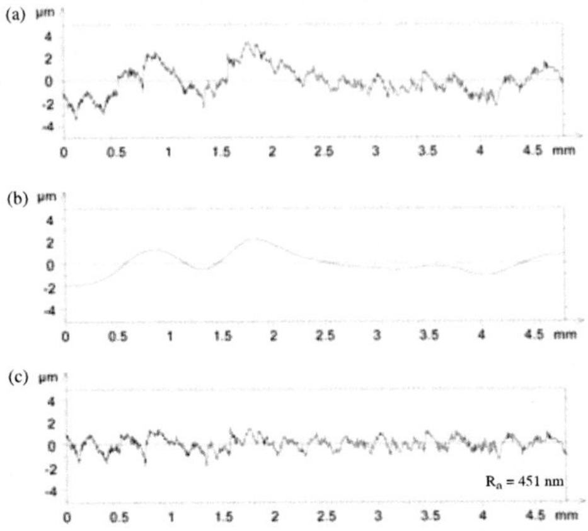

Fig. 10.16 Example of roughness profile extraction on the 460 nm *Ra* sample. (a) Raw topographic profile; (b) waviness component with longitudinal spatial cutoff at $L_{profile}/5 = \lambda c$; (c) roughness profile corresponding to (a) subtracted from (b).

Table 10.1 Comparison between stylus probe, chromatic confocal and DHM techniques on a set of calibrated roughness standards.

Sample calibrated *Ra*	Stylus instrument	Chromatic confocal	1-λ DHM	2-λ DHM
22.3 nm	-	23.7 nm	21.4 nm	-
53.4 nm	-	55.4 nm	51.7 nm	-
81.7 nm	-	90.5 nm	80.0 nm	-
229 nm	228 nm	231 nm	279 nm	-
460 nm	470 nm	470 nm	-	494 nm
876 nm	889 nm	900 nm	-	877 nm

regime' around 200 nm *Ra* when operation modes are switched from single to dual wavelength, due to the sudden amplification of noise for the synthetic wavelength (see Sect. 10.6.1). The position of this 'transition regime' can be modified by a different choice of wavelengths.

The procedure for profile roughness measurement can be transposed for two-dimensional images provided by DHM as shown in Fig. 10.17(a). Some samples with different roughness amplitudes are presented in Fig. 10.17(b,c,d).

10.6.5.3 Micro-optics Characterization

The design constraints imposed on micro-optics, such as micro-lenses, gratings, prisms, corner cubes and in particular micro-optic wafers, are extremely demanding.

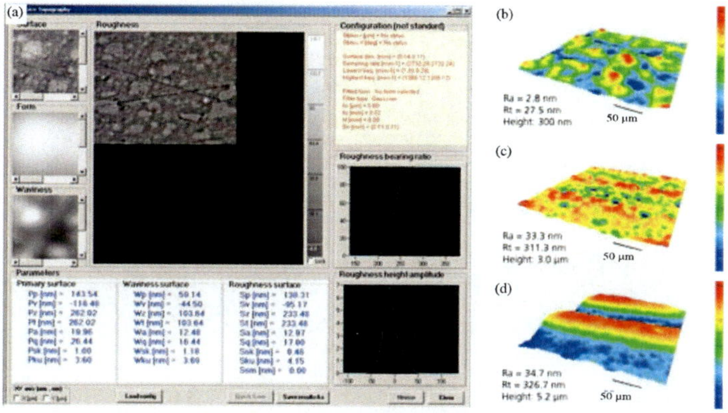

Fig. 10.17 (a) Roughness tool that decomposes the profile surface topography into form, waviness and roughness, in addition to their parameters. (b,c,d) different kind of roughness amplitude.

DHM instruments allow the accurate control of the shape, surface quality (defects detection and roughness as presented in Sect. 10.6.5.1 and 10.6.5.2, respectively), optical performance and uniformity of these parameters across a wafer (Fig. 10.18). Quantitative evaluation of micro-optic shape, radius of curvature, surface quality and aberration coefficients are also available in the instrument software (Charrière et al, 2006a; Colomb et al, 2006c; Kozacki et al, 2009; Merola et al, 2009). It should be noted that no setup adaptation is necessary to investigate different types of micro-optics. For high aspect ratio samples (for example, retroreflectors or high numerical aperture lenses), the use of immersion liquids can diminish the apparent aspect ratio (Sect. 10.4.4). For transparent micro-optics, transmission mode DHM gives access to the transmitted output optical wavefront, which is impossible to retrieve in reflection mode.

Finally, the infinite focus (Sect. 10.6.4) technique is particularly useful to retrieve the accurate shape of micro-optics and in particular micro-lenses. In PSI for example, the focus is usually placed at a point 70 % of the distance to the edge of the lens to minimize the wavefront error (Zecchino , 2003). The DHM infinite focus technique allows reconstruction of all the micro-optics in focus and the correct surface profile as demonstrated by Colomb et al (2010) and shown in Fig. 10.19.

10.6.5.4 MEMS and MOEMS

Stroboscopic measurement with DHM is well adapted for MEMS and MOEMS investigations. For example in Fig. 10.20, a frequency scan (a) is used to detect the resonance frequencies. Then, for one frequency, for example 87 kHz in (b), a reconstructed phase sequence is performed from which spatial deformation (c,d) and time displacement (e) are measured for different positions from the same acquired data. This is an advantage of full field techniques (for example, DHM or PSI) compared with other high-frequency displacement measurement techniques such as

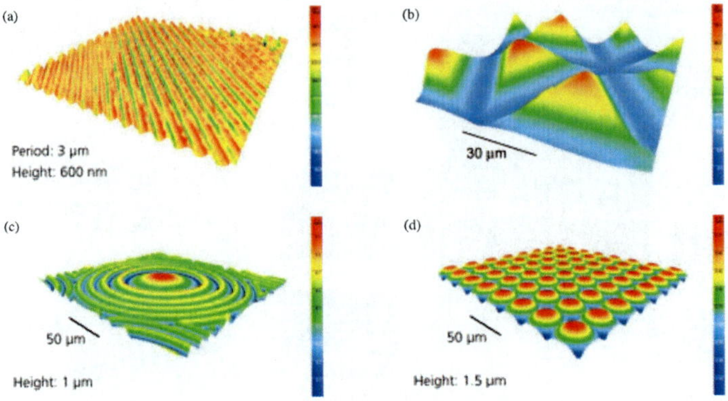

Fig. 10.18 Micro-optics applications. (a) Gratings, (b) corner cube wafer measured in an immersion liquid with a transmission DHM (c) Fresnel lens and (d) micro-lens array.

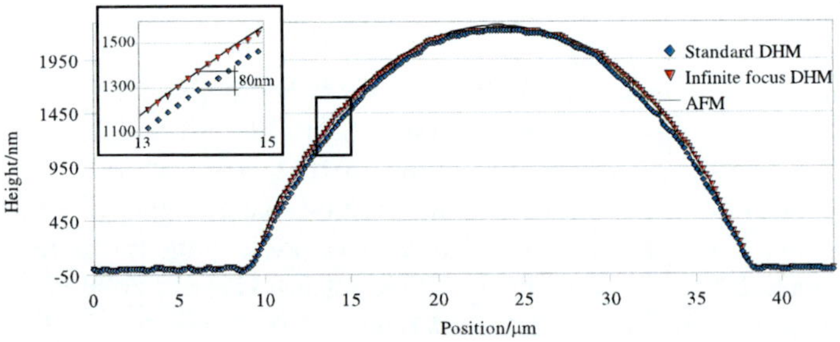

Fig. 10.19 Height profile measured on a micro-lens by standard DHM and infinite focus technique compared with an AFM measurement.

vibrometry, because multiple points are measured simultaneously. For example, in the case of an array of MEMS, it is possible, not only to determine the resonance frequency for each of the components, but also to determine whether the time delay between the excitation and the response is identical for all the components (Sénégond et al (2009)).

10.6.5.5 Semi-transparent Micro-structures

DHM reflectometry (Sect. 10.6.3) is able to accurately measure semi-transparent layered micro-structures. As an example, a crater in a SiO_2/Si wafer is sputtered with secondary ion mass spectrometry. The measured data (Fig. 10.21(a,b)) are fitted by the theoretical function defined by the model (c). Using the mean complex values in the reference area (dashed rectangle in (a,b) and in the bottom of the crater (plain line rectangle) and the theoretical refractive indexes allows the determination of the SiO_2 layer thickness ($d_{SiO_2} = 103$ nm). Then, the topographic image (d) is obtained and compares well with the measurement using a stylus instrument (e).

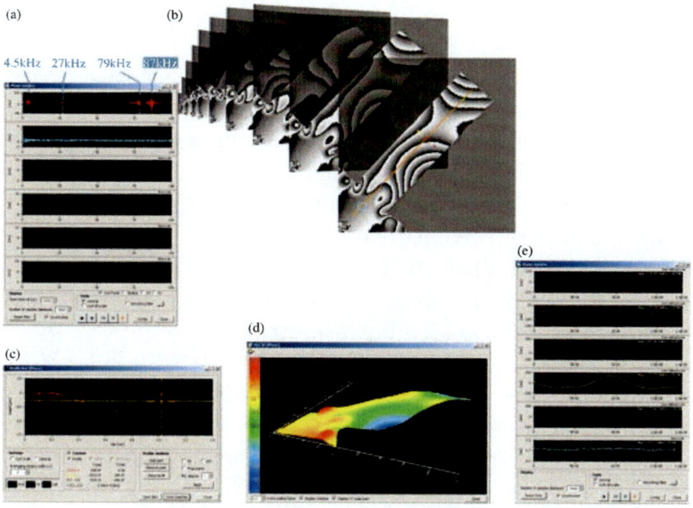

Fig. 10.20 Cantilever analysis by DHM stroboscopic module. A frequency scan (a) detects the resonance frequencies (4.5 kHz, 27 kHz, 79 kHz and 87 kHz). For an 87 kHz resonance frequency, the reconstructed phase images sequence (b) enables the measurement of the displacement profile (c) along the orange line defined in (c) and 3D perspective (d) for a given position. The time monitoring (e) measures the deformation in time for the positions defined by the small square in (b).

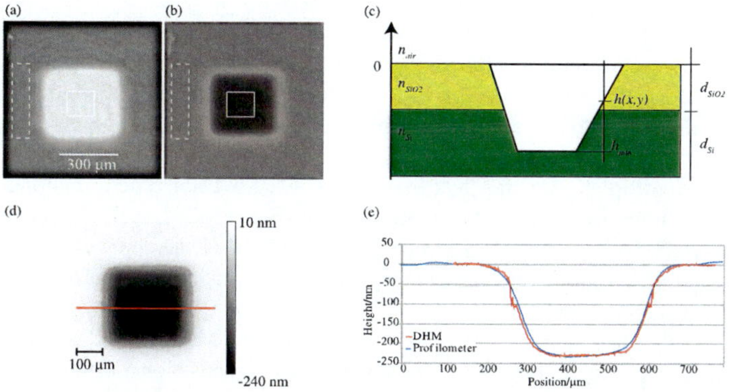

Fig. 10.21 DHM reflectometry measurement on a crater sputtered in a SiO_2/Si wafer. (a) Amplitude and (b) phase reconstructions. (c) model of sputtered crater in a two-layered wafer. The SiO_2 thickness is approximately 100 nm. The use of theoretical refractive indexes, and the average values in reference area (dashed rectangle in (a,b)) and in the bottom of the crater (plain rectangle in (a,b)) enables the computation of the SiO_2 layer thickness and crater depth determination ($d_{SiO_2} = 103$ nm, $h_{min} = 226.7$ nm). (d) Topographic image and the corresponding profile (e) computed from fitted experimental data using the theoretical function defined by the refractive indexes and the computed value d_{SiO_2}. The DHM reflectometry profile and the measurement achieved with a stylus instrument compare well.

10.7 Conclusions

Over the years, DHM has added to its inherent advantages (real-time and non-contact acquisition, vertical calibration by the wavelength, digital focusing); research developments (aberration compensation, multiple-wavelength DHM, infinite focus, DHM reflectometry) with updated equipment (motorized stages, light sources, digital cameras), and conventional modified technologies (stitching, automatic working distance research, stroboscopic acquisition). The pooling of these elements has enabled DHM to become a competitive technique in the field of optical sensors. Undoubtedly, DHM will increase its penetration in to industrial applications and research through its fast and accurate measurements.

Acknowledgements. The research leading to the DHM reflectometry and extended depth of focus has received funding from the European Community's Seventh Framework Programme FP7/2007 to 2013 under grant agreement n° 216105. The author would like to thank Frédéric Montfort for formatting the figures in this chapter.

References

Allaria, E., Brugioni, S., De Nicola, S., Ferraro, P., Grilli, S., Meucci, R.: Digital holography at 10.6 μm. Opt. Commun. 215, 257–262 (2003)

Balciunas, T., Melninkaitis, A., Vanagas, A., Sirutkaitis, V.: Tilted-pulse time-resolved off-axis digital holography. Opt. Lett. 34, 2715–2717 (2009)

Baumbach, T., Kolenovic, E., Kebbel, V., Jüptner, W.: Improvement of accuracy in digital holography by use of multiple holograms. Appl. Opt. 45, 6077–6085 (2006)

Charrière, F., Kühn, J., Colomb, T., Montfort, F., Cuche, E., Emery, Y., Weible, K., Marquet, P., Depeursinge, C.: Characterization of microlenses by digital holographic microscopy. Appl. Opt. 45, 829–835 (2006a)

Charrière, F., Montfort, F., Cuche, E., Depeursinge, C.: Shot noise perturbations in digital holographic microscopy phase images. In: Proc. SPIE, vol. 6252, p. 62521G (2006b)

Charrière, F., Rappaz, B., Kühn, J., Colomb, T., Marquet, P., Depeursinge, C.: Influence of shot noise on phase measurement accuracy in digital holographic microscopy. Opt. Express 15, 8818–8831 (2007)

Colomb, T., Cuche, E., Charrière, F., Kühn, J., Aspert, N., Montfort, F., Marquet, P., Depeursinge, C.: Automatic procedure for aberration compensation in digital holographic microscopy and applications to specimen shape compensation. Appl. Opt. 45, 851–863 (2006a)

Colomb, T., Kühn, J., Charrière, F., Depeursinge, C., Marquet, P., Aspert, N.: Total aberrations compensation in digital holographic microscopy with a reference conjugated hologram. Opt. Express 14, 4300–4306 (2006b)

Colomb, T., Montfort, F., Kühn, J., Aspert, N., Cuche, E., Marian, A., Charrière, F., Bourquin, S., Marquet, P., Depeursinge, C.: Numerical parametric lens for shifting, magnification and complete aberration compensation in digital holographic microscopy. J. Opt. Soc. Am. A23, 3177–3190 (2006c)

Colomb, T., Kühn, J., Depeursinge, C., Emery, Y.: Several micron-range measurements with sub-nanometric resolution by the use of dual-wavelength digital holography and vertical scanning. In: Proc. SPIE, vol. 7389, p. 673891H (2009)

Colomb, T., Krivec, S., Hutter, H., Akatay, A.A., Pavillon, N., Montfort, F., Cuche, E., Kühn, J., Depeursinge, C., Emery, Y.: Digital holographic reflectometry. Opt. Express 18, 3719–3731 (2010)

Colomb, T., Pavillon, N., Kühn, J., Cuche, E., Depeursinge, C., Emery, Y.: Extended depth-of-focus by digital holographic microscopy. Optics Express 13(18), 6738–6749 (2005)

Conway, J., Osborn, J., Fowler, J.: Stroboscopic imaging interferometer for mems performance measurement. J. MEMS 16, 668–674 (2007)

Coppola, G., Striano, V., De Nicola, S., Finizio, A., Ferraro, P.: Analysis of the actuation of an RF-MEMS by means of digital holography. Journal of Holography and Speckle 5, 175–179 (2009)

Cuche, E., Marquet, P., Depeursinge, C.: Simultaneous amplitude-contrast and quantitative phase-contrast microscopy by numerical reconstruction of fresnel off-axis holograms. Appl. Opt. 38, 6994–7001 (1999)

Cuche, E., Marquet, P., Depeursinge, C.: Spatial filtering for zero-order and twin-image elimination in digital off-axis holography. Appl. Opt. 39, 4070–4075 (2000)

Cuche, E., Emery, Y., Montfort, F.: Microscopy: one-shot analysis. Nature Photonics 3, 633–635 (2009)

Delacrétaz, Y., Bergoënd, I., Depeursinge, C. (2009) Digital holographic microscopy for micro-systems investigation in near infrared. In: 3rd EOS Topical Meeting on Optical Microsystems, Capri, Italy, EOS (September 26-30, 2009)

Ferraro, P., De Nicola, S., Finizio, A., Coppola, G., Grilli, S., Magro, C., Pierattini, G.: Compensation of the inherent wave front curvature in digital holographic coherent microscopy for quantitative phase-contrast imaging. Appl. Opt. 42, 1938–1946 (2003)

Ferraro, P., Nicola, S., Coppola, G., Finizio, A., Alfieri, D., Pierattini, G.: Controlling image size as a function of distance and wavelength in fresnel-transform reconstruction of digital holograms. Opt. Lett. 29, 854–856 (2004)

Ferraro, P., Grilli, S., Alfieri, D., Nicola, S., Finizio, A., Pierattini, G., Javidi, B., Coppola, G., Striano, V.: Extended focused image in microscopy by digital holography. Opt. Express 13, 6738–6749 (2005)

Gabor, D.: A new microscopic principle. Nature 161, 777–778 (1948)

Gass, J., Dakoff, A., Kim, M.: Phase imaging without 2π ambiguity by multiwavelength digital holography. Opt. Lett. 28, 1141–1143 (2003)

Jin, H., Wan, H., Zhang, Y., Li, Y., Qiu, P.: The influence of structural parameters of ccd on the reconstruction image of digital holograms. J. Mod. Opt. 55, 2989–3000 (2008)

Khmaladze, A., Restrepo-Martínez, A., Kim, M., Castañeda, R., Blandón, A.: Simultaneous dual-wavelength reflection digital holography applied to the study of the porous coal samples. Appl. Opt. 447, 3203–3210 (2008)

Kozacki, T., Józwik, M., Józwicki, R.: Determination of optical field generated by a microlens using digital holographic method. Opto-Electronics Review 17, 211–216 (2009)

Kim, M.K.: Principles and techniques of digital holographic microscopy. SPIE Reviews 1, 018005 (2010)

Kreis, T., Jüptner, W.: Suppression of the dc term in digital holography. Opt. Eng. 36, 2357–2360 (1997)

Kühn, J., Cuche, E., Emery, Y., Colomb, T., Charrière, F., Montfort, F., Botkine, M., Aspert, N., Depeursinge, C.: Measurements of corner cubes microstructures by high-magnification digital holographic microscopy. In: Proc. SPIE, vol. 6188, p. 618804 (2006)

Kühn, J., Colomb, T., Montfort, F., Charrière, F., Emery, Y., Cuche, E., Marquet, P., Depeursinge, C.: Real-time dual-wavelength digital holographic microscopy with a single hologram acquisition. Opt. Express 15, 7231–7242 (2007)

Kühn, J., Charrière, F., Colomb, T., Cuche, E., Montfort, F., Emery, Y., Marquet, P., Depeursinge, C.: Axial sub-nanometer accuracy in digital holographic microscopy. Meas Sci. Technol. 19, 074007 (2008)

Kühn, J., Solanas, E., Bourquin, S., Blaser, J.-F., Dorigatti, L., Keist, T., Emery, Y., Depeursinge, C.: Fast non-contact surface roughness measurements up to the micrometer range by dual-wavelength digital holographic microscopy. In: Proc. SPIE, vol. 7718, p. 771804 (2010)

Leval, J., Picart, P., Boileau, J., Pascal, J.: Full-field vibrometry with digital fresnel holography. Appl. Opt. 44, 5763–5772 (2005)

Liu, C., Li, Y., Cheng, X., Liu, Z., Bo, F., Zhu, J.: Elimination of zero-order diffraction in digital holography. Opt. Eng. 41, 2434–2437 (2002)

Mann, C., Yu, L., Lo, C.M., Kim, M.: High-resolution quantitative phase-contrast microscopy by digital holography. Opt. Express 13, 8693–8698 (2005)

Martìnez-Cuenca, R., Martìnez-León, L., Lancis, J., Mìnguez-Vega, G., Mendoza-Yero, O., Tajahuerce, E., Clemente, P., Andrés, P.: High-visibility interference fringes with femtosecond laser radiation. Opt. Express 17, 23016–23024 (2009)

Merola, F., Miccio, L., Paturzo, M., Nicola, S., Ferraro, P.: Full characterization of the photorefractive bright soliton formation process using a digital holographic technique. Meas Sci. Technol. 20, 045301 (2009)

Miccio, L., Alfieri, D., Grilli, S., Ferraro, P., Finizio, A., De Petrocellis, L., Nicola, S.: Direct full compensation of the aberrations in quantitative phase microscopy of thin objects by a single digital hologram. Appl. Phys. Lett. 90, 41104–41104-3 (2007)

Mico, V., Zalevsky, Z., García-Martínez, P., García, J.: Superresolved imaging in digital holography by superposition of tilted wavefronts. Appl. Opt. 45, 822–828 (2006)

de Nicola, S., Finizio, A., Pierattini, G., Ferraro, P., Alfieri, D.: Angular spectrum method with correction of anamorphism for numerical reconstruction of digital holograms on tilted planes. Opt. Express 13, 9935–9940 (2005)

Ooms, T., Koek, W., Braat, J., Westerweel, J.: Optimizing Fourier filtering for digital holographic particle image velocimetry. Meas. Sci. Technol. 17, 304–312 (2006)

Parshall, D., Kim, M.: Digital holographic microscopy with dual wavelength phase unwrapping. Appl. Opt. 45, 451–459 (2006)

Pavillon, N., Seelamantula, C., Kühn, J., Unser, M., Depeursinge, C.: Suppression of the zero-order term in off-axis digital holography through nonlinear filtering. Appl. Opt. 48, H186–H195 (2009)

Petitgrand, S., Bosseboeuf, A.: Simultaneous mapping of out-of-plane and in-plane vibrations of mems with (sub)nanometer resolution. J. Micromech. Microeng. 14, S97 (2004)

Petitgrand, S., Yahiaoui, R., Danaie, K., Bosseboeuf, A., Gilles, J.: 3d measurement of micromechanical devices vibration mode shapes with a stroboscopic interferometric microscope. Opt. Las. Eng. 36, 77–101 (2001)

Rappaz, B., Marquet, P., Cuche, E., Emery, Y., Depeursinge, C., Magistretti, P.: Measurement of the integral refractive index and dynamic cell morphometry of living cells with digital holographic microscopy. Opt. Express 13, 9361–9373 (2005)

Schnars, U., Juptner, W.: Digital recording and numerical reconstruction of holograms. Meas. Sci. Technol. 13, R85–R101 (2002)

Sénégond, N., Certon, D., Bernard, J.-E., Teston, F.: Characterization of cMUT by dynamic holography microscopy. In: IEEE International Ultrasonics Symposium (IUS), pp. 2205–2208 (2009)

Takaki, Y., Ohzu, H.: Fast numerical reconstruction technique for high-resolution hybrid holographic microscopy. Appl. Opt. 38, 2204–2211 (1999)

Zecchino, M.: Measuring micro-lens radius of curvature with a white light optical profilometer. Application note of Veeco Instruments Inc. (2003),
http://www.veeco.com/library/Application_Notes.aspx?ShowOpt=0&ID=72
(cited April 19, 2010)

Wagner, C., Osten, W., Seebacher, S.: Direct shape measurement by digital wavefront reconstruction and multiwavelength contouring. Opt. Eng. 39, 79–85 (2000)

Weng, J., Zhong, J., Hu, C.: Automatic spatial filtering to obtain the virtual image term in digital holographic microscopy. Appl. Opt. 49, 189–195 (2010)

11 Imaging Confocal Microscopy

Roger Artigas

Sensofar Tech SL
Technical University of Catalonia (UPC)
Centre for Sensors, Instruments, and Systems Development (CD6)

Abstract. Imaging confocal microscopy is a well-known technology for the 3D measurement of surface topography. A confocal microscope is used for the acquisition of a sequence of confocal images through the depth of focus of the objective. The highest signal within the images of the sequence for each pixel correlates with the height position of the topography. Confocal microscopy has many advantages over other optical techniques such as having a high numerical aperture, meaning a high lateral resolution and a high measurable local slope. The field of applications is very broad, from semiconductor, materials, paper, energy, biomedical, optics, flat panel displays, and more.

11.1 Basic Theory

11.1.1 Introduction to Imaging Confocal Microscopes

Marvin Minsky invented confocal microscopy in 1957 (Minsky 1961). However, because a confocal image is generated electronically, it did not become a practical microscopy method until computers had enough processing power. The basic field of applications of confocal microscopes is for biological studies where the samples are marked with fluorophores and excited by the illumination light source. On thick biological specimens (Brakenhoff et al. 1978), out-of-focus fluorescence that reduces the contrast of the in-focus plane is blocked by means of a confocal arrangement. A high-contrast image of the in-focus plane is recovered, unveiling details of the surfaces without the need to physically cut them. Additionally, a 3D reconstruction of the specimen is carried out by scanning the sample in the axial direction.

A confocal microscope produces optically sectioned images of the sample under inspection. The basic principle to produce an optically sectioned image is by restricting the illuminated regions on the sample by means of a structured illumination pattern and observing the reflected or backscattered light by means of a second pattern identical to the illumination pattern, which blocks the light that comes from the regions of the surface out of the focal plane of the microscope's objective. The illumination and detection patterns can be a single pinhole placed on the optical axis or a set of pinholes, slits, parallel slits, or any other pattern that effectively reduces the size of the illuminated and detection regions. Independent of the geometry of the illumination and detection patterns, in-plane scanning of such patterns is needed in order to produce a complete optically sectioned image.

The in-plane scanning can be carried out mechanically or optically and is adapted to the specific geometry of the pattern.

A number of different confocal arrangements exist. There are different techniques for in-plane scanning, for generating the illumination and detection patterns, and for detection. Each different configuration optimises a given application such as maximisation of light efficiency, optimisation of signal to noise ratio, optimisation of speed, simplification or reduction of hardware cost and adaptation to different excitation wavelengths. Despite the number of different configurations, most confocal microscopes can be classified in to one of the three following categories: laser scanning, disc scanning and programmable array scanning.

11.1.2 Working Principle of an Imaging Confocal Microscope

The simplest configuration of confocal microscopy is a laser scanning microscope (Fig. 11.1). In such a configuration, a pinhole is located on the field diaphragm

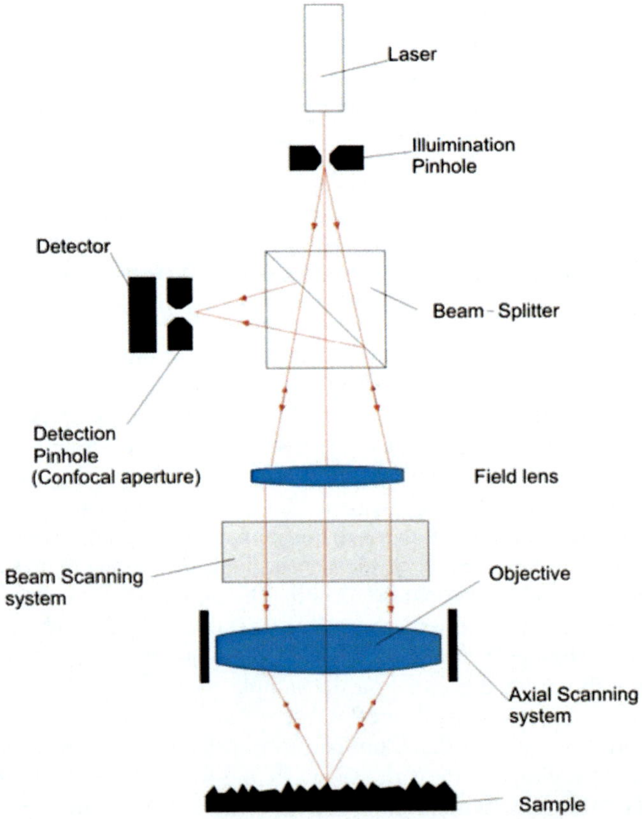

Fig. 11.1 Basic setup of a laser scanning microscope with the sample in focus position

position of the microscope and imaged onto the surface by means of the microscope's objective. The smallest illuminated spot is achieved on the focal plane of the objective and is typically a diffraction-limited spot. The light reflected from the surface passes back through the objective and is imaged onto a second pinhole, called the confocal aperture, placed on a conjugate position of the illumination pinhole. At the rear of the confocal aperture, there is a photo-detector recording the intensity signal reflected from the surface.

When the surface is placed exactly on the focal plane of the objective, the reflected light is imaged onto the confocal aperture and thus the recorded signal on the photodetector is high. On the contrary, when the surface is placed away from the focal plane of the objective (Fig. 11.2), the reflected light is imaged away from the confocal aperture. In this case, the confocal aperture filters out the light that is not coming from the focal plane of the objective and the photo-detector records a lower signal.

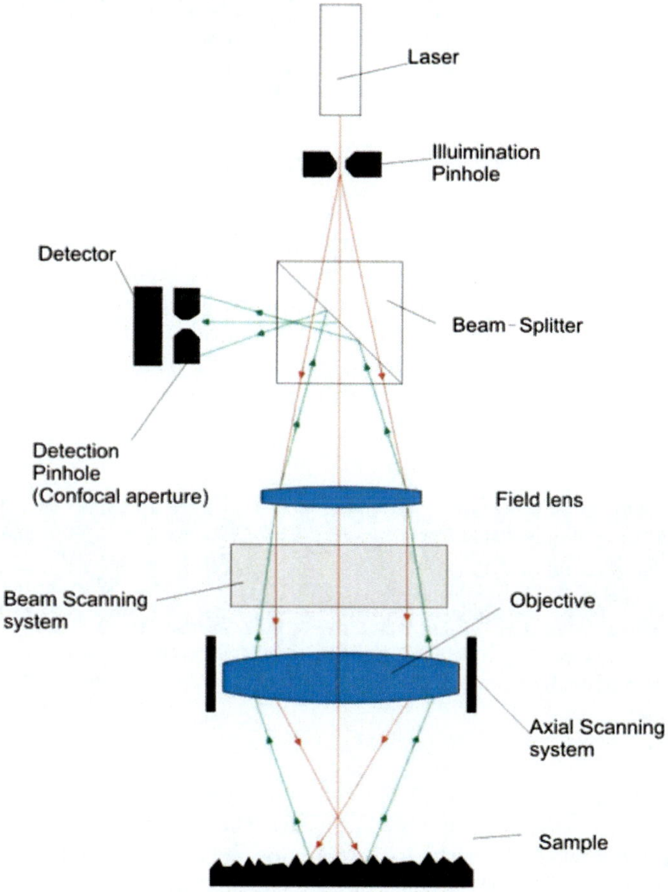

Fig. 11.2 Basic setup of a laser scanning microscope with the sample out-of-focus position

A confocal image from the focal plane of the objective is recovered by scanning the beam onto the surface point by point. In a bright field image, there is signal all along the image, showing high frequency details only on those regions that are close to the focal plane of the objective. In contrast, in a confocal image the regions in focus have a high signal, while the regions out-of-focus tend to be dark, being completely dark for those regions very far from focus. Fig. 11.3 shows the difference between bright field and confocal images.

Fig. 11.3 Bright field (left) and confocal (right) images of laser irradiated silicon

The basic principle of confocal profiling relies on storing a sequence of confocal images in the memory of a computer taken from different z axis planes along the depth of focus of the microscope's objective. A sequence of confocal images is shown in Fig. 11.4. An optically sectioned image shows bright grey pixel levels for those regions of the surface that lie within the depth of focus of the objective, and dark grey pixel levels for the rest of the parts of the surface that are out-of-focus.

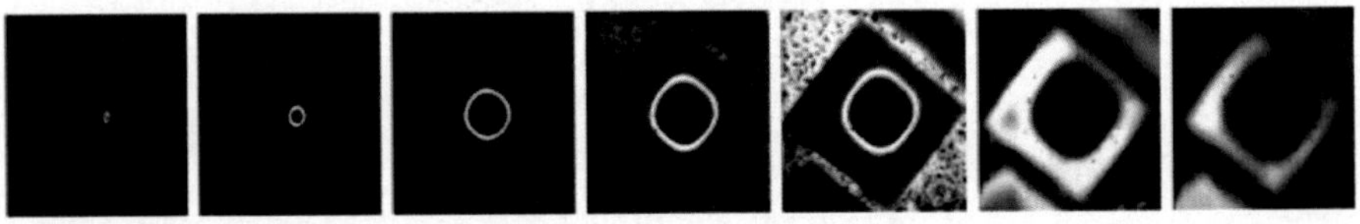

Fig. 11.4 A series of confocal images through the depth of focus of a confocal microscope's objective

Each pixel of the image contains a signal along the z direction called the axial response similar to that shown on Fig. 11.5 (left). The maximum signal of the axial response is reached when the surface is exactly located on the focal plane of the microscope's objective. Different pixels will have the axial response maximum located on different z axis positions according to the 3D surface shape. By locating the z axis position of the maximum of the axial response for each pixel, the 3D surface is reconstructed. Fig. 11.5 (right) shows the 3D surface calculated from the sequence of images in Fig. 11.4.

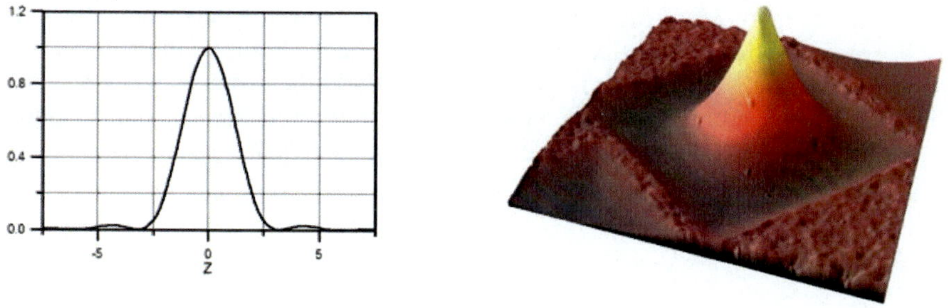

Fig. 11.5 Left: axial response of a single pixel along the z direction. Right: the 3D surface of the series of images of in Fig. 11.4

A confocal image is built up by scanning in plane (on the x and/or the y direction) the illumination and detection patterns. Depending on the kind of confocal microscope, the in-plane scanning can be very slow or very fast. A laser scanning microscope needs to scan point by point, meaning that the scanning time is slow. Commercial instruments at this time reach between one to five images per second. In contrast, disc scanning and micro-display scanning confocal microscopes are much faster, reaching more than ten images per second, and even more than one hundred with some designs. In both confocal designs, an in-plane scanning is needed, meaning that the sample and the focal plane of the microscope's objective should not be moving during this time. For areal measurement of surfaces this means that the in-plane scanning and the axial scanning need to be synchronised in order not to disturb each other.

11.1.3 Metrological Algorithm

The z axis location of the maximum of the axial response relates to the height location of the 3D surface (Conchelo and Hansen 1990). The fastest way to calculate the z axis position is to assign it to the discrete position of the scanner. This is a low-resolution method because the smallest typical step on a confocal microscope is on the order of 0.05 μm. More advanced mathematical methods are used in order to locate more precisely the metrological data. These methods can be classified as real-time fitting algorithms and off-line fitting algorithms. A real-time algorithm calculates significant mathematical data for each z plane during the axial scan, discarding the images each time. The real-time algorithm has the advantage of being not dependent on the computer's memory, but it has the disadvantage of not being able to deal with multiple peaks, such as those appearing on thick transparent materials, among other optical artefacts. The most well-known real-time algorithm is the centre of mass algorithm. In contrast, an off-line algorithm stores the entire series of confocal images in the computer's memory and calculates the 3D surface afterwards. The main advantage of an off-line algorithm is that it can deal with multiple peaks and optical artefacts, but it has the disadvantage of being limited to the computer's memory and processing time. There are several off-line algorithms. Paraboloid fitting (Fig. 11.6) uses some points around the maximum and is one of the most utilised algorithms (Wilson 1994). Paraboloid fitting is fast, requires little data, and can provide resolution of 1/100 the distance between z axis steps. An advanced algorithm is Gaussian fitting to the full axial response. Gaussian fitting is more accurate but requires more processing time.

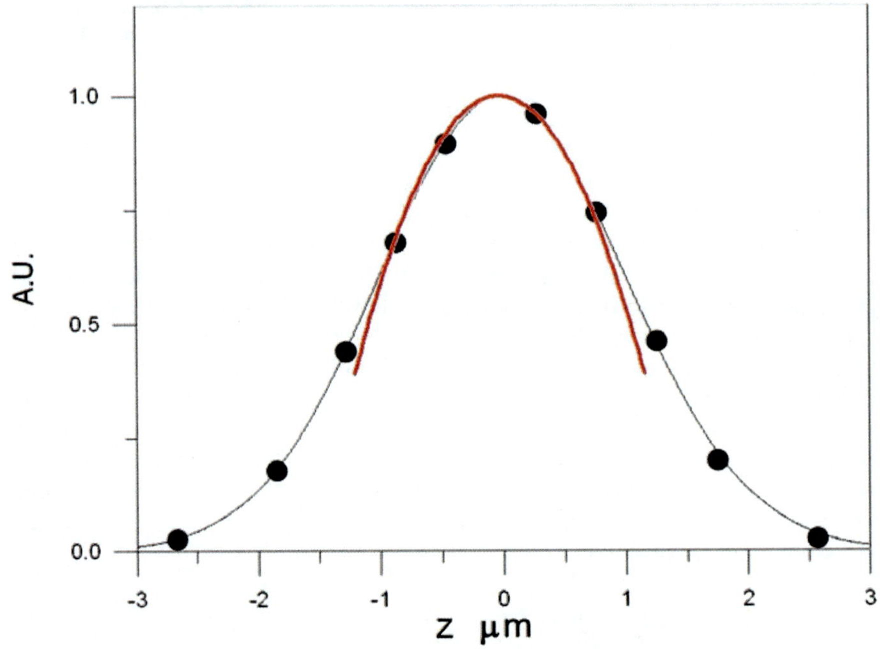

Fig. 11.6 Paraboloid fitting to some points around the maximum of the axial response

11.1.4 Image Formation of a Confocal Microscope

11.1.4.1 General Description of a Scanning Microscope

A schema of typical scanning optical microscope is shown in Fig. 11.7.

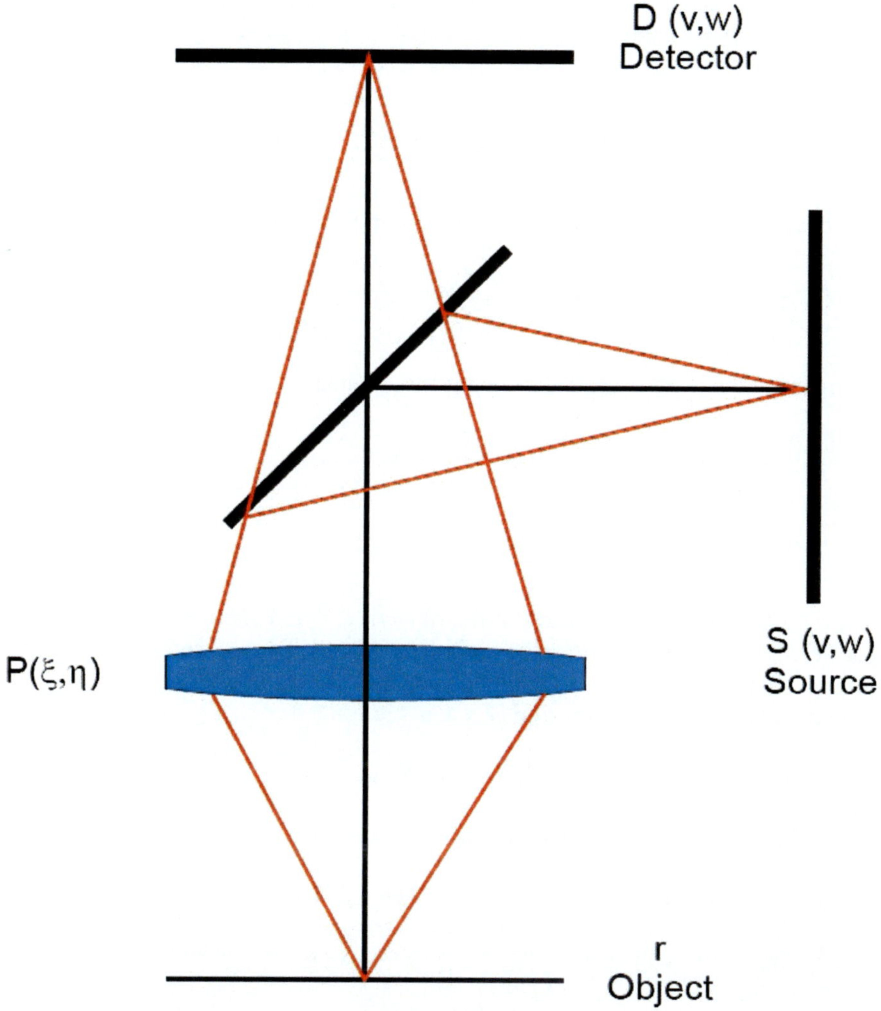

Fig. 11.7 General layout of a scanning optical microscope

A light source pattern of distribution $S(v,w)$ illuminates a microscope's objective with a pupil function $P(\xi,\eta)$. The light is focused onto the sample with reflectance distribution r, reflected back to the objective with the same pupil function and focused onto a detector with sensitivity $D(v,w)$. The intensity on a single point of the detector with incoherent propagation is given by (Sheppard and Shotton 1987, Sheppard and Moa 1988)

$$I(u,v,w) = \int_{\infty}^{\infty}\int_{-\infty}^{\infty} D(v,w)\,t(u,v',w')\left|h(0,v-v',w-w')\right|^2 dv'dw' \qquad (11.1)$$

where $t(u,v',w')$ is the projection pattern out-of-focus by a quantity u, diffraction limited on the object plane, $u/2$ being the distance between the surface and the object plane, given by

$$t(u,v',w') = \int_{-\infty}^{\infty}\int_{-\infty}^{\infty} S(v'',w'')\left|h(u,v'-v'',w'-w'')\right|^2 dv''dw'' \tag{11.2}$$

where $h(u,v,w)$ is the Fourier transform of the pupil distribution

$$h(u,v,w) = \int_{-\infty}^{\infty} P(\xi,\eta)e^{i(\xi v+\mu w)}e^{i\Delta W(u,v,w)}d\xi d\eta\,. \tag{11.3}$$

The terms v and w are the normalised optical coordinates

$$\begin{aligned} v &= \frac{2\pi}{\lambda}x\sin\alpha \\ w &= \frac{2\pi}{\lambda}y\sin\alpha \end{aligned} \tag{11.4}$$

and $\sin\alpha$ is the numerical aperture of the objective, ξ and η are the pupil coordinates normalised to the aperture radius of the pupil, $\xi = x/a$ and $\eta = y/a$, and $\Delta W(u,v,w)$ is the wavefront including focusing and aberration terms.

Equation (11.2) can be expressed as a convolution by

$$t(u,v,w) = S(v,w)\otimes\left|h(u,v,w)\right|^2, \tag{11.5}$$

resulting in a simplified expression for equation (11.1) as

$$I(u,v,w) = \left(S(v,w)\otimes\left|h(u,v,w)\right|^2\right)\left(D(v,w)\otimes\left|h(0,v,w)\right|^2\right). \tag{11.6}$$

On a circular pupil, such as those present on a typical microscope objective

$$\xi = \eta = \rho\sin\theta \tag{11.7}$$

and equation (11.3) simplifies to

$$h(u,v) = \int_0^1 P(\rho)J_0(v\rho)e^{i\Delta W(u,\rho)}\rho d\rho \tag{11.8}$$

where $J_0(x)$ is a first-order Bessel function of the first type. The defocus term for the wavefront in the pupil region is expressed as

$$\Delta W(u,\rho) = -\frac{1}{2}u\rho^2 \tag{11.9}$$

where

$$u = \frac{8\pi}{\lambda} z \sin^2(\alpha/2)\,. \tag{11.10}$$

Equation (11.8) simplifies to

$$h(u,v) = \int_0^1 P(\rho) J_0(v\rho) e^{-i\frac{1}{2}u\rho^2} \rho d\rho\,. \tag{11.11}$$

11.1.4.2 Point Spread Function for the Limiting Case of an Infinitesimally Small Pinhole

From equation (11.6), a single-point illumination *S* and extended detector *D* will give an intensity distribution of a bright field microscope given by

$$I(v,w) = \left\{ \frac{2J_1\left(\sqrt{v^2+w^2}\right)}{\sqrt{v^2+w^2}} \right\}^2 . \tag{11.12}$$

In contrast, on a confocal microscope, for the limiting case of infinitesimally small pinholes, the light source distribution *S* and detector distribution *D* approximate a Dirac delta function, giving

$$I(v,w) = \left\{ \frac{2J_1\left(\sqrt{v^2+w^2}\right)}{\sqrt{v^2+w^2}} \right\}^4 . \tag{11.13}$$

Equations (11.12) and (11.13) are depicted in Fig. 11.8 where it can be seen that both equations have the same zero location, but in the confocal case the width is 1.4 times narrower than the bright field case.

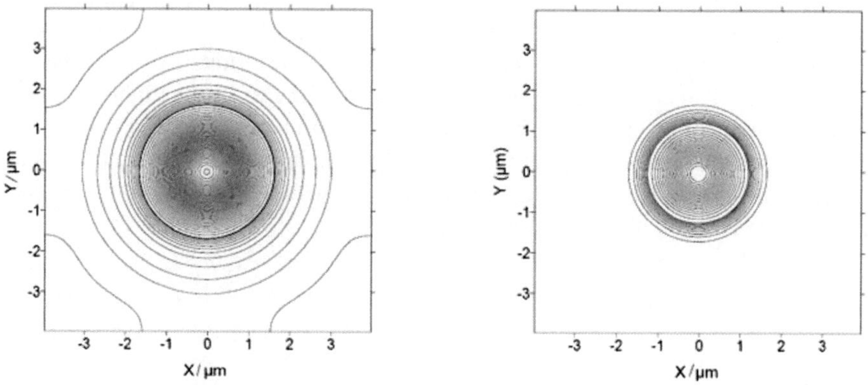

Fig. 11.8 Intensity distribution of the point spread function for a bright field microscope (left) and confocal microscope (right) for the limiting case of infinitesimally small pinholes

Analogously, the defocus distribution of equations (11.12) and (11.13) are depicted in Fig. 11.9.

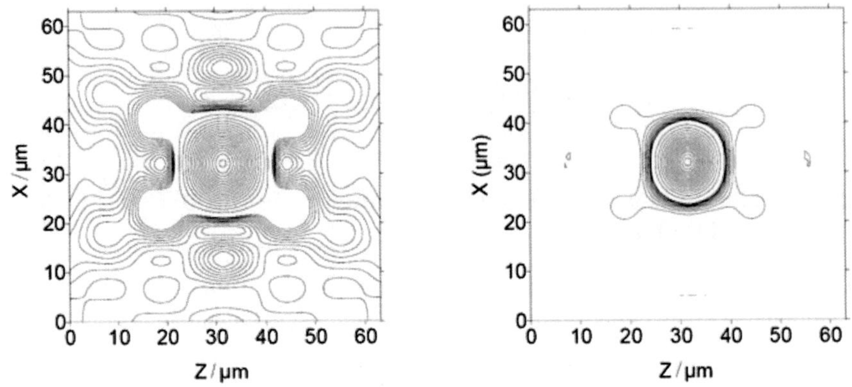

Fig. 11.9 Intensity distribution of a bright field microscope (left) and confocal microscope (right) for the limiting case of infinitesimally small pinholes along the *xz* plane. Note the optical sectioning strength of a confocal microscope, which reduces the signal for the out-of-focus regions

11.1.4.3 Pinhole Size Effect

11.1.4.3.1 Pinhole Size Effect on Lateral Resolution

It is intuitive to think that increasing the size of the confocal aperture will cause the signal distribution to broaden from a purely confocal microscope to a bright field microscope. Fig. 11.10 shows the half-width intensity of the point spread function (PSF) ($v_{1/2}$) against the normalised radius of the detection pinhole (v_p). The pinhole diameter is normalised to a reference unit called the airy unit (AU). The airy unit corresponds to the diameter of a diffraction limited spot on the objective's plane. The confocality is lost by increasing the size of the pinhole and the half-width intensity degrades to that of a bright field microscope. Fig. 11.10 also shows how the half intensity is kept constant for a pinhole radius smaller than 0.5, which means that 0.5 AU is a good pinhole size to keep good confocality and maximise the detector signal.

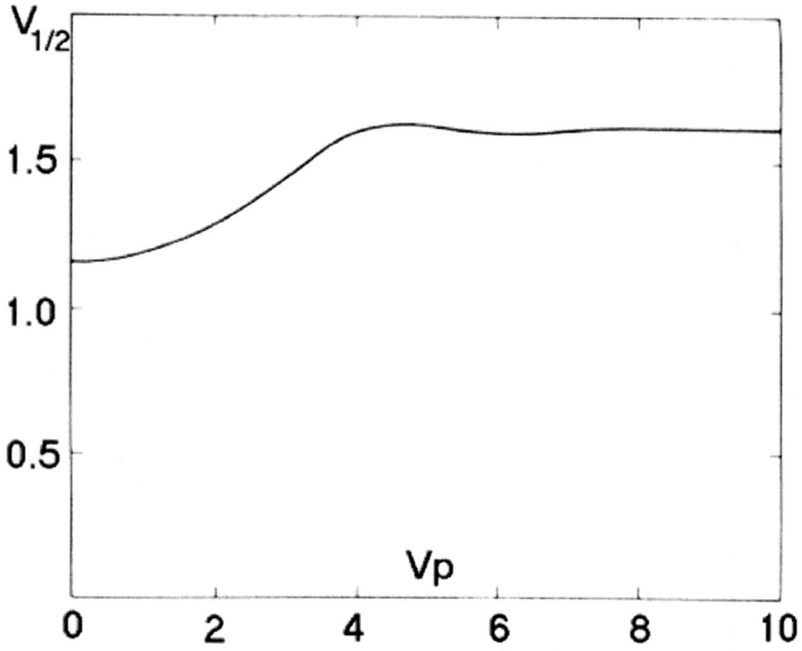

Fig. 11.10 Half-width intensity of the PSF against normalised pinhole radius

On confocal microscopes it is theoretically possible to close down the pinhole to very small diameters. Nevertheless, for pinhole sizes smaller than 0.25 AU additional diffraction effects on the pinhole should be taken into account. These effects of diffraction are called wave-optical confocality. Additionally, under the conditions for wave-optical confocality, the optical propagation of the system (the coherence and the phase of the wavefront) should be taken into account. For pinhole sizes larger than 0.25 AU the propagation is incoherent and the diffraction can be avoided. In this case the behaviour is totally geometric and called geometric-optical confocality.

The lateral resolution of a confocal microscope is determined by

Geometrical-optical confocality	Wave-optical confocality	
$\dfrac{0.61\lambda}{A_N}$	$\dfrac{0.37\lambda}{A_N}$	(11.14)

where A_N is the numerical aperture.

11.1.4.3.2 Pinhole Size Effect on the Axial Response

The size of the detection pinhole determines the optical sectioning strength of a confocal microscope. The larger the pinhole size, the lower the optical sectioning. The intensity along the defocus is given by

$$I(u) = \left(\frac{sin(u/2)}{u/2} \right)^2 \tag{11.15}$$

where the normalised defocus coordinate u is related to the real defocus z by

$$u = \frac{8\pi}{\lambda} z\, sin^2(\alpha/2). \tag{11.16}$$

Fig. 11.11 shows the effect of increasing the pinhole size on the axial response.

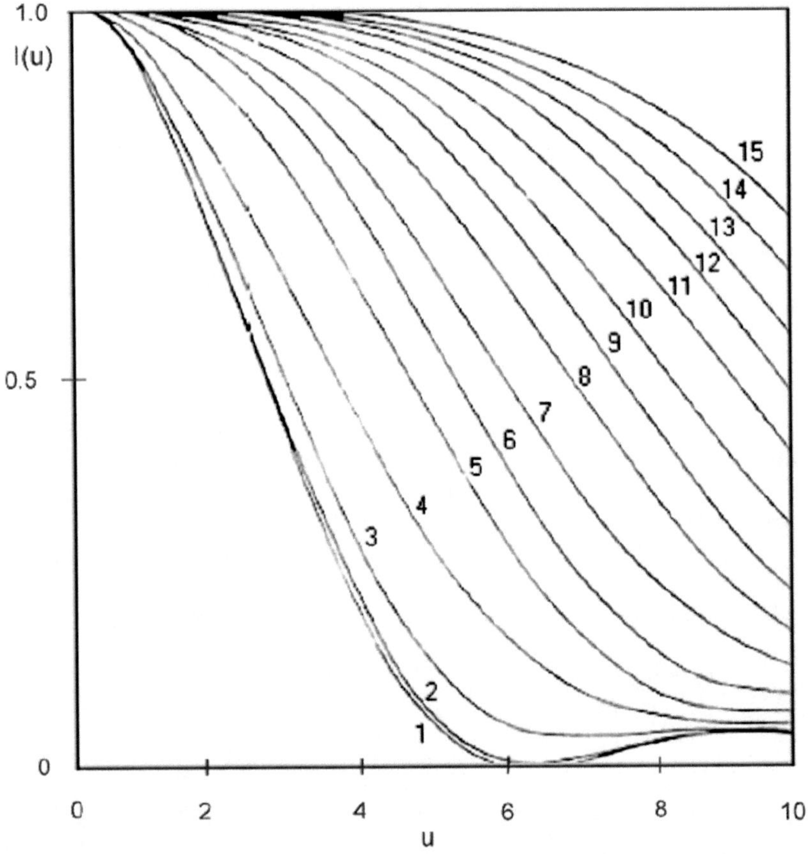

Fig. 11.11 Axial response for different pinhole sizes. The number identifying each curve indicates the pinhole diameter expressed in airy units

It is worth noting that the pinhole size d is described in the object plane (Wilson 1987, Kimura 1994). To obtain the v_p value is multiplied by the magnification between the optical system and the detector. This means that in order to have pinhole sizes less than 0.5 AU the following condition must be accomplished

$$d \leq \frac{0.61\lambda}{NA}\left(\frac{M}{2}\right) \tag{11.17}$$

where M is the magnification of the objective and d the real diameter of the pinhole. It is also worth noting that changing the objective on the system changes the magnification and the numerical aperture, and thus changes the degree of confocality.

Fig. 11.12 shows theoretical results for the axial response of different pinhole sizes. The objective used had a magnification of 50× and numerical aperture of 0.8.

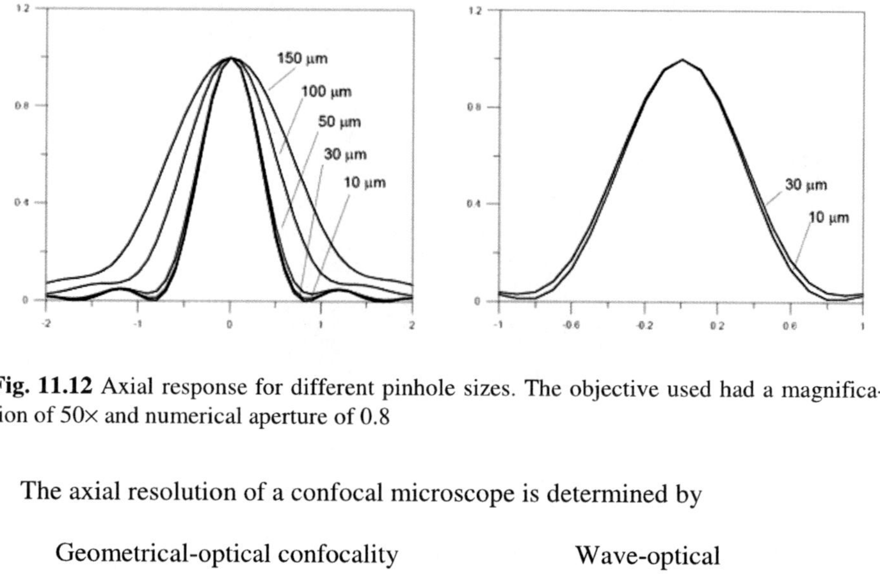

Fig. 11.12 Axial response for different pinhole sizes. The objective used had a magnification of 50× and numerical aperture of 0.8

The axial resolution of a confocal microscope is determined by

Geometrical-optical confocality	Wave-optical confocality

$$\sqrt{\left(\frac{0.88\lambda}{1-\sqrt{1-NA^2}}\right)^2+\left(\frac{\sqrt{2}\,PH}{NA}\right)^2} \qquad \frac{0.64\lambda}{1-\sqrt{1-NA^2}} \tag{11.18}$$

where PH is the size of the detection pinhole.

11.2 Instrumentation

There are many different commercial confocal microscopes on the market for biological applications. However, only a few of them are purposely designed for 3D measurement of surfaces. Fig. 11.13 shows some available commercial instruments based on different technologies.

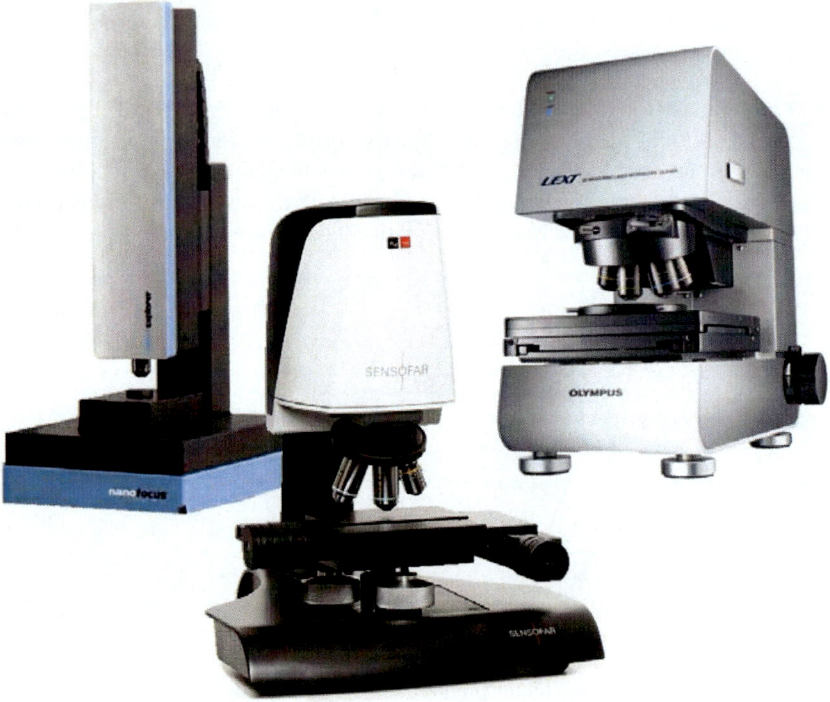

Fig. 11.13 Some commercial confocal instruments. Clockwise: Nanofocus (disc scanning), Sensofar (microdisplay scanning) and Olympus (laser scanning)

11.2.1 Types of Confocal Microscopes

11.2.1.1 Laser Scanning Confocal Microscope Configuration

Laser scanning confocal microscopes (LSCMs) are based on the basic idea of Marvin Minsky. The illumination and detection patterns are single pinholes placed on optically conjugate planes. The beam of the illumination pinhole is raster scanned over the sample in order to build up a confocal image. Fig. 11.14 shows the basic characteristics of a LSCM. A laser beam illuminates a pinhole placed on the field diaphragm position of the objective by means of a field lens. The image of the pinhole is formed onto the sample on the focal plane of the objective. The light reflected or backscattered from the sample passes back through the objective and is imaged onto a second pinhole called the confocal aperture placed on a conjugate position to the illumination pinhole. A detector on the rear of the confocal aperture records the signal reflected from the surface. Light reflected from the out-of-focus positions reaches the confocal aperture plane as an out-of-focus image, leading to a low-intensity signal at the detection plane.

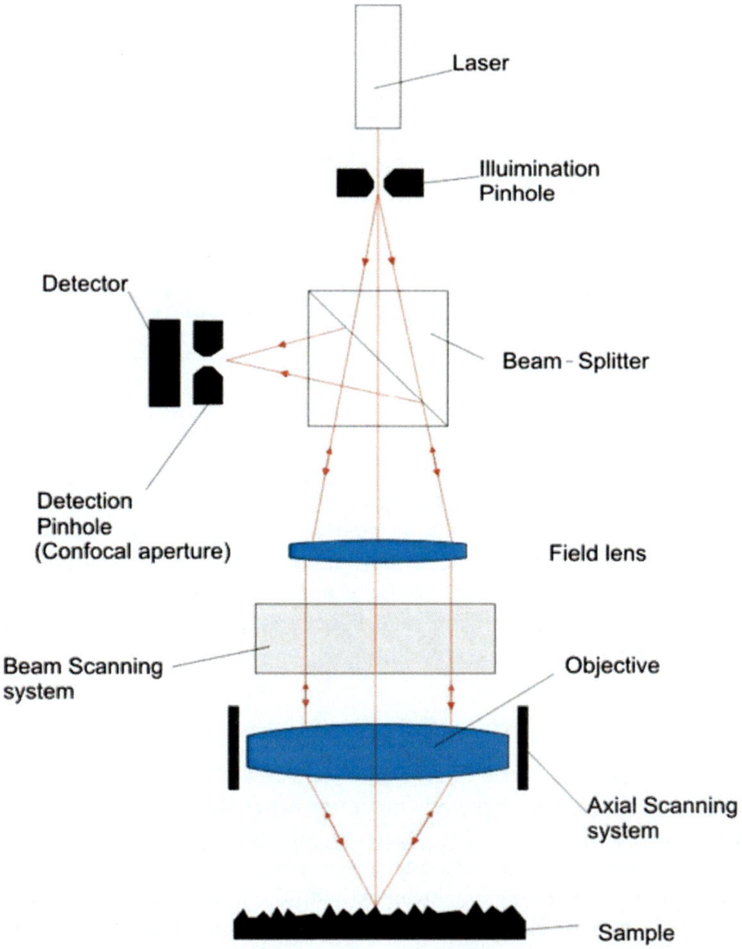

Fig. 11.14 Typical arrangement of a LSCM

11.2.1.1.1 Illumination Pinhole

The illumination pinhole is located on a conjugate image plane of the microscope's objective. The size of the pinhole determines the illuminated area on the surface, which is typically a diffraction-limited spot and adapted to the numerical aperture and magnification of the objective.

Some LSCM microscopes use slit scanning instead of pinhole scanning. Slit scanning has the advantage of increasing the signal intensity and speed, although there is a loss in sectioning strength in the direction parallel to the slit (Sheppard 1988, Artigas 1999, Botcherby 2009).

11.2.1.1.2 Light Source

The light source on a LSCM is a laser. For biological confocal microscopes, the wavelength of the laser is adapted to the excitation of a given fluorophore. In contrast, for 3D measurement of engineering surfaces, the wavelength is typically chosen to be as short as possible. Solid state diode lasers operating in the violet region of the spectrum are increasingly common.

During the beam scanning to reconstruct a confocal image, each pixel is illuminated for only a few microseconds. Therefore, laser light sources are needed on a LSCM.

11.2.1.1.3 Confocal Aperture

The confocal aperture is a second pinhole placed on a conjugate position from the illumination pinhole. The size of the confocal aperture determines the optical sectioning strength of a confocal microscope. The smaller the diameter of the confocal aperture, the higher the optical sectioning strength and the smaller the received signal. Increasing the size of the confocal aperture allows more signal to be detected, but at the cost of allowing out-of-focus signal to reach the detector, thus reducing the optical sectioning strength.

11.2.1.1.4 Detector

Photomultipliers are the most commonly used detector devices in confocal microscopes. Photomultipliers have high quantum efficiency and a wide responsivity spectrum. Other detectors used on commercial instruments are photo-diodes and avalanche photo-diodes. The analogue signal from the photon detector is fed into a computer frame grabber and synchronised with the beam scanner. In order that the frame grabber records an image, the scanner generates pixel clock pulses related to its x and y locations.

11.2.1.1.5 Beam Scanning

The beam of the illumination pinhole is scanned in a raster fashion along the x and y axes in order to generate a confocal image. Generally, the beam is bent by two mirrors that move in perpendicular directions. The beam scanning can fall in to two categories: conventional scanners and resonant scanners.

Conventional scanners are comprised of two galvanometric mirrors moving in discrete steps. The x axis scanner moves in a quasi-continuous movement and generates the lines on the confocal image, while the y axis scanner steps at a lower frequency line by line. Conventional scanners have high positioning repeatability meaning low pixel jitter and high confocal image quality. In contrast, the low-frequency movements of the mirrors result in low frame rates.

Resonant scanners generate video rate confocal images by moving one of the mirrors in a resonant manner. There are resonant galvanometer scanners that achieve some tens of kilohertz with sinusoidal motion, and MEMS-based scanners that can achieve similar frequencies with simplification of the optical setup. Resonant scanners move in a sinusoidal manner, meaning that time integration for each pixel must be adjusted to the speed of the scanner. Acousto-optical devices are also used for continuous high-speed beam displacement along the x direction. The y direction scanner is typically a galvanometric mirror moving in a saw tooth pattern.

11.2.1.2 Disc Scanning Confocal Microscope Configuration

Fig. 11.15 is the basic schematic diagram of a disc scanning confocal microscope (DSCM). A light source is collimated and directed to a disc. The disc contains an illumination pattern (and detection pattern at the same time) similar to those shown in Fig. 11.16. DSCM uses a Nipkow disk on which a set of equally sized pinholes are arranged on an Archimedean spiral (Xiao 1988). Each one of the pinholes is imaged onto the surface by means of a field lens and the microscope's objective. The light reflected or backscattered from the surface for each illuminated spot passes back through the objective and the field lens and is focused on the disc onto the same pinhole.

Light arising from the focal plane is accurately focussed on to the disc surface while light from out of the focus regions is focused on to planes before or after the disc. Each one of the pinholes acts as an illumination and detection element at the same time. Light transmitted through the pinholes is focused onto a two-dimensional detector, such as a CCD camera. The Nipkow disc is rotated at high speed, illuminating and filtering out-of-focus light sequentially and producing an optically sectioned image.

The main advantage of DSCMs is the fact that confocal images are taken at high frame rates. Some commercial instruments deliver up to one thousand images per second. In contrast, light efficiency is very low in comparison with other confocal arrangements. This is due to the low pinhole fill factor on the disc (typically between 1 % and 5 %). Additionally, light reflected from the first surface of the disc is directed to the imaging sensor, increasing the background level and reducing the confocal contrast. To avoid this unwanted light, a polarizer is placed in front of the disc and an analyser in front of the detector. A quarter-wave polarizer is placed on the back side of the disc in order to allow the reflected light from the surface to pass through the analyser. This combination of polarization significantly reduces the amount of signal reaching the detector.

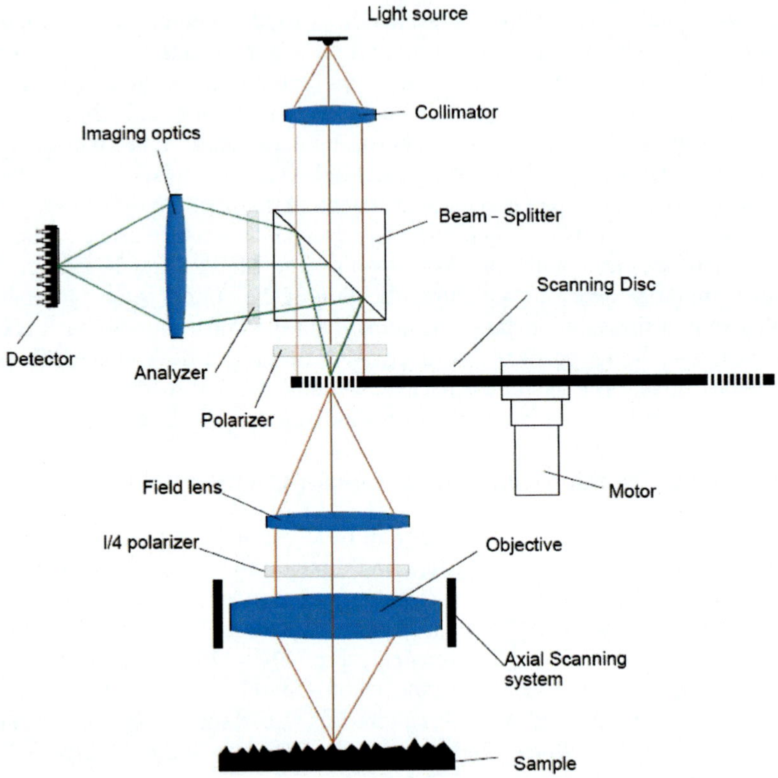

Fig. 11.15 Typical arrangement of a DSCM

To increase the light throughput of the Nipkow disc, some configurations of DSCM include a disc with a microlens array placed on top of the pinhole disc (Tanaami 2002, Tiziani 2000). Each one of the microlenses focuses the light onto each one of the pinholes. The two discs are perfectly matched, and rotate simultaneously at high speed. This confocal arrangement can raise the light efficiency up to 70 %, but at the price of high manufacturing complexity. Other configurations to increase light efficiency are the use of parallel slits, such as those showed in Fig. 11.16. This slit disc configuration has the advantage of signal increase at the price of some loss of optical sectioning strength in the direction parallel to the slits. To overcome this loss, the slits can be arranged as shown in Fig. 11.16 (right). During the rotation of the disc, each one of the pixels on the imaging detector averages all possible slit directions, averaging at the same time the loss of confocality parallel to the slit.

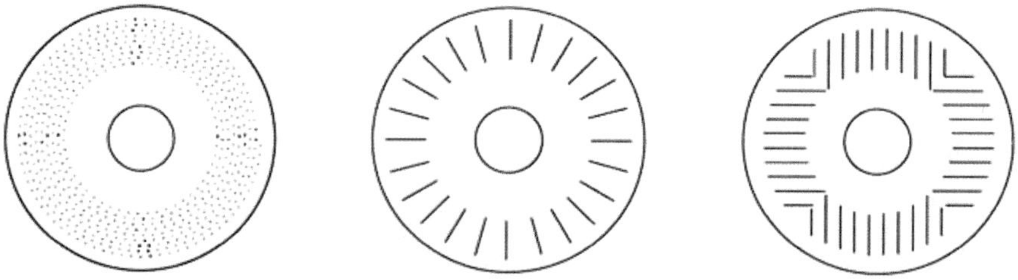

Fig. 11.16 Three different disc patterns for DSCMs. Left to right: Nipkow disc, parallel slits, point rotating slits

11.2.1.2.1 Light Source

Light sources on DSCMs can be white light sources, for example, xenon or mercury lamps, and monochromatic LEDs. Laser light sources are avoided with disc scanning systems because the high coherence of a laser introduces out-of-focus speckle that is imaged through neighbouring pinholes, increasing the noise of the confocal image.

White light sources have the benefit of producing a colour confocal image in real time with colour depth coding due to the chromatic aberration of the objective that focuses different wavelengths at different focal planes. This is useful for surface inspection, but for 3D areal measurements this effect reduces the accuracy and increases the noise. The use of apochromatic objectives (see Sect. 2.3) reduces this effect.

High-power LEDs are commonly used. A LED delivers an appropriate amount of power, has a nearly monochromatic spectrum and the short coherence length reduces the speckle contrast (Fewer 1997).

11.2.1.2.2 Scanning Disc

The most common scanning disc is a Nipkow disc. The disc is made of glass with a layer of chrome or aluminium and a patterned pinhole arrangement. The pinholes are arranged on a spiral. The disc can have two annular regions with different pinhole sizes or different pinhole spacing. This permits the selection between optical sectioning strength against signal strength. The optimum distance between pinholes to avoid crosstalk is a multiple of the pinhole diameter.

Light efficiency is increased with a microlens disc. For such an arrangement a disc is made with a similar spiral pattern but with microlenses instead of pinholes. The focal plane of the microlenses is used as illumination spots and the focusing through the microlens is used as a filter for the out-of-focus light. The imaging detector images the pupils of the microlenses, reducing the lateral resolution.

11.2.1.2.3 Polarization

Polarization elements are used to suppress reflected light from the upper surface of the disc and prevent it from falling onto the imaging detector.

11.2.1.2.4 Imaging Detector

The most common imaging detector for DSCM is a CCD camera, producing either an analogue or a digital signal. The frame rate of the CCD camera is matched with the rotation speed of the disc in order to give a confocal image. The higher the frame rate, the higher the rotation speed and the lower the overall signal strength. For systems to achieve high frame rates it is typical to use cooled cameras, where the noise of the camera itself is lower than the detected signal.

11.2.1.3 Programmable Array Scanning Confocal Microscope Configuration

As with DSCM, programmable array confocal microscopes (PACMs) use parallel illumination to increase the scanning speed, the signal strength or both (Verveer 1998). The active element on a PACM is a micro-display placed at the field diaphragm position of the microscope. The micro-display is used to generate illumination and/or detection patterns (Karadaglic 2008).

A PACM can be arranged in illumination only mode, and in illumination and detection mode (Artigas 2004). In illumination only mode the pixels of the micro-display are used to restrict the light impinging onto the surface while the optical sectioning is achieved by the use of the pixels of a CCD camera. In contrast, in illumination and detection mode the pixels of the micro-display are used to illuminate the surface and at the same time to filter out the light that falls out-of-focus.

Fig. 11.17 shows a typical configuration of a PACM in illumination only mode. The light source is collimated and directed to a micro-display by means of a polarizing cube beam-splitter. The micro-display is formed by a matrix of active elements that could be turned ON or OFF at high speed. The light reflected from an ON element passes through the microscope's objective and is focused onto the sample. The light reflected or backscattered from the surface passes back through the objective and is directed to an imaging detector by means of a beam-splitter. Each element of the micro-display corresponds to one pixel of the CCD camera. To generate an optically sectioned image a set of elements of the micro-display are turned ON, creating an illumination pattern. Each correlated pixel on the CCD records its signal. Light from out-of-focus regions falls on neighbouring pixels that are not taken into account, and the signal on correlated pixels is smaller than when the light is focused on them. A confocal image is reconstructed from a sequence of recorded images on the CCD camera correlated by shifting the ON elements on the micro-display.

A PACM in illumination and detection mode is very similar to a DSCM, where the disc is replaced by a micro-display. Each pixel of the micro-display acts as an illumination and detection element at the same time. PACMs in illumination and detection mode are used for high speed imaging, but suffer from low light efficiency. The main benefit of a PACM is the fact that the illumination and detection pattern can be adapted to the surface under inspection. The illumination pattern can be a series of equally spaced elements (acting like pinholes), simulating a Nipkow disc, or a series of parallel slits, or any other pattern that effectively reduces the amount of illuminated regions. A simple Nipkow disc can be constructed from widely spaced elements to avoid cross-talk or closely spaced elements to increase signal strength. A slit pattern can be adapted to surface lay structures in the perpendicular direction, increasing the detection capability for trenches.

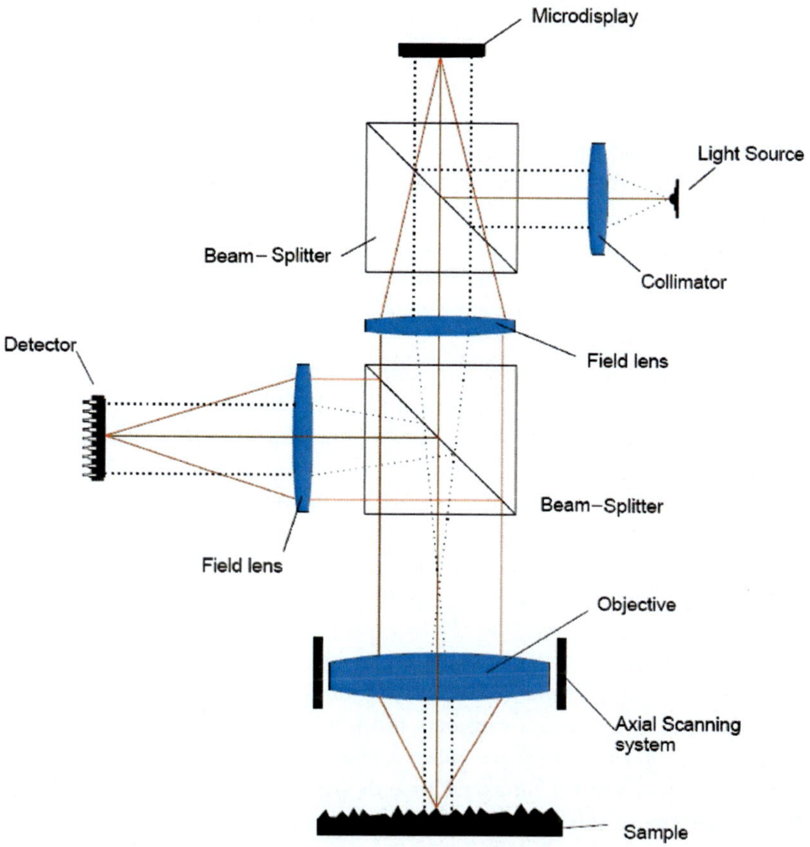

Fig. 11.17 Typical arrangement of a PACM

11.2.1.3.1 Light Source

The most common light sources used on PACM are monochromatic LEDs. As with DSCM, laser light sources introduce too much speckle noise due to their high degree of coherence, making them unsuitable for confocal scanning.

11.2.1.3.2 Micro-display Technologies

There are many technologies for micro-display manufacturing, although only two have been proven to be effective for confocal microscopy: ferroelectric liquid crystal on silicon (FLCoS) and digital micro-mirror devices (DMDs) (Smith 2000, Cha 2000). The reason for the limited choice of micro-display technologies is that image scanning has to be carried out at high speed, both when the micro-display is used in an illumination and detection arrangement, or when it is only used in an illumination arrangement. Micro-display technology can fall basically in two modes: transmission and reflection. In transmission mode, the most typical is a liquid crystal device (LCD). With LCD technology each pixel can change the amount of direction of polarization according to the applied voltage. There is an entrance polarizer and output analyser that converts the polarization state to a grey level. The biggest disadvantage of LCD technology is that it takes a few milliseconds to change from pure white to pure black, making the device rather slow for confocal microscopy. In reflection mode there are basically liquid crystal on silicon (LCoS), FLCoS and DMD. LCoS micro-displays are LCDs with a silicon backplane acting as a mirror. The working principle is the same as LCD but the light enters and exits from the same surface. LCoS suffers from the same low speed problem as with LCD making them unsuitable for confocal microscopy.

FLCoS are the most appropriate micro-displays for confocal microscopy. The working principle of FLCoS is very similar to that of a LCD, but the main difference is that the device is only stable in two polarization states. This makes the device binary by nature. The zero angle polarization state is manufactured as the black point of the device, while the 90° polarization state is the white point. Switching time between images is a few nanoseconds, making the device very fast and able to project several thousand images per second.

DMD is a MEMS-based micro-display where each pixel is a bistable mirror. DMD are also known simply as micro-mirrors. Each pixel has a built-in mirror that can be positioned at two stable tilt positions. The optical arrangement is set up in a way that the positive stable tilt position directs light to the microscope, while the negative stable tilt position directs the light out to an absorbent material. As with FLCoS, DMD is binary by nature. The switching time is very fast — it is possible to project several thousand images per second. The optical fill factor of DMD is lower than FLCoS.

11.2.1.3.3 Imaging Detector

The imaging detector on a PACM is a CCD camera. As with DSCMs, the camera is electronically synchronised with the micro-display in order to achieve a confocal image.

When the micro-display is used in illumination and detection mode, each frame of the CCD contains a set of shifted images of the micro-display. This means that the micro-display is synchronised with the camera in order to show all the images within few milliseconds.

In contrast, when the micro-display is used only for illumination, each frame of the CCD camera contains one image projected on to the micro-display. The CCD and the micro-display are synchronised in a way that each shifted image corresponds with each recorded image, creating a sequence of images for synthetic generation of the confocal image.

11.2.2 Objectives for Confocal Microscopy

Metallographic objectives are the most utilised for 3D applications because they are designed to operate in air. There are many different types of objectives intended for different applications, but the most common way to choose an objective is to try to select the usable magnification with the highest possible numerical aperture. The most common objective used is the bright field plan apochromatic (APO) objective, which has the highest aberration correction, including wavefront and colour correction. Even if a confocal microscope is using a single wavelength for imaging, it is common to use a colour camera for sample positioning and 3D rendering with the real texture. Plan APO objectives are available from many manufacturers, can have magnifications from 1× to 200× and numerical apertures from 0.045 to 0.95. There are extra long (ELWD) and super long (SLWD) working distance objectives that are very useful for the measurement of very difficult to access surfaces. For example, a 20× magnification SLWD objective can have up to 24 mm of working distance in comparison with the 4.5 mm for the plan APO counterpart. The disadvantage of a SLWD objective is that the numerical aperture reduces to 0.3, which leads to less confocal depth discrimination and thus less signal strength, a higher noise level and a lower maximum local slope capability on smooth surfaces.

When measuring 3D surfaces with an imaging confocal microscope, it is important to select the appropriate objective. The highest possible numerical aperture should be selected in order to reduce the instrument noise and increase accuracy. In general, 50× or higher magnifications are available with 0.95 numerical aperture, the highest possible on a dry objective with an instrument noise of the order of 1 nm or less. For lower magnifications the noise is higher and the maximum local slope capability is lower. Also for lower magnifications, it is common to strike a balance between a lower magnification objective and field stitching with higher magnification. Fig. 11.18 shows laser-irradiated silicon measured with a 20× objective (0.45 numerical aperture) and field stitching (625 single fields) spanning a total field of 16 mm by 16 mm. Such relatively large fields could not be measured with lower magnification due to the high local slopes on the surface, which reach more than 20°.

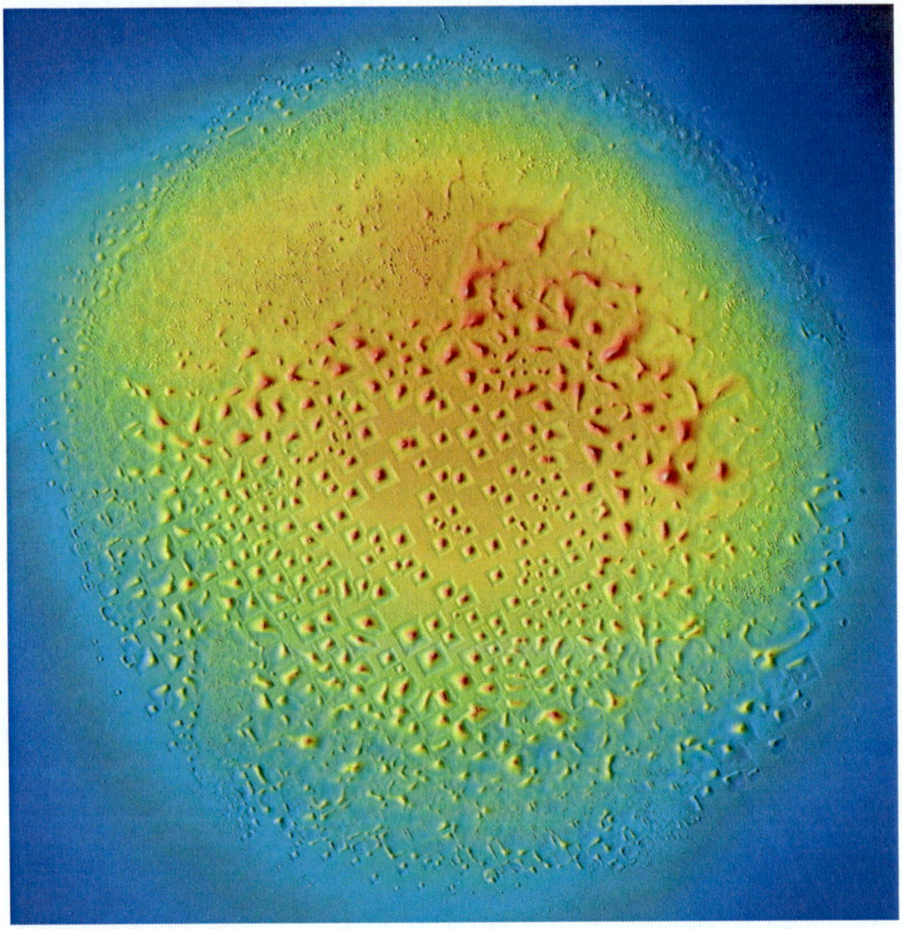

Fig. 11.18 Laser-irradiated silicon measured with a 20× objective and field stitching spanning a total field of 16 mm by 16 mm

Fig. 11.19 shows a part of the surface shown in Fig. 11.18 with a 20× magnification single field (left) and the measurement on a peak with a 150× magnification (right).

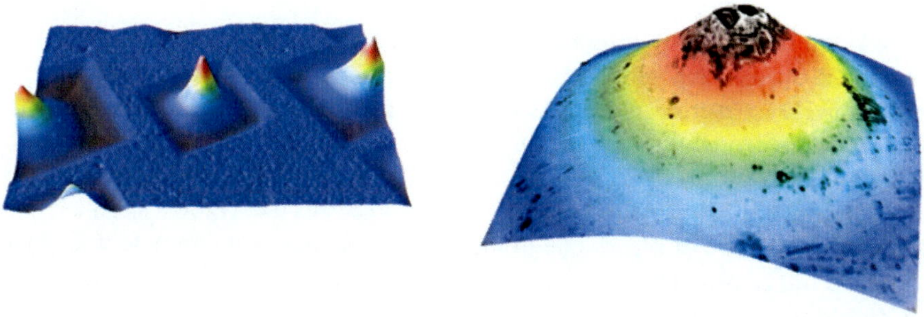

Fig. 11.19 (Left) single field measurement with a 20× objective of laser irradiated silicon. (Right) measurement of the tip of a peak with a 150× objective

Water immersion objectives are also available for confocal microscopes. These objectives are constructed in a way that it is possible to dip them into water and image a surface that is completely immersed in water. There are many applications for the characterisation of surface texture evolution under wet conditions. For example, in the paper industry for paper roughness change under water, in the food packaging industry for the texture evolution of film depositions of the envelopes under wet conditions or in the semiconductor industry for the surface change in wet treatments.

Collar ring (CR) objectives are also available and very useful for some applications in imaging confocal microscopes. These objectives are designed to image a surface that is located on the rear of an optical window. They have a ring with marks indicating the thickness of the glass. By turning the ring to the appropriate number, the objective corrects the spherical aberration introduced by the optical window. A typical application of these objectives is the measurement of MEMS in cryogenic conditions, where the sample is located in a freezing chamber and the device is seen through a very small window. In the paper industry there is also a need to measure paper roughness under pressure. In this case, the paper is pressed against glass and the roughness is measured from the glass side.

IR and UV objectives are also used with confocal microscopes. Nevertheless, the optical design of the microscope should be adapted to the use of such wavelengths reducing the broad offer of different objectives. UV wavelengths have the advantage of improving the lateral resolution close to 0.2 μm, while IR wavelengths are needed for those applications imaging through silicon structures.

Biological objectives are less common for 3D measurement of surfaces. The main benefit of biological objectives is the fact that they are oil immersion objectives. The numerical aperture is the highest possible, but at the cost of dropping oil onto the surface, loosing the non-contact nature of confocal microscopy.

11.2.3 *Vertical Scanning*

The vertical scanner on a confocal microscope determines the height values of the constructed surface topography image and, therefore, is the most important component for 3D measurements. The location in height value of the maximum of the axial response for each pixel of the image is directly related to the positioning accuracy of the z axis stage itself. Any non-linearity of the vertical scan will be embedded in the 3D surface measurement. Typically, the vertical scanner on a confocal microscope displaces one of the three components: the sample, the objective or the full sensor. The vertical scanner can be closed-loop or open-loop.

Open-loop scanners are used for 3D measurements on many commercial instruments, reducing the overall cost. Although there are high-quality open-loop scanners on the market, they are limited in accuracy.

In contrast, closed-loop scanners increase the accuracy of the measurement by incorporating a measuring device on the stage itself. On motorised stages, it is typical to incorporate an optical linear encoder with a period of a few micrometres. Alternatively, piezoelectric scanners use capacitive or piezo-resistive network sensors.

11.2.3.1 Motorised Stages with Optical Linear Encoders

There are many types of linear stages and actuation motors. One of the highest quality stages for confocal microscopy uses two parallel linear guides with recirculation ball sliders and a central ball screw. The two side sliders define the straightness and smoothness of the motion. Any deviation of straightness (pitch and yaw) will introduce relative tilt between the measurement axis and the optical axis. The central ball screw grade will have a direct relationship with the accuracy of the instrument. The screw is typically actuated with a stepper motor or a servomotor with a high-resolution on-axis encoder.

Optical linear encoders are the most common devices used for closed-loop control of stages on confocal microscopes. Optical linear encoders are used for large travel ranges from a few millimetres up to one metre. Manufacturing tolerances of the grating and the mechanical assembly of the optical sensor on the linear encoder limit the accuracy to about 1/50 of its grating period. The resolution for such measuring devices depends on the interpolation electronics and can be as high as $1/4000^{\text{th}}$ of the grating period. For a 2 μm period grating the resolution can be less than 1 nm. Despite such good resolution, a double-frequency error, called a quadrature error, appears within a single period of the grating, limiting the accuracy deliverable from the sensor. Fig. 11.20 shows the error present on an optical encoder and a stage showing the typical maximum error dependence on the signal period.

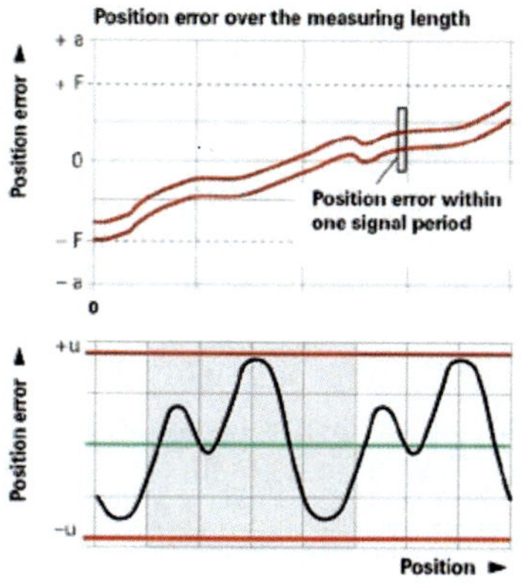

Scale signal period (μm)	Maximum error (μm)
2	± 0.02
10	± 0.1
20	± 0.2

Fig. 11.20 Optical linear encoder error

The accuracy of the measurement of a step height artefact will be related to the value of the height and to the starting position of the scanner relative to the linear scale. Fig. 11.21 shows two cases for a 500 nm step height located at two different positions of the vertical scanner. In Fig. 11.21 the signal period is 2 μm and each box on the figure is equal to a 0.5 μm height.

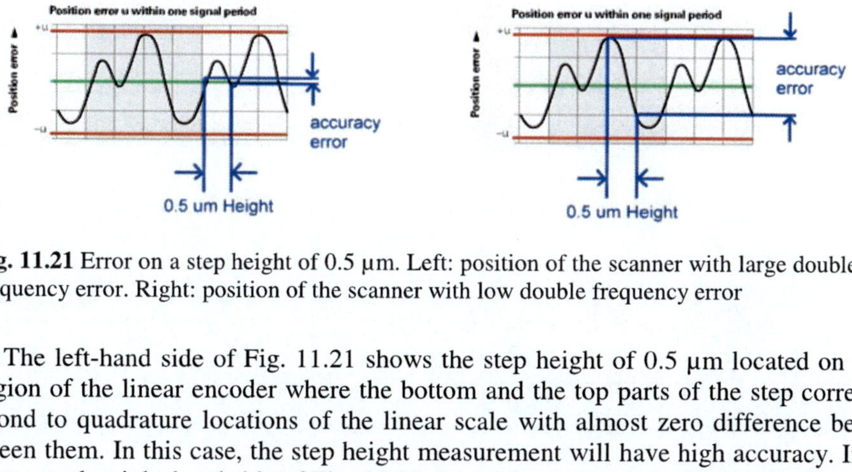

Fig. 11.21 Error on a step height of 0.5 μm. Left: position of the scanner with large double-frequency error. Right: position of the scanner with low double frequency error

The left-hand side of Fig. 11.21 shows the step height of 0.5 μm located on a region of the linear encoder where the bottom and the top parts of the step correspond to quadrature locations of the linear scale with almost zero difference between them. In this case, the step height measurement will have high accuracy. In contrast, the right-hand side of Fig. 11.21 shows the same step height located on a different position within the linear encoder. In this case, the top and the bottom of the step height are located on nearly the peak to valley error of the quadrature signal error, which has a maximised error.

11.2.3.2 Piezoelectric Stages

Piezoelectric stages used for confocal microscopy use a flexure arrangement to produce linear motion and are non-contact. There is a piezoelectric actuator on one face of the flexure and a position sensor on the opposing face. Typical travel ranges of the stage range from a few micrometres up to half a millimetre.

Capacitive sensors and piezo-resistive sensors are able to deliver resolutions below 1 nm and accuracies of less than 0.05 % along their full measuring range. The noise of such sensors can be as low as 0.1 nm, but it is impossible for them to operate in closed-loop mode at high frequency at this resolution. Furthermore, the maximum permissible closed-loop frequency depends on the inertia of the actuator, which should be avoided by the control electronics.

Piezoelectric stages have the highest position accuracy and repeatability, but they suffer from the maximum permissible weight. Commercial stages are able to deal with a maximum 1 kg mass before the flexure breaks. Special designs discussed with the manufacturers of the stages are needed to withstand the mass of a full confocal sensor or a nosepiece with up to six objectives. There are also single objective piezoelectric stages that screw directly onto the position of one objective on the nosepiece. The performance of these single-objective actuators is good, although they are not practical if the intention is to have several objectives installed on the instrument.

11.2.3.3 Comparison between Motorised and Piezoelectric Scanning Stages

Fig. 11.22 shows the difference between measuring a flat mirror with a closed-loop motorised linear stage with an optical linear encoder of 2 μm and a piezoelectric stage. The measurement with the linear stage has some non-linearities of the order of a few nanometres that cannot be corrected due to the inherent error of the sensor. The measurement with the piezoelectric stage clearly shows higher quality results meaning better accuracy performance and higher linearity.

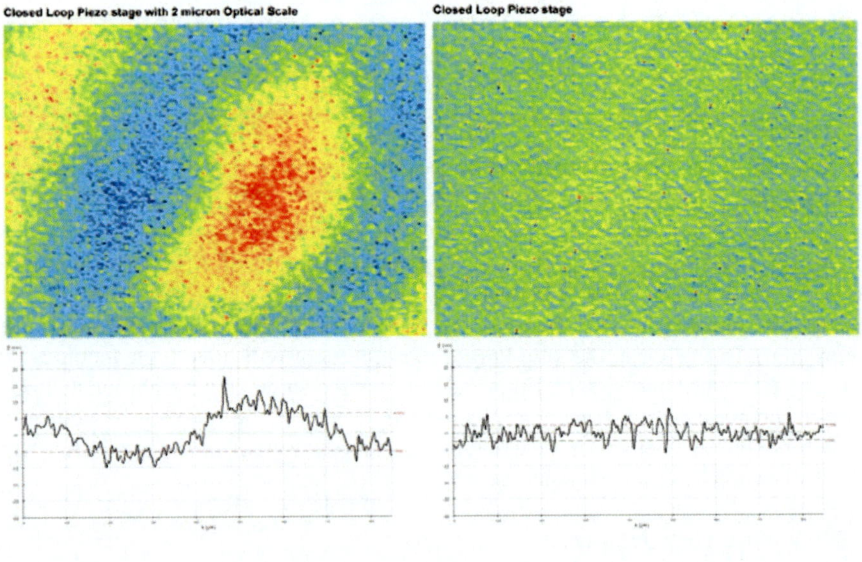

Fig. 11.22 Measurement of a high-quality mirror with a closed-loop linear stage incorporating a 2 μm optical encoder (left) and a piezoelectric stage with piezo-resistive sensor (right)

11.3 Instrument Use and Good Practice

11.3.1 Location of an Imaging Confocal Microscope

A confocal microscope is a highly robust instrument that requires very low maintenance. Nevertheless, the environment in which the instrument is located can seriously affect its performance.

The least environment requirements are for those measurements that take only a few seconds within a single field of view and at low magnifications. For large area stitching measurements and the use of high magnification, the external vibration should be minimised with the use of vibration isolation tables. For measurements of very smooth and flat surfaces the stability of the instrument is also very important. In that case, it is recommended that the instrument is installed with the highest possible stable conditions such as on top of a vibration isolation table isolated in a cabinet, and preferably on the ground floor of the building; avoid installing instruments on higher levels where the floor itself is acting like a membrane, introducing low frequency movement.

Cabling is also one of the main sources of instabilities on optical profilers. The cables should exit the instrument without introducing any stress. Stiff cables tend to introduce external vibration.

11.3.2 Setting Up the Sample

Most commercial instruments have the possibility of showing a bright field image to allow easier sample set up. For rough surfaces the focusing is done very quickly and easily. In contrast, for smooth surfaces focusing can be more difficult.

Also, for rough surfaces or surfaces with significant texture, it is not usually problematic if there is tilt between the sample and the optical axis. For smooth surfaces, however, it is recommended to use a tilting stage to place the sample as perpendicular to the optical axis as possible. This assures a good correspondence between the measurement and the calibration of the instrument.

11.3.3 Setting the Right Scanning Parameters

Before beginning to take measurements the appropriate scanning parameters must be chosen. Some commercial instruments have automatic features that search for the optimum parameters such as scanning range, focusing and light level. For instruments in control of production this is necessary, but for instruments in research institutions it is much better to train an expert user.

Light and Gain Adjustment

The light level should be adjusted avoiding saturation on smooth surfaces. On rough surfaces, a small percentage of saturated pixels are acceptable, avoiding always having large saturated regions. On some surfaces, a single light level is not enough to deal with high-reflectivity and low-reflectivity regions at the same time. Some commercial instruments have the option to illuminate at the same focal region with different light levels. A series of several light level sequences of confocal images are stored and the sequence with the highest signal with no saturation is used for the topographical calculation.

Objective

On instruments with automatic nosepieces the software recognises the objective in use and automatically adjusts for the optimum step size between planes, and the number of planes necessary to cover the desired range. With manual nosepieces the user needs to inform the software of which objective is in use. When the objective used for the measurement does not match with the software, many errors can result; the most significant being the mismatch of flatness error calibration (see Sect. 11.4.3.1).

Area

With single-field measurements it is normal to choose the largest area available. On the contrary, for stitching the desired area will determine how many single fields are needed. In this case, it is typical to reduce the resolution for each single field in order to keep a reasonable amount of measured pixels. For example, suppose a confocal microscope with a single field measures with one million pixels. If the desired area needs ten by ten fields, this will result in a 100 million pixel measurement. For an instrument with a confocal stack image and colour image, 32 bits are needed for the topography, 8 bits for the stack and 24 bits for colour. This means 64 bits per pixel, or 8 bytes per pixel, or 800 Mbytes of data for the full area. Now suppose a fast Fourier transform (FFT) needs to be performed on the full area file. A FFT needs auxiliary memory by a factor of four, meaning 3.2 Gbyte of memory. Reducing the resolution by a factor of two in each direction will result in 200 Mbyte, much easier to handle by the software.

During field stitching some errors can occur. Alignment of the *xy* stage with the pixels of the image should be carried out as carefully as possible. Some overlapping is used between measurements and the measured data are correlated to correct for misalignments. Typical overlapping values are between 10 % and 25 % of the field. For highly textured surfaces a small amount of overlapping is enough. On the contrary, for smooth surfaces with very little surface features, it is recommended to increase the overlapping to the maximum.

Threshold

The most common use of a threshold value is to avoid the noise present on dark regions of a surface that will lead to spike-like data. The threshold value varies from one manufacturer to another. Some instruments refer to the threshold as a grey level, while some use a relative number as a percentage of a fixed number.

Data Interpolation

Data interpolation is used to replace non-measured points for measured data with the use of interpolation algorithms. Some analysis software is able to show the profile data in a different colour for those pixels that has been interpolated. Fig. 11.23 shows a measurement of a steel roll surface measured with low magnification. The crater region has a local slope that cannot be measured with the numerical aperture of the objective used, (non-measured regions are shown in black). The image on the right shows the restored topography.

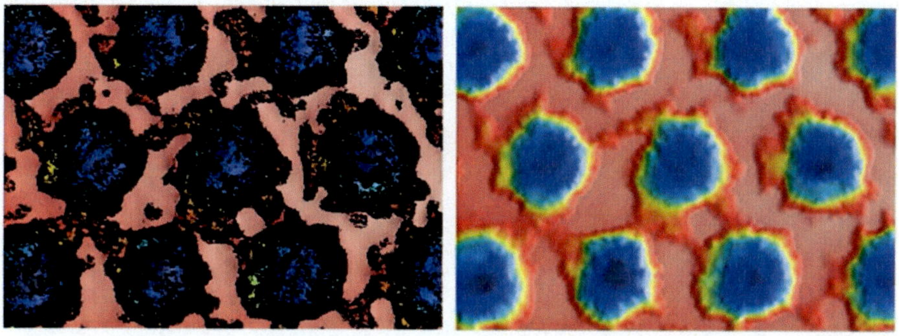

Fig. 11.23 Measurement of a steel roll with non-measured points (left) and interpolated points (right)

11.3.4 Simultaneous Detection of Confocal and Bright Field Images

During the axial scanning, the sequence of confocal images is stored on the computer's memory for further processing of the topography data. In addition to the 3D information, the confocal stack is also computed by assigning to each pixel a grey level corresponding to the highest signal within the sequence. This image is also known as the infinite focus image, which has all the areas in perfect focus removing the blur images familiar with a bright field microscope. In order to get a good stack image it is very important to be sure that the image is not saturated at any point all along the axial scan. Fig. 11.24 shows a single confocal image (left) taken with a 50× objective on an aluminium plate and the stack image (right) after processing the confocal sequence.

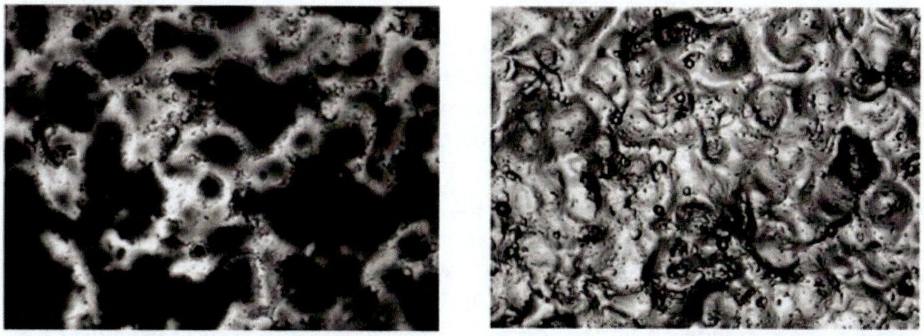

Fig. 11.24 Confocal image (left) and confocal stack (right) of an aluminium plate

Some confocal microscopes incorporate an additional colour camera. The colour camera is used for sample positioning and inspection as well as for real colour texturing of the 3D images (see Fig. 11.25). During the axial scanning, a sequence of colour images is stored in parallel with the sequence of confocal images. The colour assigned for each pixel corresponds to the colour image of the z axis plane where the maximum signal of the axial response is found.

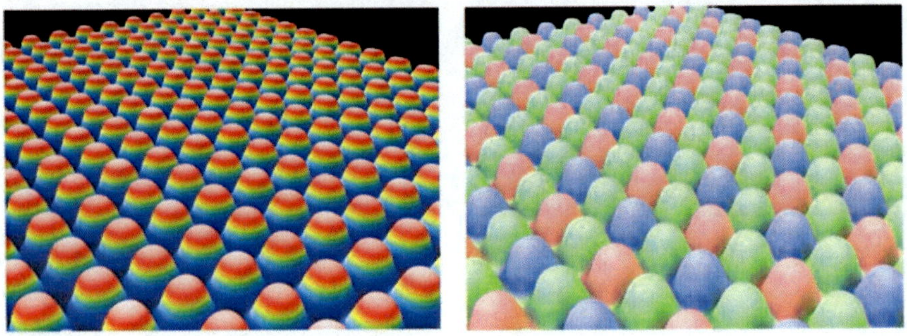

Fig. 11.25 Topography image of a CCD sensor microlens array with colour height codification (left) and real texture colour (right)

11.3.5 Sampling

The sampling is the size of a single pixel projected onto the surface. On a confocal microscope, the optical design is a compromise between the lateral resolution of the instrument and the sampling. For medium magnification objectives (around 50×) the lateral resolution and the pixel size are more or less identical. Lower magnifications have smaller pixel size than resolution and higher magnifications have larger sampling. A typical pixel size for a 150× magnification is 0.1 μm, while the optical resolution for such an objective with a mean wavelength of 0.46 μm is 0.3 μm, which corresponds to sampling three pixels per resolving area. Nevertheless, for critical dimension measurements on the lateral resolution limit it is desirable to increase the sampling.

Some commercial instruments have the option to zoom on the confocal image. This has several benefits, although the most important is to correctly sample on the resolution limit. On laser scanning instruments, it is very common to have an optical zoom. The laser is scanned on a shorter area by means of the amplitude of the galvanometric mirrors or by means of a bending lens on the scanning path. Optical zoom has the advantage of keeping the final number of pixels and scanning speed. With low magnification objectives the pixel size can be increased by a factor between 2× and 3× still being below the resolution limit. A 10× objective can be zoomed to emulate the magnification of a 20× or 30× objective while still having a sharp image, although the numerical aperture is kept constant. On the contrary, applying zoom to a high magnification objective, which already is sampled beyond the resolution limit, will make the image larger but at the same time fuzzy. This is useful for the correct sampling beyond the resolution limit. Disc scanning and programmable array microscopes cannot use an optical zoom. Instead, they use a digital zoom on the confocal image. A digital zoom has less quality than an optical zoom, especially for low magnifications. On high magnifications a digital zoom, already zooming on an over-sampled image, has nearly the same quality as an optical zoom.

Fig. 11.26 shows the topographical image of an AFM calibration artefact with 0.66 µm period. The image on the left is under sampled with 0.1 µm per pixel, while the image on the right has a digital zoom of 6× and 0.015 µm per pixel.

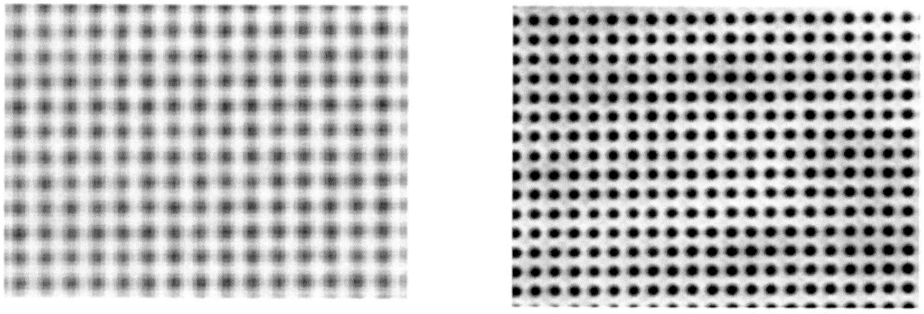

Fig. 11.26 Under-sampled (left) and correctly sampled (right) measurement of an AFM calibration artefact (0.66 µm pitch) with 150× magnification. The image on the right has a 6× digital zoom factor applied to the confocal images during the scan

11.3.6 Low Magnification against Stitching

Sometimes it is important to compromise between using low magnification objectives and field stitching. The use of low magnification gives a large measurement area, but it inherently comes with low numerical aperture and thus high noise in confocal instruments. Additionally, low numerical aperture also limits the maximum local slope measurable on smooth surfaces, as well as on moderately rough surfaces with high wall angles. Most of the commercial instruments are equipped with *xy* stages for sample positioning, multiple repetitive measurements and topography stitching. In general, the use of topography stitching with higher magnifications improves the result, especially when using magnifications higher than 20×. However, with stitching the measurement time is longer and the instrument has to operate under stable conditions.

Fig. 11.27 shows the profile of a stainless steel structure with 60° wall slope. The black line is a single field measurement with a 20× magnification, 0.45 numerical aperture with outlier pixels on the slope due to low numerical aperture. The threshold has been lowered to zero in order to use any signal for the calculation even on the wall region. If the threshold is set to higher levels it is possible to clean out all the spikes at the cost of having non-measured points in that region. Data interpolation is sometimes useful for geometrical reconstruction, but the restored points cannot be used for further data analysis such as for roughness. The red line in Fig. 11.27 shows a measurement of the same sample with a 50×, 0.95 numerical aperture objective, which can measure up to 72° local slopes on smooth surfaces. In order to cover the full sample profile it is necessary to measure three fields and stitch the results.

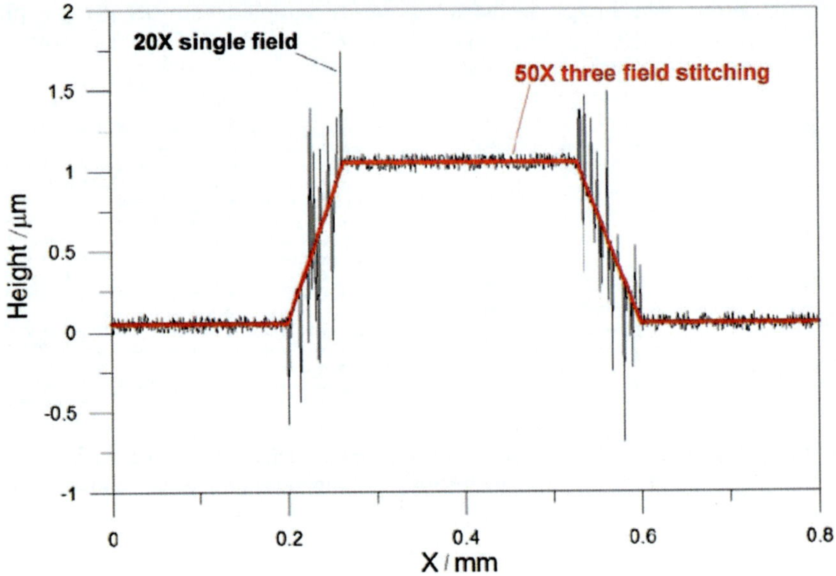

Fig. 11.27 Measurement of a stainless steel structure with 60° wall slope. The black line is a single-field measure with a 20× magnification with outlier pixels on the slope due to low numerical aperture. The red line is a three-field stitching with a 50× magnification, 0.95 numerical aperture objective

11.4 Limitations of Imaging Confocal Microscopy

11.4.1 Maximum Detectable Slope on Smooth Surfaces

The numerical aperture of the objective largely determines the maximum local slope measurable on a non-scattering surface. Fig. 11.28 shows two typical cases where light illuminates the surface, reflects and goes back to the objective (green rays), and light that does not reach the objective (red rays). Fig. 11.28 (right) shows the limiting case where the surface tilt equals the numerical aperture of the objective and only the marginal ray of the objective illuminates the surface and goes back on itself. Surfaces with higher tilt do not provide any signal back.

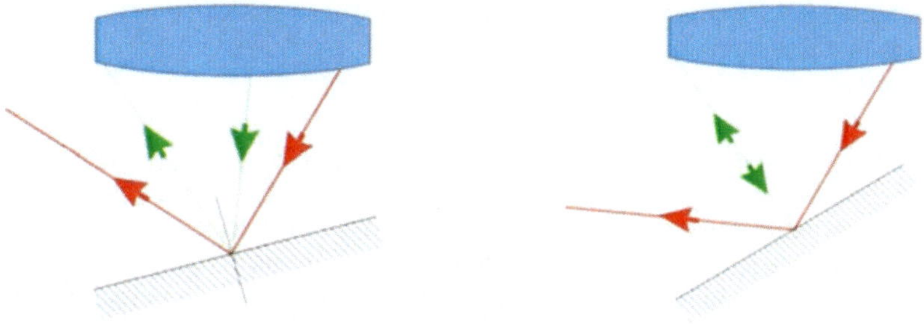

Fig. 11.28 Surface providing signal (left) and surface with tilt equal to the numerical aperture (right)

One of the biggest advantages of a confocal microscope is the use of high numerical aperture objectives. It is possible to use dry objectives up to 0.95 numerical aperture providing a maximum measurable local surface slope up to 72°. Larger slopes can be measured with higher numerical aperture objectives such as water immersion or oil immersion, although they are not practical for 3D measures of technical surfaces (see Sect. 11.2.2). Fig. 11.29 shows the dependence of maximum local slope on numerical aperture, assuming that the roughness of the surface and its propensity for scattering light in different directions is not significant.

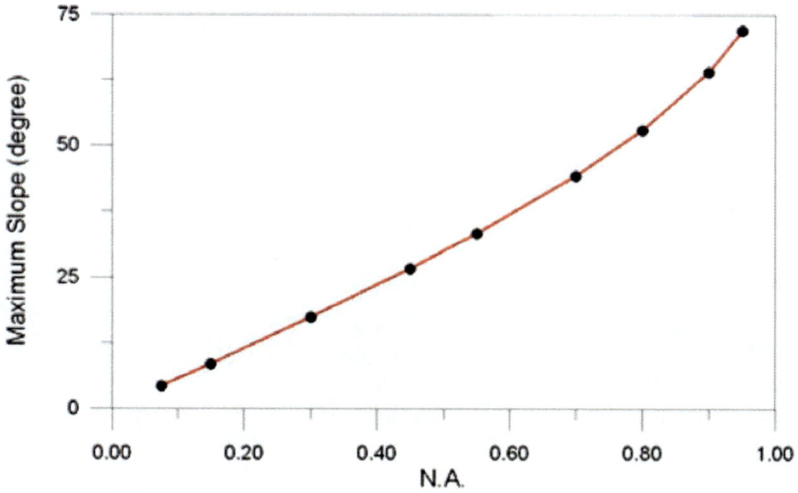

Fig. 11.29 Maximum local slope against an objective's numerical aperture

The slope limitation is only applicable to optically smooth surfaces or for those surfaces with very low light scatter (Häusler 1994). For optically rough surfaces the maximum local slope measurable can be much higher, even reaching close to a vertical step of 90°.

11.4.2 Noise and Resolution in Imaging Confocal Microscopes

Although the metrological algorithm used in confocal microscopy increases the resolution of the z axis discrete steps during the axial scan, the resolution is limited by the numerical aperture of the objective. The lower the numerical aperture, the larger the axial response. The optical slice thickness (FWHM) of a confocal microscope is approximately given by

$$FWHM = \left(\frac{0.88\lambda}{1-\sqrt{1-A_N{}^2}} \right) \tag{11.19}$$

where λ is the wavelength and A_N the numerical aperture of the objective. Fig. 11.30 shows different axial responses for different numerical apertures with a mean wavelength of 0.55 μm.

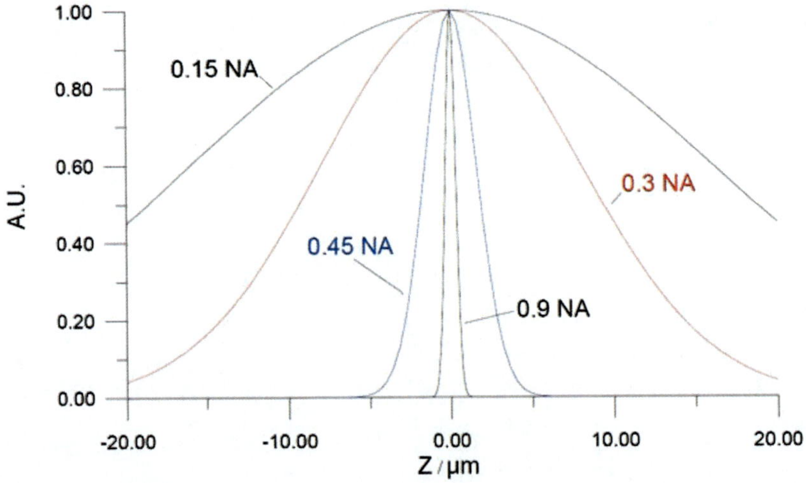

Fig. 11.30 Axial response for different numerical apertures

For those objectives with low numerical aperture, the maximum position is more difficult to locate because the signal decreases slowly. Objectives with high numerical aperture have sharper peaks and a much more accurately located maximum. The z axis step between planes may be matched to the resolution of the objective. Table 11.1 shows some typical values.

Table 11.1 Optimal and conventional ranges of z axis steps between confocal images during an axial scan for some numerical apertures

Numerical aperture	Noise limited resolution optimal step /μm	Conventional optimal step /μm
0.3	2.0 - 8.0	8.0 – 16.0
0.45	0.5 - 1.0	1.0 – 4.0
0.8	0.1 - 0.4	0.4 – 1.6
0.9	0.05 - 0.2	0.2 – 0.8

Making the optimal step smaller than the optimum range does not improve the noise of confocal microscopes. Fig. 11.31 shows the dependence of the resolution of a confocal microscope on the numerical aperture of the objective. Low magnification objectives have typically low numerical aperture, meaning that the noise of a confocal microscope is high. For objectives reaching 0.75 numerical aperture or larger, the noise approaches 1 nm.

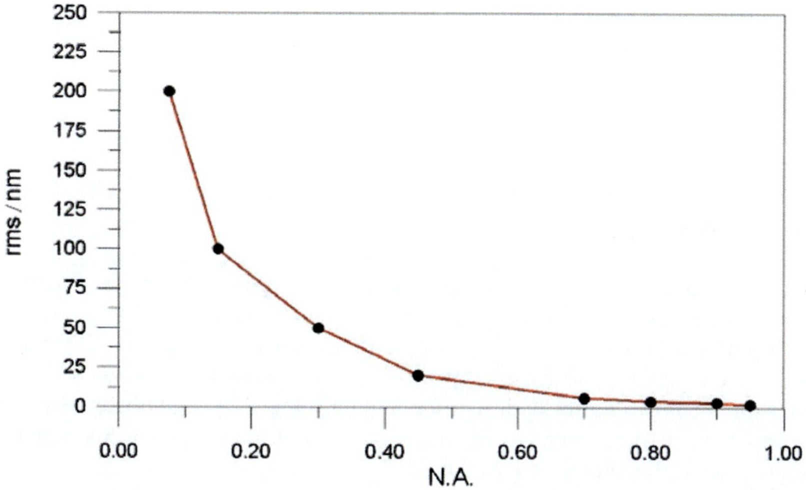

Fig. 11.31 Dependence of the z axis resolution of a confocal microscope with the numerical aperture

The height resolving power is directly linked to the noise. The smallest roughness that can be measured is the noise of the instrument itself, while the smallest step height should be typically three times larger than the noise. Typical noise values for an objective with 0.95 numerical aperture are on the order of 1 nm. This should mean that the smallest step height, and thus the axial resolution, is on the order of 3 nm. Nevertheless, confocal image averaging can significantly improve the noise level at the cost of larger measuring time. Fig. 11.32 shows the result of the measurement of a 10 nm step height with one confocal image per plane (black line) and ten confocal images averaged per plane (red line).

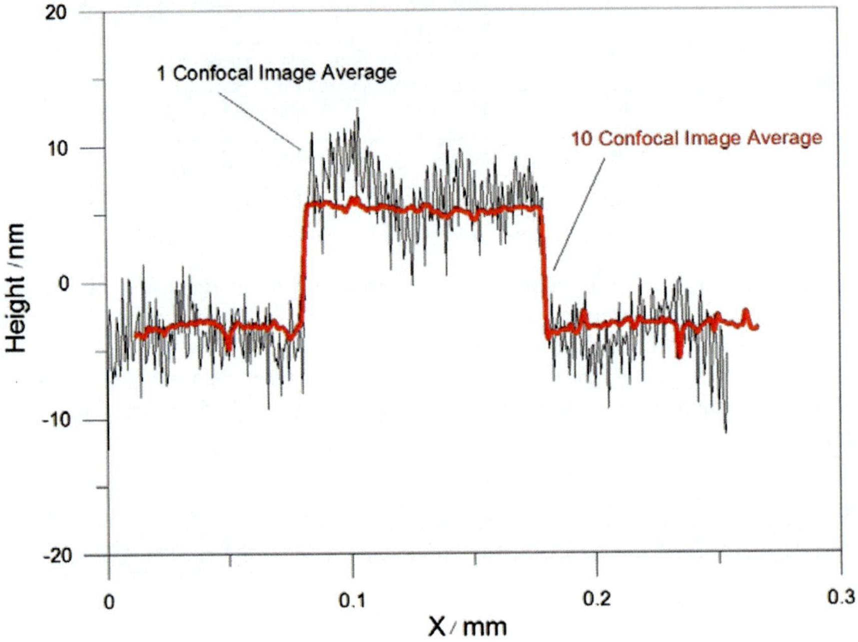

Fig. 11.32 Measurement of a 10 nm step height with a 50× magnification, 0.95 numerical aperture objective. Black line with one confocal image per plane. Red line with ten confocal images averaged per plane

Averaging too many images will not indefinitely improve the noise. Mechanical stress, small thermal variations, external vibration, etc. limit the low-spatial frequency components that affect the calibration of the instrument. These low-frequency components appear on the measurement as waviness, and they are stable on a stabilised instrument within around one minute. This practically limits the noise of a confocal instrument to the order of 0.3 nm, meaning that the smallest step height resolved is on the order of 1 nm.

11.4.3 Errors in Imaging Confocal Microscopes

11.4.3.1 Objective Flatness Error

Objective manufacturers design the optics in a way so as to keep all the aberrations within the depth of field of the objective itself. This ensures that normal use of the microscope, visually or with a camera, will not show small residual optical aberrations in the design such as field curvature, spherical aberration, astigmatism or coma (Gu 1992, see Sect. 2.2). Most of the optical correction is carried on the objective itself, but some manufacturers combine the lenses on the objective, the field lens, and the binocular to get better optical correction.

A 3D imaging confocal microscope pushes the design of an objective to its limits. The height resolution is close to 1 nm for the highest numerical aperture of 0.95, with a depth of field of 0.7 μm. This value is hundreds of times higher than the resolution, which will show any non-corrected optical aberration present within the depth of field. The most influential optical aberration is spherical aberration. Fig. 11.33 shows the spherical aberration of three typical objectives (20×, 50× and 100×) before instrument calibration. The decrease of the profile length due to the field of view of each objective is worth noting.

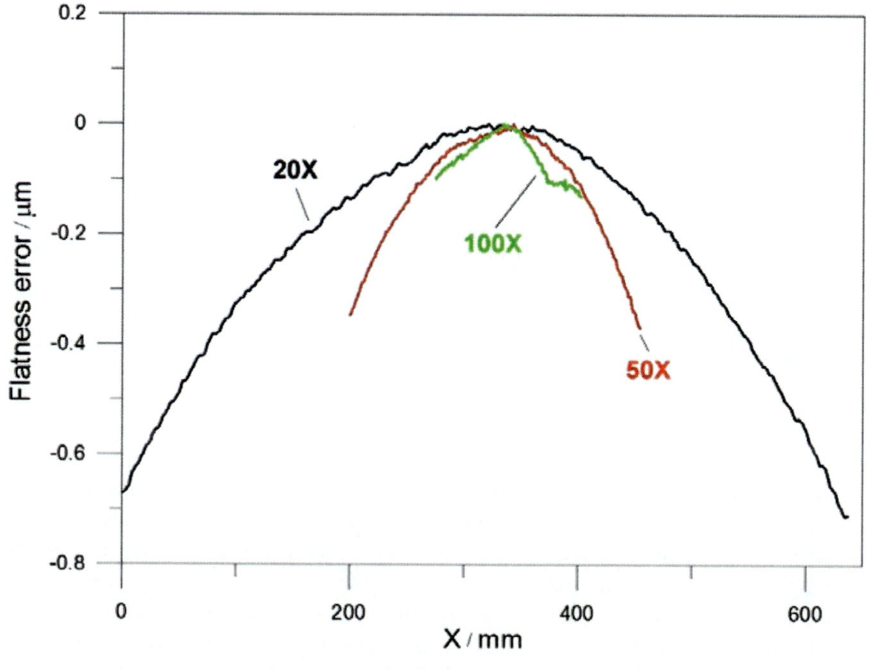

Fig. 11.33 Optical aberrations before instrument calibration

11.4.3.2 Calibration of the Flatness Error

In order to correct for optical aberrations the flatness of confocal microscope has to be calibrated (see Chap. 4). The most typical way to do such calibration is with a reference flat surface such as a good-quality mirror. The mirror is placed under the microscope and tilted to a position as perpendicular to the optical axis as possible. The software takes a reference topography image that is stored and automatically subtracted from subsequent measurements. The flatness calibration process is repeated for all objectives. Fig. 11.34 shows the measurement of a flat calibration mirror after the flatness calibration procedure. The effect of noise decrease with the magnification is also shown.

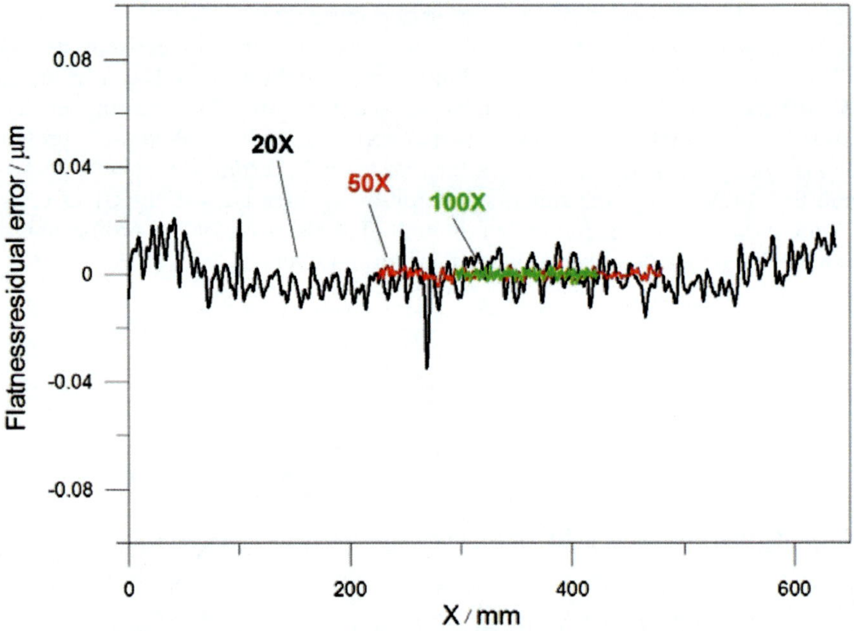

Fig. 11.34 Residual profile of the optical aberrations after flatness calibration

11.4.3.3 Measurements on Thin Transparent Materials

Confocal microscopes are insensitive to dissimilar materials. This is only true with optically opaque materials or for transparent materials that are optically resolved with the objective in use. When measuring step heights of transparent materials with thicknesses smaller than the depth of focus of the objective and the coherence length of the light source, an additional effect appears: the light reflected from the top and the bottom parts of the transparent materials interferes with an interference state that depends on the wavelength, the thickness, the refractive index and the numerical aperture of the objective. This interference shifts the peak of the axial response and consequently, the measured topography is not equal to the mechanical height of the step.

This is a typical effect when measuring step height of silicon oxide on silicon, and leads to a mismatch of measurements with other instruments (for example, a stylus instrument – see Sect. 1.5.1), even with the same instrument with different objectives due to the change of the numerical aperture.

11.4.4 Lateral Resolution

The lateral resolution of an imaging confocal microscope is the highest that can be achieved with an optical instrument. As discussed in Sect. 11.1.3.3.1, the lateral resolution is equal to $0.61\lambda / A_N$, where λ is the mean wavelength of the source.

The highest numerical aperture objective in air is 0.95 and typical wavelength is in the violet to blue region (between 0.4 μm and 0.46 μm). This leads to a lateral resolution between 0.25 μm and 0.29 μm.

Despite the lateral resolution as given by the Rayleigh criteria, confocal microscope manufacturers often discuss half of this limit. They use the line and space (L&S) criteria often used in the semiconductor industry. In that case, the minimum feature size during wafer inspection is the observation of gratings and is equal to half of the Rayleigh lateral resolution. If a grating is a sinusoid, the cut-off frequency (highest grating frequency resolved) will be equal to the diffraction limit corresponding to the optics in use. At the same time, the grating is composed by the L&S, meaning that it is possible to resolve a feature size half of the grating frequency. With the L&S criteria, the lateral resolution of an confocal microscope decreases from 0.12 μm to 0.15 μm, depending on the mean wavelength in use.

Fig. 11.35 shows the confocal image of a periodic structure of 0.4 μm period. The circular arc has an average diameter of 0.2 μm.

Fig. 11.35 Confocal image of a periodic structure of 0.4 μm period and 0.2 μm diameter

Despite the ability to revolve features around 0.15 μm with a confocal microscope it is much more difficult to accurately resolve the height of such features (see Sect. 2.5).

11.5 Measurement of Thin and Thick Film with Imaging Confocal Microscopy

11.5.1 Introduction

One advantage of confocal microscopes is their ability to correctly measure surfaces containing dissimilar materials without the need for any correction of the measurements (see Sect. 11.4.3.3). This feature makes confocal microscopes useful for applications such as biomedics, materials testing, chemistry and microelectronics.

However, one measurement task, that is considered to be very difficult to carry out with most optical techniques, is the measurement of stratified media. These media show refractive index variations in the axial direction. Typical examples of stratified media are integrated circuit architectures, integrated optics structures and optoelectronic devices.

11.5.2 Thick Films

Axial imaging is obtained in a confocal microscope by scanning the sample through the confocal depth of focus. For a thick film the instrument measures peaks in the

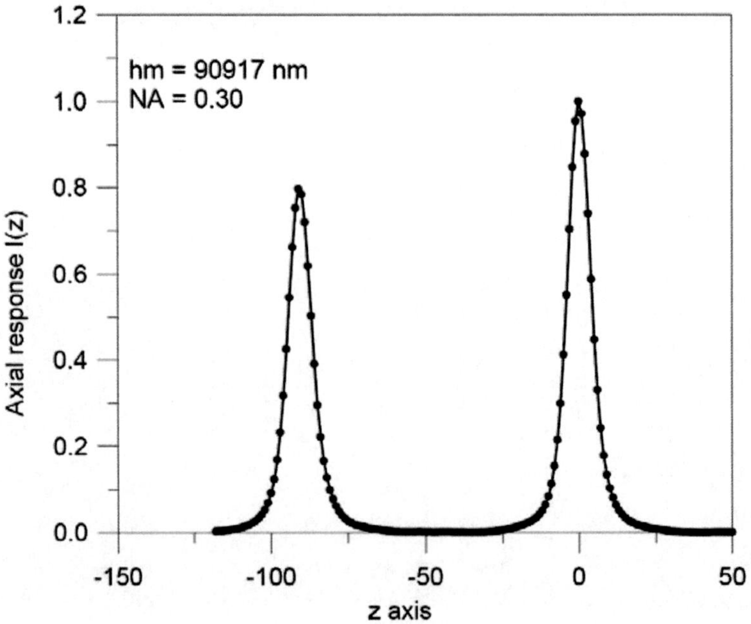

Fig. 11.36 Axial response of a 140 μm thick glass sheet

axial response arising from reflection at the two parallel reflecting surfaces. The widths of the peaks are reduced when the numerical aperture of the objective increases and the distance between the peaks increases with the thickness of the film (Sheppard 1994). Fig. 11.36 shows the experimental axial responses of a 140 μm thick glass sheet ($n_1 = 1.52$) obtained with a 10×, 0.3 numerical aperture objective.

The measured separation of the peaks h_m is very different from the real thickness h of the sheet because of two important factors: (i) depth distortion due to the refractive index n of the medium, and (ii) the spherical aberration caused by focusing with high numerical aperture optics through a refractive medium. The relationship between h_m and h can be predicted using a simple geometrical model (Sheppard 1994)

$$\frac{h_m}{h} = \frac{\int_0^{\theta_0} \sin\theta d\theta}{\int_0^{\theta_0} \frac{\left(n^2 - \sin^2\theta\right)^{1/2}}{\cos\theta} \sin\theta d\theta} \tag{11.20}$$

where θ is the angle of incidence of a ray impinging on the layer surface and $\sin\theta$ is the numerical aperture of the objective.

Fig. 11.37 shows the results of calculating the correction factor h_m/h for the thick glass sheet. Some experimental values obtained from the axial responses of

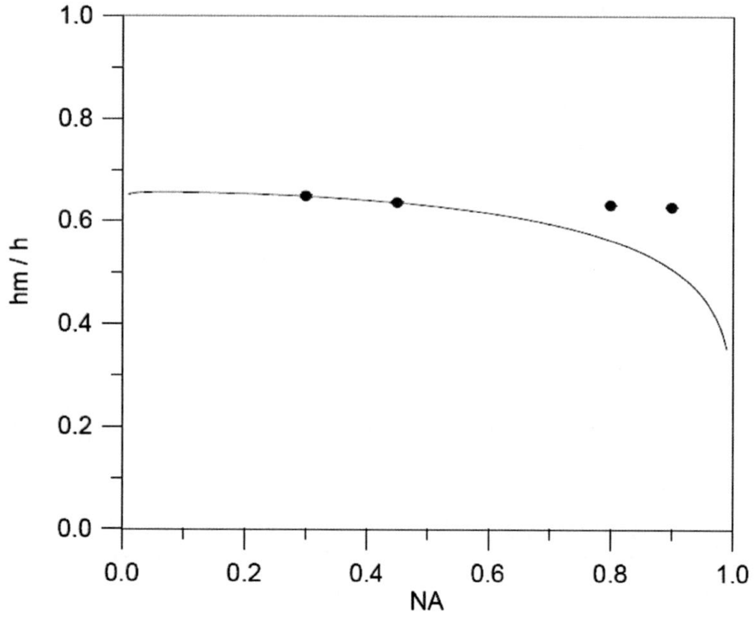

Fig. 11.37 Relationship h_m/h as a function of the numerical aperture of the objective predicted by the simple geometrical model for a 140 μm thick glass sheet (continuous line). Experimental values obtained from the axial responses are also plotted

Fig. 11.37 are also plotted. The agreement is fairly good for the lower numerical aperture objectives but is less satisfactory for numerical aperture values above 0.5. Differences between experimental and theoretical results may also be due to apodisation and residual aberrations in the objectives.

The geometrical model is also incorrect when axial imaging is used to measure the thickness of thin layers (Cadevall 2003). In these samples, the peaks obtained in the axial scanning are much closer, and effects such as multiple reflections and interference between wavefronts reflected on the interfaces become significant (see Sect. 11.5.3).

11.5.3 Thin Films

According to wave theory, in a confocal microscope the sample is illuminated by an angular spectrum of plane waves (Flagello 1996, Stamnes 1986). Each of these components is reflected by a structured surface with an amplitude reflection coefficient $R(\theta)$. The effect of the confocal aperture is to integrate over the amplitudes of the reflected angular spectrum. When the sample is scanned along the z axis, the detected intensity is given by

$$I(z) = \left| \int_0^{\theta_0} R(\theta) P(\theta) \exp(2ikn_0 z \cos\theta) \sin\theta \cos\theta d\theta \right|^2 \tag{11.21}$$

where $P(\theta)$ is the pupil function, k is the wave number and n_0 is the refractive index of the medium surrounding the surface of the sample. Considering a layer of refractive index n_1 on a substrate of refractive index n_s, the reflection coefficient $R(\theta)$ for linearly polarized illumination is given in terms of the reflection coefficients for parallel and perpendicular polarization by

$$R(\theta) = (R_\sigma - R_\pi)/2 \tag{11.22}$$

where

$$R_\sigma = \frac{R_{01,\sigma} + R_{1s,\sigma} \exp(2i\beta)}{1 + R_{01,\sigma} R_{1s,\sigma} \exp(2i\beta)} \tag{11.23}$$

and

$$\beta = kn_1 h \cos\theta_1 \,. \tag{11.24}$$

Similarly for R_π, where:

$$R_{01,\sigma} = \frac{n_0 \cos\theta - n_1 \cos\theta_1}{n_0 \cos\theta + n_1 \cos\theta_1}, \qquad R_{01,\pi} = \frac{n_1 \cos\theta - n_0 \cos\theta_1}{n_1 \cos\theta + n_0 \cos\theta_1}$$

$$R_{1s,\sigma} = \frac{n_1 \cos\theta_1 - n_s \cos\theta_s}{n_1 \cos\theta_1 + n_s \cos\theta_s}, \qquad R_{1s,\pi} = \frac{n_s \cos\theta_1 - n_1 \cos\theta_s}{n_s \cos\theta_1 + n_1 \cos\theta_s} \tag{11.25}$$

Fig. 11.38a and Fig. 11.39b show the axial response $I(z)$ calculated from equation (11.21) for two layers of silicon dioxide ($n_1 = 1.46$) deposited on a silicon substrate ($n_s = 3.88$) with thicknesses h of 1.5 μm and 1 μm respectively. The pupil function is $P(\theta) = 1$ (aberration-free objective), the numerical aperture is 0.9 and the source wavelength is 632.8 nm. In Fig. 11.38a the axial response shows two peaks: the smaller one, which appears in the position $z = 0$, comes from the reflection on the air-silicon dioxide interface; the larger one, which appears in the position $z = -0.847$ μm, comes from the reflection on the silicon dioxide-silicon interface. Because of the multiple reflections, some additional weaker peaks appear deeper in the z axis scanning. The two main peaks are partially overlapped, but it is still possible to measure the distance h_m between them. The situation is very different in Fig. 11.38b, where the two peaks cannot be resolved and it is, therefore, not possible to assign any h_m value to this axial response.

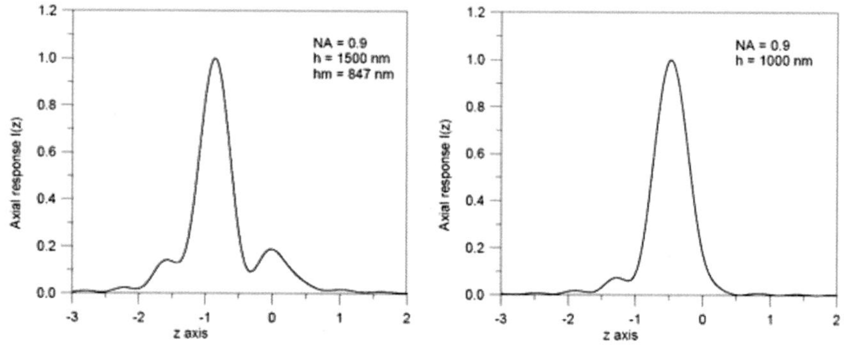

Fig. 11.38 Axial responses $I(z)$ calculated from equation (11.21) for a layer of silicon dioxide deposited on a silicon substrate

Fig. 11.39 shows the relationship between the separation of the peaks h_m predicted by equation (11.21) and the real thickness of the layer h for a numerical aperture of 0.9. For comparison, the results obtained from equation (11.20) are also plotted (dashed lines). Three main factors can be noted:

- The wavy behaviour of the h_m/h relationship caused by interference between wavefronts reflected at the layer interfaces.
- The lack of continuity of the h_m/h relationship for h values smaller than 2 μm. From the axial responses calculated from equation (11.21), it is observed that the z axis position of the larger peak coming from the reflection on the silicon dioxide-silicon interface oscillates slightly when the thickness of the layer h increases. However, the small peak coming from the reflection on the air-silicon dioxide interface oscillates with much larger amplitude in such a way that sometimes it becomes embedded in the

larger peak. In these conditions it is not possible to assign any h_m value to the axial response.

- The existence of a limit of resolution in the layer thickness measurement, which is reached when the two peaks of the axial response become unresolved. Even s microscope objectives with high numerical aperture is used, this limit of axial resolution turns out to be very close to layer thicknesses h of 1.5 µm or above. This is the situation reproduced in Fig. 11.38a, where the overlapping of the two peaks is very close to the limit given by the Rayleigh criterion.

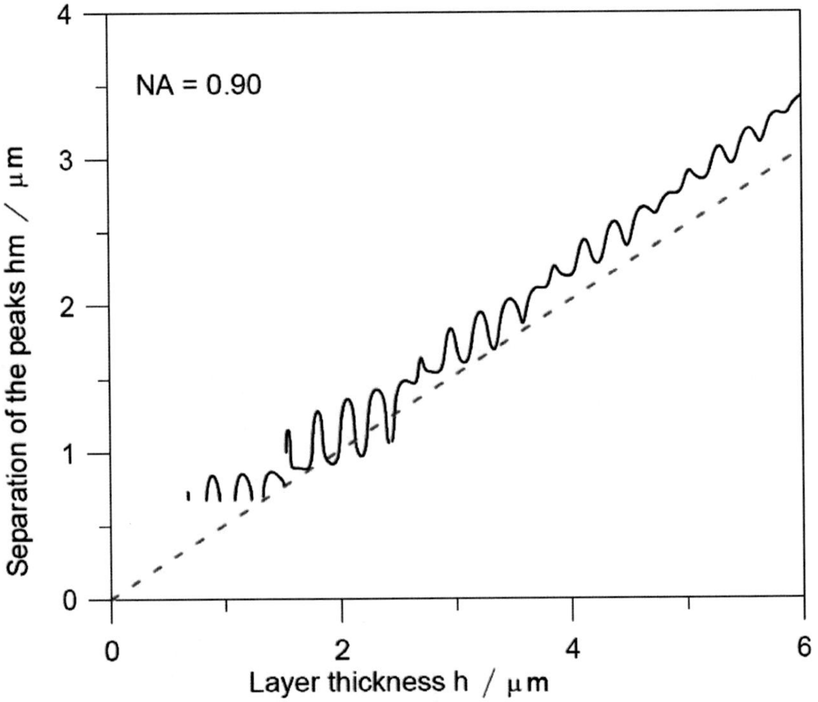

Fig. 11.39 Relationship between the separation of the peaks h_m predicted by equation (11.21) and the real thickness h of a layer of silicon dioxide deposited on a silicon substrate. Red line: linear regression

Unfortunately, the limit of resolution seems to mean that confocal technology is unable to measure the shape of structured surfaces such as those obtained when developing integrated circuits, integrated optics, MEMS and optoelectronic devices. These samples are obtained by the growth or the deposition of various layers of dissimilar materials (silicon, silicon dioxide, silicon nitride, photoresists, etc.) with thicknesses close to or well under 1 µm. After the deposition, the layers are patterned by well-known photolithographic processes, thus providing different layouts.

11.6 Case Study: Roughness Prediction on Steel Plates

This example will illustrate how an imaging confocal microscope can be used for the assessment of key parameters in the steel industry. During the manufacturing of steel plates for automotive and aerospace applications, the flat and smooth plate is processed to change its surface texture. One of the main processes is the increase of the roughness in a very well-controlled manner. Roughness will have a direct impact with paint adhesion. Too smooth surfaces will have low paint adhesion, while higher roughness surfaces will have better adhesion. Nevertheless, random roughness can create a waviness effect on the upper layers of the paint, creating a visual effect that makes the final customer of a car feel that the surface finish is of low quality. The steel plate is pressed against a roll. The roll has a manufactured surface texture in a way that after pressing the steel plate will imprint on it the desired texture. Such texture increases the effective adhesion area by creating craters and keeping a flat upper region.

Fig. 11.40 shows the surface texture of one of such rolls manufactured with electron beam discharge (EBD) technology. A steel roll is introduced into a vacuum chamber and each one of the electron discharges creates a crater-like structure. The measurement of the surface was carried out with a 20× objective and field stitching. Approximately, each field of view of the objective was measuring four craters. Six by six single fields were needed spanning a total area of 3.5 mm by 3.5 mm.

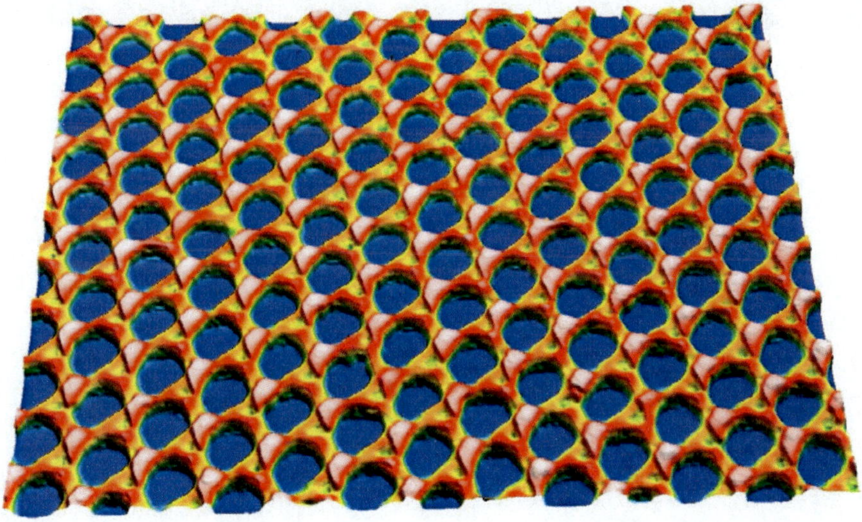

Fig. 11.40 Texture of steel roll after an EBD texturing process

Under a certain amount of pressure between the roll and the steel plate, only 70 % of the highest texture points of the surface are imprinted. Fig. 11.41 shows the bearing area curve of the previous surface. The predicted roughness on the final steel plate will be calculated by cutting the surface at 70 % of its bearing area point and inverting the result.

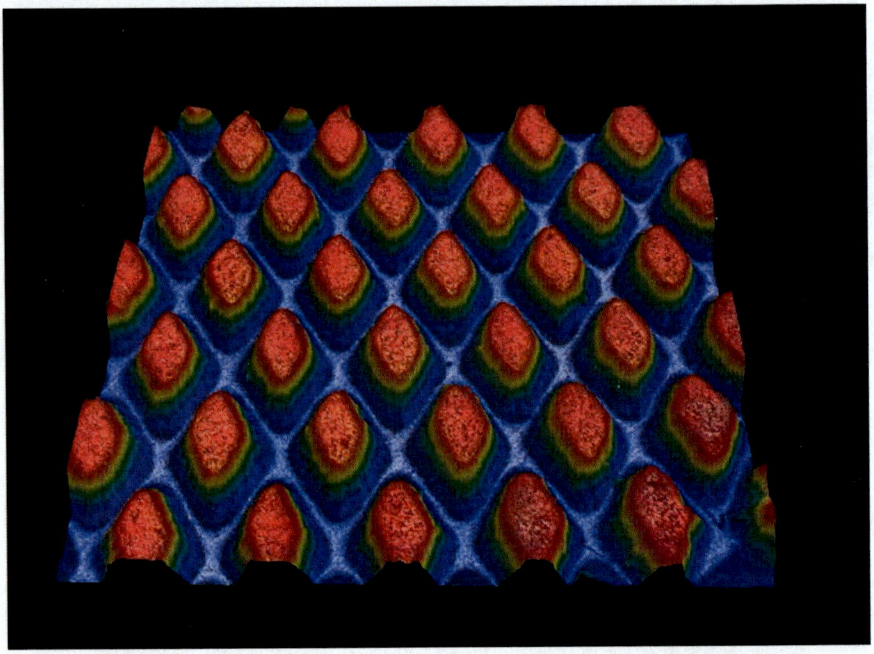

Fig. 11.41 Bearing area curve of the surface of Fig. 11.40. The cutting point for the prediction of the surface is located at 70 % of the highest point

Fig. 11.42 shows the result of cutting the surface at the mentioned point and inverting the result. The surface roughness of the original roll surface and the predicted surface are also shown.

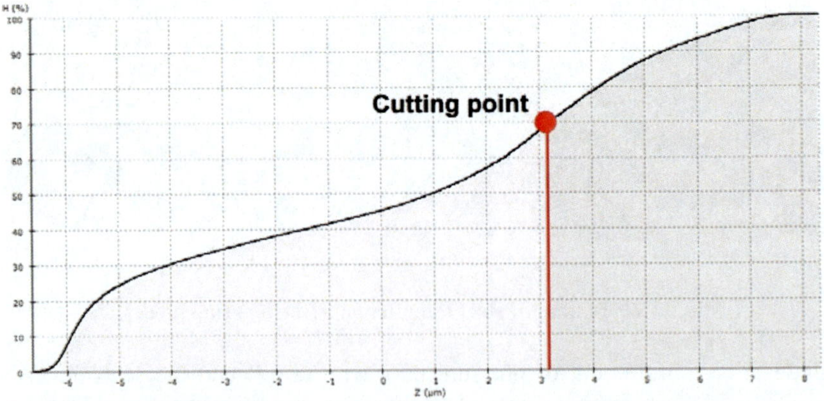

Fig. 11.42 Predicted surface after pressing a steel roll on a steel plate

References

Artigas, R., Pintó, A., Laguarta, F.: Three-dimensional micromeasurements on smooth and rough surfaces with a new confocal optical profiler. Optical Measurement Systems for Industrial Inspection 3824, 93–103 (1999)

Artigas, R., Laguarta, F., Cadevall, C.: Dual-technology optical sensor head for 3D surface shape measurements on the micro and nano-scales. In: Proc. SPIE, vol. 5457, pp. 166–174 (2004)

Bennett, J.M., Mattson, L.: Introduction to surface roughness and scattering. Optical Society of America, Washington (1989)

Brakenhoff, G.J., van der Voort, H.T.M., van Spronsen, E.A., Nanninga, N.: 3-Dimensional imaging of biological structures by high resolution confocal scanning laser microscopy. Scanning Micros. 2, 33–40 (1978)

Botcherby, E., Booth, M., Juškaitis, R., Wilson, T.: Real-time slit scanning microscopy in the meridional plane. Opt. Lett. 34, 1504–1506 (2009)

Cadevall, C., Artigas, R., Laguarta, F.: Development of confocal-based techniques for shape measurements on structured surfaces containing dissimilar materials. In: Proc. SPIE, vol. 5144, pp. 206–217 (2003)

Cha, S., Lin, P.C., Zhu, L., Sun, P., Fainman, Y.: Nontranslational three-dimensional profilometry by chromatic confocal microscopy with dynamically configurable micromirror scanning. Appl. Opt. 39, 2605–2613 (2000)

Conchelo, J.-A., Hansen, E.W.: Enhanced 3-D reconstruction from confocal scanning microscope images. 1: deterministic and maximum likelihood reconstructions. Appl. Opt. 29, 3795–3804 (1990)

Dorsch, R.G., Häusler, G., Herrmann, J.M.: Laser triangulation: fundamental uncertainty in distance measurement. Appl. Opt. 33, 1306–1314 (1994)

Fewer, D.T., Hewlett, S.J., McCabe, E.M.: Influence of source coherence and aperture distribution on the imaging properties in direct view microscopy. J. Opt. Soc. Am. A14, 1066–1075 (1997)

Flagello, D.G., Milster, T., Rosenbluth, A.E.: Theory of high NA imaging in homogeneous thin films. J. Opt. Soc. Am. 13, 53–64 (1996)

Gu, M., Sheppard, C.J.: Effects of defocus and primary spherical aberration on three-dimensional coherent transfer functions in confocal microscopes. Appl. Opt. 31, 2541–2549 (1992)

Karadaglic, D.: Image formation in conventional brightfield reflection microscopes with optical sectioning property via structured illumination. Micron 39, 302–310 (2008)

Kimura, S., Munakata, C.: Depth resolution of the fluorescent confocal scanning optical microscope. Appl. Opt. 29, 489–494 (1994)

Minsky, M.: Microscopy apparatus. US patent 3.013.467 (1961)

Sheppard, C.J.R., Mao, X.Q.: Confocal microscopes with slit apertures. J. Mod. Opt. 35, 1169–1185 (1988)

Sheppard, C.J.R., Shotton, D.M.: Confocal laser scanning microscopy. BIOS Scientific Publishers, Oxford (1997)

Sheppard, C.J.R., Connolly, T.J., Lee, J., Cogswell, C.J.: Confocal imaging of a stratified medium. Appl. Opt. 33, 631–640 (1994)

Smith, P.J., Taylor, C.M., Shaw, A.J., McCabe, E.M.: Programmable array microscopy with a ferroelectric liquid-crystal spatial light modulator. Appl. Opt. 39, 2664–2669 (2000)

Stamnes, J.: Waves in focal regions. IOP Publishing, Boston (1986)
Tanaami, T., Otsuki, S., Tomosada, N., Kosugi, Y., Shimizu, M., Ishida, H.: High-speed 1-frame/ms scanning confocal microscope with a microlens and Nipkow disks. Appl. Opt. 41, 4704–4708 (2002)
Tiziani, H., Wegner, M., Steudle, D.: Confocal principle for macro- and microscopic surface and defect analysis. Opt. Eng. 39 (2000)
Verveer, P.J., Hanley, Q.S., Verbeek, P.W., van Vliet, L.J., Joven, T.M.: Theory of confocal fluorescence imaging in the Programmable Array Microscope (PAM). J. Microscopy 189, 192–198 (1998)
Xiao, G.Q., Corle, T.R., Kino, G.S.: Real time confocal scanning optical microscope. Appl. Phys. Lett. 53, 716–718 (1988)
Wilson, T., Carlini, A.R.: Depth discrimination criteria in confocal optical systems. Optik 76, 164–166 (1987)
Wilson, T.: Confocal microscopy. Academic, London (1990)
Wilson, T., Masters, B.: Confocal microscopy. Appl. Opt. 33, 565–566 (1994)

12 Light Scattering Methods

Theodore V. Vorburger[1], Richard Silver[1], Rainer Brodmann[2], Boris Brodmann[2], and Jörg Seewig[3]

[1] Precision Engineering Division, National Institute of Standards and Technology
100 Bureau Drive, Stop 8212
Gaithersburg, MD 20899, USA
[2] OptoSurf GmbH
Nobelstrasse 9-13, 76275 Ettlingen
Germany
[3] Technische Universitat Kaiserslautern
Gottlieb-Daimler-Strasse
67663 Kaiserslautern
Germany

Abstract. Light scattering belongs to a class of techniques known as area-integrating methods for measuring surface texture. Rather than relying on coordinate measurements of surface points, light scattering methods probe an area of the surface and yield parameters that are characteristic of the texture of the area as a whole. The specular beam intensity, the angle-resolved scatter and the angle-integrated scatter are examples of measurands from light scattering that can yield useful parameters of the surface texture. Uses of light scatter for inspecting the surfaces of mechanical and optical components as well as surfaces produced in semiconductor manufacturing are primarily reviewed here. Several documentary standards describing best practice are also briefly reviewed.

12.1 Introduction

Since the beginnings of optical technology in ancient times (Flinn 2010), people have understood that smooth surfaces are mirror-like and that rough surfaces scatter light in many directions. This phenomenon of light scattering has been used since the early 1960s (Bennett and Porteus 1961, Bennett and Mattsson 1989, Griffiths et al. 1994) to help quantify surface roughness. The theory is well founded and has been extensively developed, and the designs of light scattering instrumentation have greatly varied. In this chapter the theory of light scattering from surfaces will be briefly outlined, and a number of different ways in which light scatter can be used to measure surface roughness will be described.

As indicated in Chap. 1, methods for measuring rough surfaces may be classified into three types (see Fig. 1.1): line profiling, areal topography and area-integrating (Vorburger et al. 2007, ISO 25178-6 2010). Line profiling and areal topography methods are coordinate based and are similar to one another.

Area-integrating methods are quite different. These methods sense an area of the surface as a whole and provide a measure of the overall surface roughness of that area, perhaps even with a single measured parameter. The most widely used area-integrating methods are based on light scattering. Two important types of light scatter methods are illustrated in this chapter: angle-resolved scatter (Fig. 12.1), where the light scattered in different directions is measured and analysed, and total integrated scatter (Fig. 12.2), where essentially all of the light that is not specularly reflected is captured.

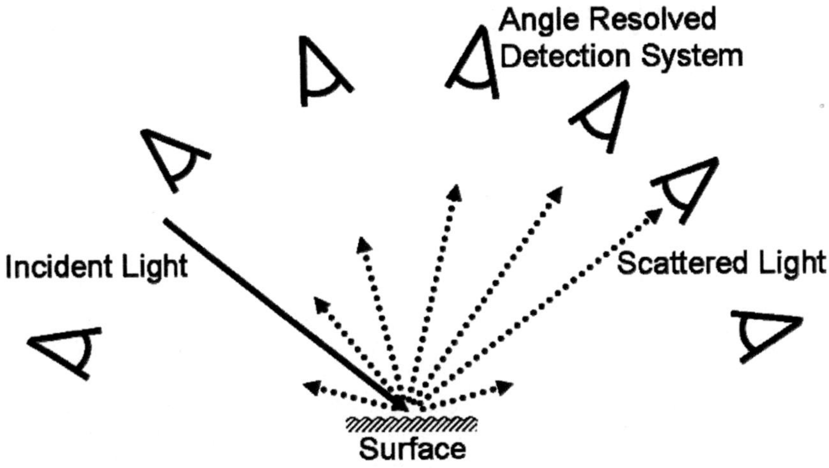

Fig. 12.1 Schematic diagram of light scattering from a rough surface and an angle-resolved detection system consisting here of a number of detectors. Alternatively, the system could have a moving detector

When a beam of laser light is incident on a smooth surface most of the reflected light travels in the specular direction such that the angle of reflection equals the angle of incidence, as shown in Fig. 12.1. As the surface roughness increases, more light is scattered into different directions and the reflected specular beam loses intensity. The theory (for example, Beckmann and Spizzichino 1987) that describes these phenomena comes from electromagnetism. Specifically, the theory describes the behaviour of the electric and magnetic fields that compose propagating beams of light and their behaviour when boundaries, such as reflecting surfaces, are encountered in the medium of propagation. The focus of this chapter is on the specular beam and the overall angle-resolved scatter distribution. Laser speckle effects (Asakura 1978) and changes in polarization upon reflection (Azzam and Bashara 1977) are also sensitive to surface roughness, but such phenomena will not be discussed here.

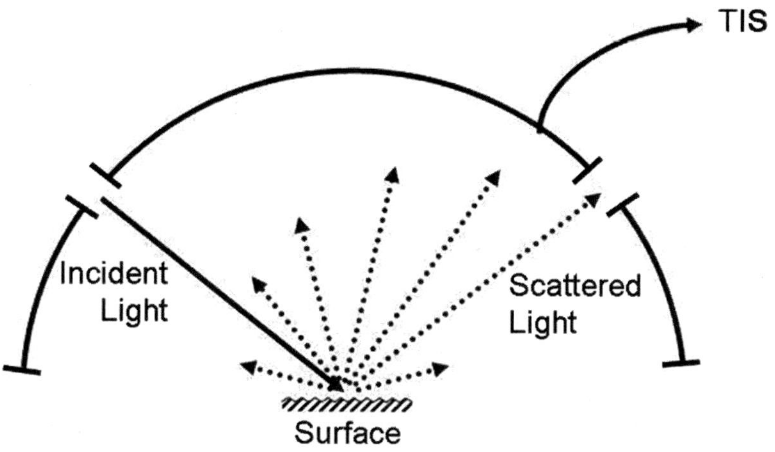

Fig. 12.2 Schematic diagram of light scattering from a rough surface and a device for collecting the total integrated scattered light

12.2 Basic Theory

A relatively simple theory for an electromagnetic boundary problem is illustrated in Fig. 12.3. The light is incident from the upper left with direction vector $\vec{\mathbf{K}}_i$, and it illuminates an area of an undulating (rough) surface. To simplify the geometrical arguments here, it is assumed that the surface is smooth along the *y* direction and that the light is incident in the *xz* plane as shown. The reflected light is scattered into a wide range of angles θ_s in the circle above the surface. If the surface is

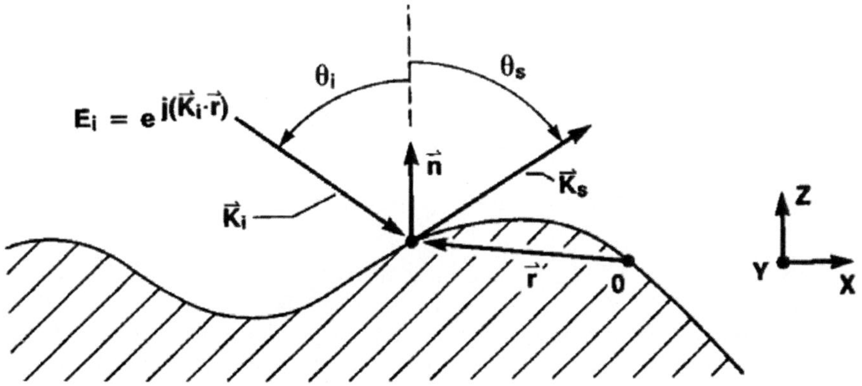

Fig. 12.3 Schematic diagram of key quantities associated with the electromagnetic boundary value problem describing light scattering from a rough surface. The incoming plane wave has direction vector $\vec{\mathbf{K}}_i$ with polar angle θ_i with respect to the normal $\vec{\mathbf{n}}$ of the mean plane of the surface. The theory aims to calculate the electric field in any direction θ_s from knowledge of the incident electric field E_i and the surface roughness *z*(*x*)

only moderately rough most of the reflected light proceeds in the specular direction associated with reflection from the mean plane of the surface. For the 2D schematic diagram shown here, the specular direction is such that $\theta_{\text{spec}} = \theta_i$. In addition, if the surface is transparent, some of the light will travel through the material, but that situation will not be discussed here, and only the reflected light will be described.

The boundary value problem that models the light-surface interaction may be approximated and described by the following equation if the surface is highly reflecting and has moderate slopes (Beckmann and Spizzichino 1987)

$$E(\theta_s) = F(\theta_i, \theta_s) \int_0^L \exp(j\vec{\mathbf{V}} \bullet \vec{\mathbf{r}})\, \mathrm{d}x \tag{12.1}$$

where $E(\theta_s)$ is the electric field of the scattered light, expressed here as a scalar, F contains a known trigonometric function of the angles and is also proportional to the incident electric field, 0 to L indicates the extent of the illuminated region, j is equal to $(-1)^{1/2}$, $\vec{\mathbf{V}} = \vec{\mathbf{K}}_\mathrm{i} - \vec{\mathbf{K}}_\mathrm{s}$, that is, $\vec{\mathbf{V}}$ is the vector difference between the wave vector of the incident light travelling in the direction θ_i and the wave vector of the scattered light for a particular direction θ_s, the magnitude of both wave vectors is $2\pi/\lambda$, where λ is the wavelength of the incident light, and $\vec{\mathbf{r}}$ $(= x\hat{\mathbf{e}}_\mathrm{x} + z(x)\hat{\mathbf{e}}_\mathrm{z})$ is the vector from the coordinate origin (O) to the points on the illuminated surface over which the integration is taken.

Furthermore, the dot product $\vec{\mathbf{V}} \bullet \vec{\mathbf{r}}$ contains the mathematical description of the surface profile $(x, z(x))$, and the flux density of the scattered light is proportional to $|E(\theta_\mathrm{s})|^2$ and is termed $I(\theta_\mathrm{s})$. Therefore, equation (12.1) indicates how the flux density of scattered light may be related to the surface roughness profile of the illuminated region and vice versa.

Under a number of conditions, including especially the assumption of a Gaussian distribution function of surface heights, equation (12.1) leads to a relationship (Beckmann and Spizzichino 1987) for the magnitude P_{spec} of the light flux reflected into the specular direction given by

$$P_{\text{spec}} = P_0 \exp\{-[4\pi(Rq)\cos\theta_i / \lambda]^2\}, \tag{12.2}$$

where P_0 is the total reflected flux, and Rq is the root mean square (RMS) roughness of the surface. For now, equation (12.1) for the electric field along a non-specular direction θ_s will be examined. A useful simplification results if the surface roughness is much smaller than the wavelength of the incident light. Then, the exponential term $\exp(j\vec{\mathbf{V}} \bullet \vec{\mathbf{r}})$ in equation (12.1) may be approximated by its first-order equivalent $\exp(jV_x x)[1 + jV_z z]$. The flux density $I(\theta_s)$ as a function of scatter angle may then be described approximately by the following (Elson and Bennett 1979, Church et al. 1979)

$$I(\theta_s) = [F_1(\theta_s)/\lambda^4]\left|\int_0^L \exp(jV_x x)\, z(x)\, dx\right|^2, \tag{12.3}$$

where F_1 is another known trigonometric function. The flux density in a non-specular direction is still a function of the roughness profile; however, the integral on the right-hand side of equation (12.3) is a close approximation to the power spectral density function (PSD) of the surface profile.

Equations (12.2) and (12.3) give two approaches to quantify the surface roughness in terms of scattered light. Equation (12.2) can be inverted to yield

$$Rq = (\lambda / 4\pi \cos\theta_i)\,[\ln(P_0 / P_{\text{spec}})]^{1/2}, \tag{12.4}$$

which indicates that the RMS roughness Rq can be determined from measurement of the light flux scattered into the specular direction (Bennett and Porteus 1961). This relationship is especially useful when the surface is moderately rough and the specular intensity is decreasing exponentially fast with increasing roughness, so that an accurate determination of roughness can be made from a measurement of the specular beam. Equation (12.2) is not at all useful when the surface is so rough that the specular beam is so small as to be un-measurable. It is also not useful when the surface is very smooth and the specular intensity comprises nearly all the reflected light. For smooth surfaces it is more useful to measure the small fraction of integrated scattered light ($P_{\text{scat}} = P_0 - P_{\text{spec}}$) and use the approximate relationship (Davies 1954),

$$Rq = (\lambda / 4\pi \cos\theta_i)(P_{\text{scat}} / P_{\text{spec}})^{1/2}. \tag{12.5}$$

This is the principle of the technique of total integrated scatter (TIS) (Bennett and Mattsson 1989, Stover 1995), an approach for which a documentary standard was developed (SEMI MF 2009).

For smooth surfaces the angular distribution of scattered light is closely related to the PSD of the surface, which describes the decomposition of the surface profile into its component spatial frequencies f (= $1/d$), where d is a component spatial wavelength. So a measurement of the angular distribution of scattered light can yield the PSD. The spatial wavelength d is related to the scattering angle by

$$2\pi / d = V_x = (2\pi / \lambda)\,(\sin\theta_s - \sin\theta_i), \tag{12.6}$$

which reduces to the relationship

$$\sin\theta_s = \sin\theta_i + \lambda / d. \tag{12.7}$$

Equation (12.7) shows the mapping between the spatial frequency (1/*d*) and the scattering angle θ_s. Small surface spatial wavelengths *d* scatter the light into large angles θ_s with respect to the specular direction θ_{spec} (= θ_i), and large spatial wavelengths of the surface scatter light into angles very close to the specular direction.

Equation (12.7) has been used a great deal (Elson and Bennett 1979, Church et al. 1979, Stover 1988) to describe and measure the quality of optical surfaces, where the RMS roughness is much smaller than the wavelengths of visible light. For slightly rougher surfaces these relationships become more complex. Equation (12.7), in particular, is generalised to

$$\sin\theta_s = \sin\theta_i + m\lambda / d, \quad (12.8)$$

where *m* is an integer (0, ±1, ±2, …). Equation (12.8) is the diffraction equation (Beckmann and Spizzichino 1987), which is useful to describe the scattered light from periodic gratings of any amplitude and any combination of spatial wavelengths. A particular spatial wavelength *d* scatters light into specific directions given by different diffraction orders *m*.

For ordinary rough surfaces, with a continuum of spatial wavelengths and heights, equation (12.3) is no longer valid, but for a limited range of slightly rough surfaces, it is still possible to obtain the surface autocorrelation function from a measurement of the angle-resolved light scatter (Chandley 1976).

One more relationship is important. Given the assumption that the surface has moderate slopes, it has been shown (for example Rakels 1989) that the RMS width of the distribution of scattered light (Γ) is proportional to the RMS width of the distribution of slopes on the surface *Rdq*

$$Rdq = 0.5\,\Gamma. \quad (12.9)$$

where, say, both quantities are expressed in radians or as tangents. Therefore, the RMS surface slope can be obtained from measurement of the RMS width of the scattered light (Cao et al. 1991). This relationship is derivable both from the theory of physical optics and from geometrical optics using a facet model to describe the scattered light.

Now the limitations of these relationships in terms of measurable amplitude and spatial wavelength will be investigated. Fig. 12.4 (Vorburger et al. 1993) shows approximate limitations in amplitude for the validity of the four approaches described above. These limitations are expressed in terms of the ratio of RMS roughness *Rq* to the wavelength λ of the incident light. Note that the boundaries shown here are only approximate.

For the smoothest surfaces the angle-resolved scatter (ARS) distribution directly yields the surface PSD. A criterion for the upper limit of $Rq/\lambda \approx 0.05$ was proposed elsewhere (Vorburger et al. 1993). If the surface is slightly rougher, one can still measure the angle-resolved scatter distribution and calculate the autocorrelation function (Chandley 1976). At approximately $Rq/\lambda = 0.14$, this operation is no longer mathematically feasible (Marx et al. 1993).

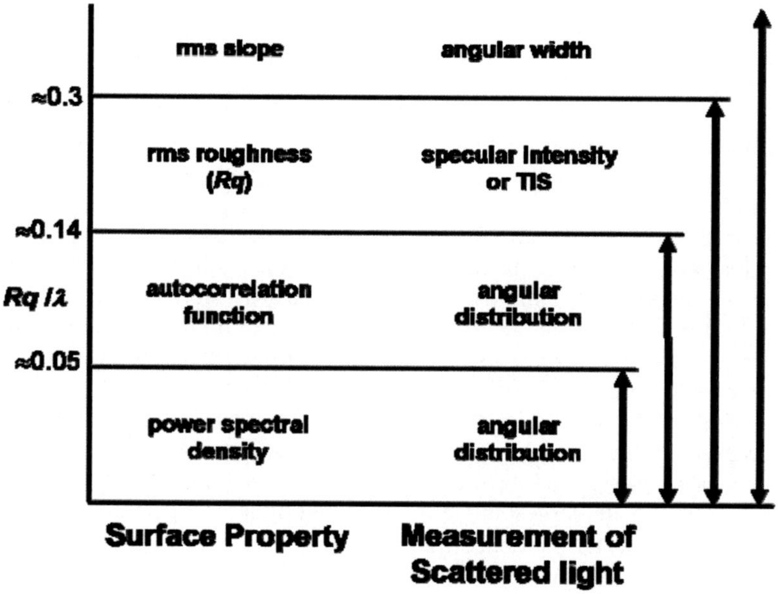

Fig. 12.4 Schematic diagram showing regimes of roughness for which measurable properties of the scattered light are paired with surface parameters or functions that may be derived from them (Vorburger et al. 1993)

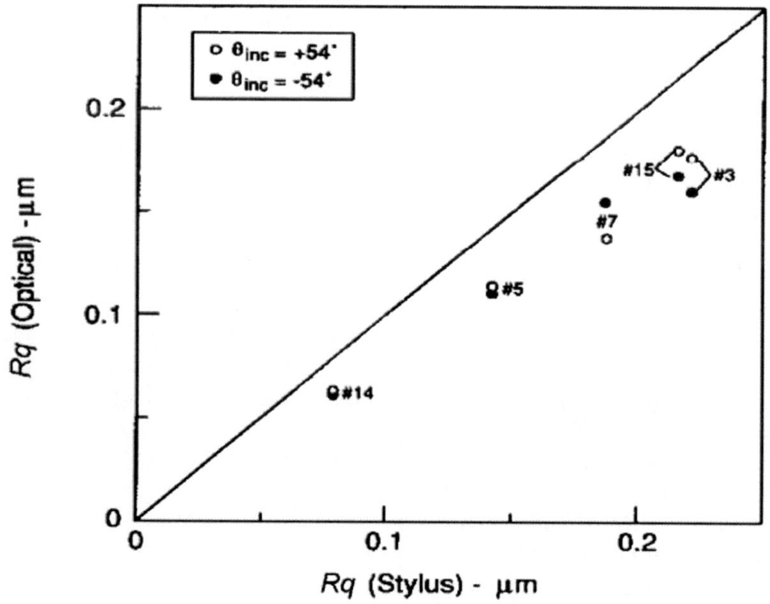

Fig. 12.5 Comparison of *Rq* values calculated from optical scattering and from stylus profiles for a set of five hand lapped surfaces. Data are shown for light incident at both +54° and -54° from the surface normal. The diagonal line indicates a slope of unity (Marx and Vorburger 1990)

Rq can still be calculated so long as there is a measurable specular spot (that is, the surface still appears a little glossy), but the approximate limitation there is $Rq/\lambda \approx 0.3$, and only if the light is incident at a high angle θ_i where $\cos\theta_i$ is small (see equation (12.2)). An example is shown in Fig. 12.5 (Marx and Vorburger 1990), where the Rq values calculated from measurement of the specular beam using equation (12.4) are plotted against the Rq values obtained from profiles measured with a stylus instrument for a set of five hand lapped surfaces of differing roughness. The correlation coefficient between the two sets of Rq values is 0.986. The optical values are systematically smaller than the stylus values, likely because the spatial bandwidth limits of the two instruments are different.

Finally, the width of the angle-resolved scatter can be used as an estimator of the RMS slope of the surface using the relationship derivable both from physical optics and geometrical optics and represented here by equation (12.9).

As with surface profiling instruments, there are limitations in the surface spatial wavelengths (d) that may be assessed by light scattering (Church 1979). The shortest measurable spatial wavelengths are determined, in light of the diffraction equation (12.8), by the largest angles of scattered light that can be collected by the instrument. For normal incidence light, the limit is $d = \lambda$, assuming that the instrument can capture light scattered right at the horizon. Conversely, long spatial wavelengths scatter light into small angles very close to the specular beam, so the long wavelength limit is determined in principle by the size of the specular spot in the plane of detection or by apertures in the instrument used to detect the specular beam, or conversely, by apertures allowing the specular beam to pass through the instrument. For smooth surfaces, where the index m only takes the value of one and higher orders are insignificant, this limit is easily determined by the diffraction equation with $m = \pm 1$. For example, if the specular spot has a diameter of 1 mm at the detector and the detector is located at a radius (R) of 100 mm from the sample, then the angular width of the specular spot ($\Delta\theta$) is approximately 0.005 rad. Assuming that the wavelength of the incident light is 0.6 μm and is normally incident on the surface, the largest measurable spatial wavelength is approximately given by

$$d \approx \lambda / \sin(\Delta\theta) = 0.6\ \mu\text{m} / \sin(0.005\ \text{rad}) = 120\ \mu\text{m}. \qquad (12.10)$$

For moderately rough surfaces higher values of m become significant in equation (12.8), and the limit is not so clear cut. In principle, rough surfaces with very long spatial wavelengths can produce measurable scattered light at high orders m outside the specular spot. Therefore, long spatial wavelengths, whose first-order scattered light may lie inside the specular spot and remain undetectable, may also have sufficient amplitude to produce higher scattering orders that lie outside the specular spot and are detected. Hence the long spatial wavelength cut-off is difficult to characterise for rough surfaces.

12.3 Instrumentation and Case Studies

12.3.1 Early Developments

From the 1960s, a number of light scatter instruments were custom developed by manufacturers of optical surfaces and by surface metrology researchers but were not commercialised. One notable example is the goniometric optical scatter instrument (GOSI) developed at NIST (Germer and Asmail 1997) for calibrated measurements of scattered light flux. However, several companies have also produced light scatter products*. Commercial glossmeters, for example, are fairly common, and physical standards for glossmeters have been developed (Nadal and Thompson 2000).

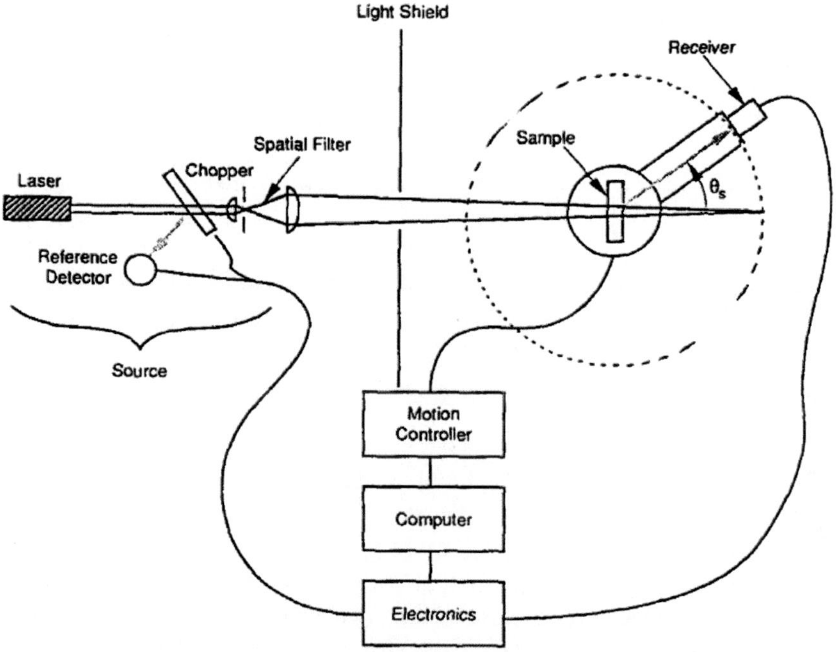

Fig. 12.6 Schematic diagram of an apparatus to measure angle-resolved scattering (Stover 1995, reprinted with permission)

* Certain commercial equipment, instruments or materials are identified in this chapter to specify adequately certain experimental procedures. Such identification does not imply recommendation or endorsement by the National Institute of Standards and Technology, nor does it imply that the materials or equipment identified are necessarily the best available for the purpose.

Several other examples of commercial instruments intended for quantitative measurement of surface texture parameters are cited here. Stover et al. have developed a line of angle-resolved scatterometers intended mainly for the study of smooth optical and semiconductor surfaces. A typical scatterometer design (Stover 1995) is shown in Fig. 12.6.

To assess the quality of optically smooth surfaces, these instruments operate over a wide dynamic range of scattering angles and flux levels. These capabilities are important for the study of very smooth surfaces, such as laser gyro mirrors, where small amounts of light scatter affect the function of the device and need to be specified. Typically, high-quality scatterometers can detect scattered light flux ten orders of magnitude smaller than the specular beam flux as well as light scattered within 0.1° of the specular beam by long spatial wavelengths of the surface.

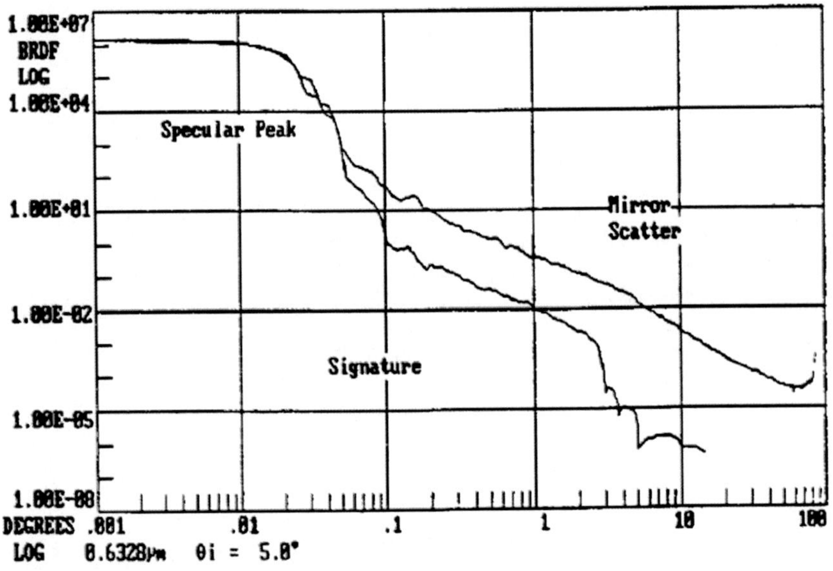

Fig. 12.7 BRDF from a molybdenum mirror measured as a function of scattering angle. The scale is logarithmic along both axes. Separation from the instrument signature and meaningful data are obtained beyond about 0.1° (Stover 1995, reprinted with permission)

Figure 12.7 shows a typical measured angle-resolved scattering distribution function, in this case for a molybdenum mirror (Stover 1995). The distribution is expressed as the bidirectional reflectance distribution function (BRDF), a quantity that will be discussed further in Sect. 12.4.2. Note that both the BRDF and the scatter angle are plotted logarithmically. The upper curve is the measured BRDF as a function of angle in the plane of incidence. The lower curve is the instrument signature, which is the measured function when there is no sample in place. The instrument signature represents scattering from various surfaces in the instrument and along the air path. Figure 12.7 shows BRDF measurements over roughly twelve orders of magnitude and shows good separation between the mirror scatter, the quantity to be measured, and the instrument signature to within 0.1° of the specular peak.

A portable commercial instrument (Fig. 12.8), developed by Brodmann et al. (Brodmann et al. 1985, Brodmann and Allgäuer 1988), was mainly intended for the study of moderately rough surfaces in a manufacturing environment and even during the machining process. The angle of incidence is near normal. Because a single housing contains both the light source and the detection system, the instrument is useful (Zimmerman et al. 1988) for automated alignment with respect to the surface being measured, an important function for quality control of parts during manufacture.

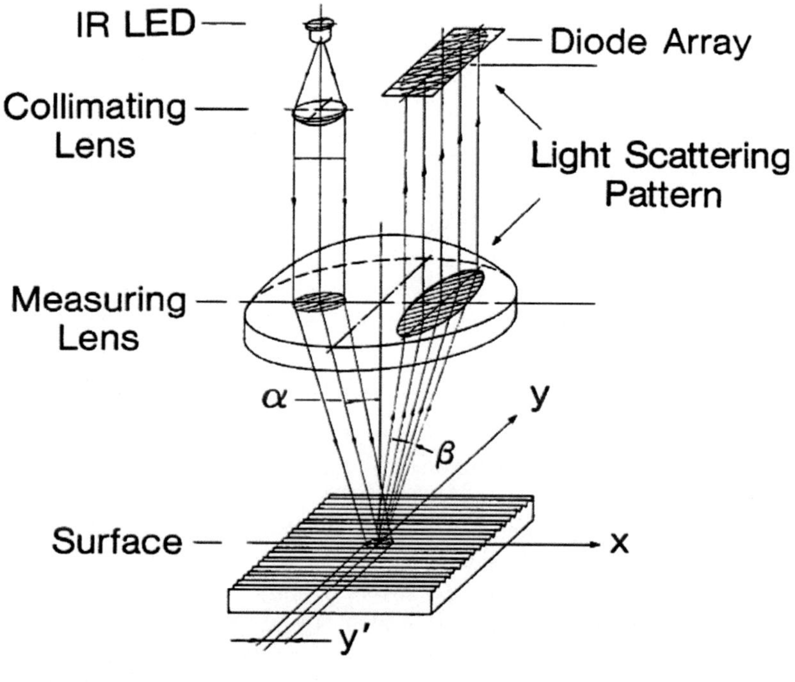

Fig. 12.8 Schematic diagram of an optical scatterometer. IR LED stands for infrared light-emitting diode (Brodmann and Allgäuer 1988, reprinted with permission)

Subsequently, Kiely et al. (Kiely 1991) experimentally showed for a set of hand lapped components the expected correlation between the RMS width of the light scattering distribution measured with the instrument and the RMS slope of the surface as measured with a stylus instrument.

A different commercial instrument for control of surfaces in manufacturing environments was developed by Valliant (Valliant et al. 2000) and is shown in Fig. 12.9. Once again both the source and the receiver components are located in a single housing. The angle of incidence is approximately 75° here. Mechanisms for monitoring and compensating for vibrations during measurements in the manufacturing operation make integration of the scatterometer technology into the plant environment possible. The instrument shown in Fig. 12.9 is one of several having different sizes and degrees of portability.

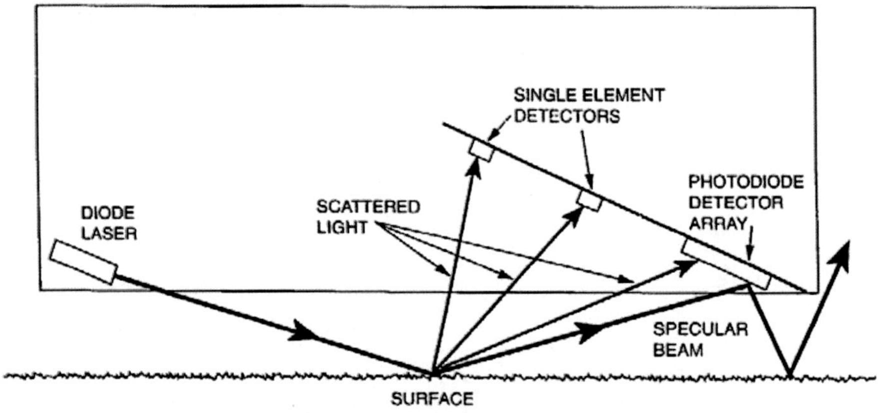

Fig. 12.9 Schematic diagram of Lasercheck optical scatterometer (Valliant et al. 2000, reprinted with permission)

12.3.2 Recent Developments in Instrumentation for Mechanical Engineering Manufacture

The commercial light scattering technologies of Brodmann et al. (1988) and Valliant (2000) have been developed further for in-line measurement on the shop floor (Brodmann et al. 2009, Optosurf 2010a, Schmitt Industries 2010). The sensor recently developed by Brodmann et al. has high dynamic range (16 bit), a measurement time of 0.5 ms, and software packages for roughness and form analysis. Following a recent guideline developed for scattered light measurement (Sect. 12.4.5), a roughness parameter called Aq is calculated, which describes the variance in angle of the scattered light distribution recorded by the sensor. The focus of the applications is the assessment of roughness and form of fine machined mechanical surfaces including automotive parts, such as finished bearings of crankshafts (Fig. 12.10), piston bolts, steering gearing components and gear shafts.

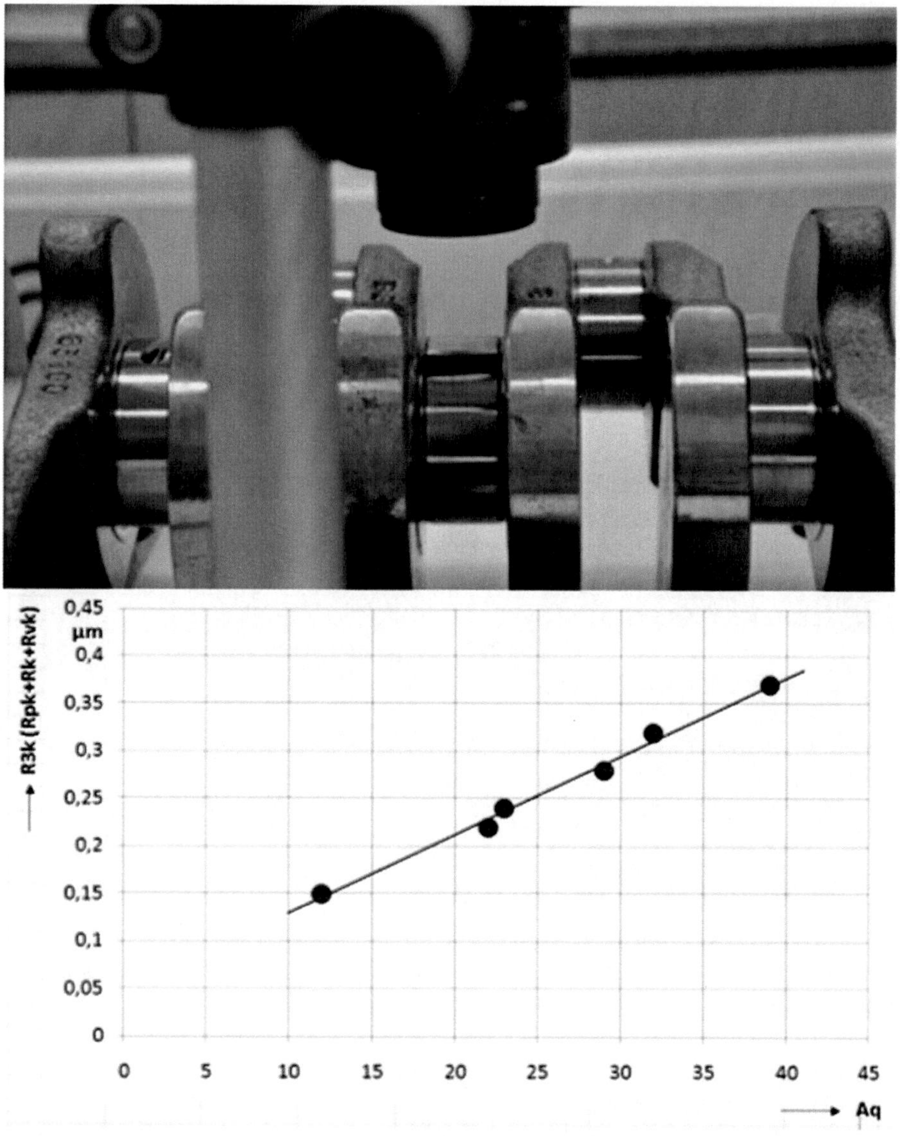

Fig. 12.10 Scattered light sensor shown measuring the bearings of crankshafts (top). The scattered light roughness value *Aq* correlates with the *R3k*-roughness parameter (bottom), which equals *Rpk* + *Rk* + *Rvk* (ISO 13565-1 1996), courtesy of Hercke T, Daimler AG (Optosurf 2010b)

Scattered light sensors can be used in an oil-vapour environment close to the manufacturing process. Based on the specific capability of the scattered light method to measure polished surfaces, scatterometry has also found application in biomedical device manufacturing for 100 % measurement of artificial femoral heads. The sensor may be used with two precision rotary stages. One stage rotates the sample and the other pivots the sensor across the entire surface. Such a system is shown in Fig. 12.11. Sensor outputs are used for scratch detection and roundness measurement of the ball. The sensor detects scratches with depths of 0.2 μm or less and a few micrometres in width. The uncertainty of the roundness measurement is approximately 0.2 μm (peak to valley).

Fig. 12.11 Automated surface measurement of artificial femoral head (left). Scratch detection and form measurement along the equator and across the pole of the ball are obtained. Measurement time of the surface is less than 10 s

The measurement of form with scattered light may be achieved by a triangulation technique whereby the centre of mass of the measured angle distribution is detected and integrated along the scan direction. The sensitivity of the method as a function of surface spatial wavelength has been derived by Seewig et al. (2009).

As an example, a roundness measurement artefact with different sinusoidal waves was measured by Seewig et al. (2009). Fig. 12.12 shows the corresponding roundness profiles in polar coordinates. A reference measurement obtained with a tactile roundness instrument is shown on the left side. The reconstructed roundness profile obtained with the light scattering sensor is shown on the right side. The *Pt* (peak to valley) values differ by approximately 3 %.

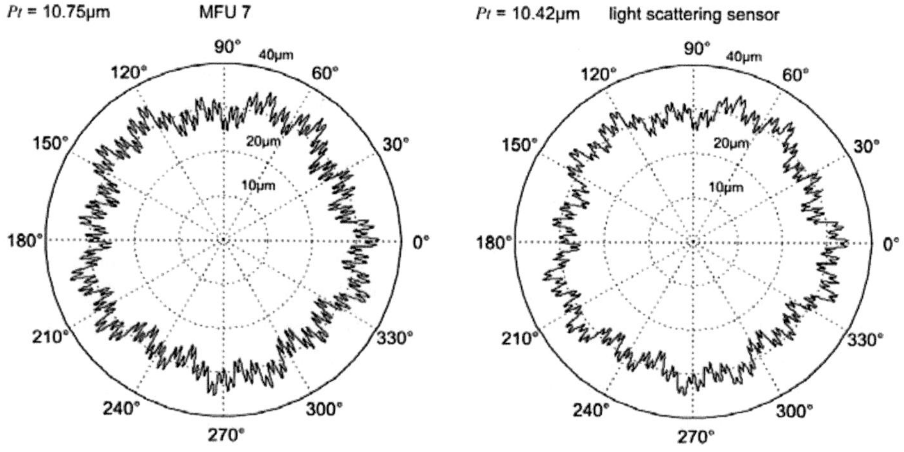

Fig. 12.12 Left: tactile profile measurement of a roundness artefact with a *Pt* value of 10.75 µm. Right: reconstructed profile of the optical sensor data with a *Pt* value of 10.42 µm (Seewig et al. 2009)

As mentioned earlier, the surface roughness can be characterised by the parameter *Aq*, which is proportional to the second statistical moment, the variance, of the angle distribution. A mirror-like surface has a small *Aq* value, and *Aq* increases depending on the RMS slope of the surface roughness. However, not only the roughness influences the *Aq* value. Any additional curvature caused by the workpiece form increases the value of the parameter. The influence of curvature can be suppressed using a correction term. For a cylindrical workpiece the correction term is proportional to $\left(D'/R_{zyl}\right)^2$, where D' is the diameter of the incident light spot and R_{zyl} is the radius of the cylinder. Correction terms for arbitrary form components can be calculated according to a mathematical formalism given by Seewig et al. (2009).

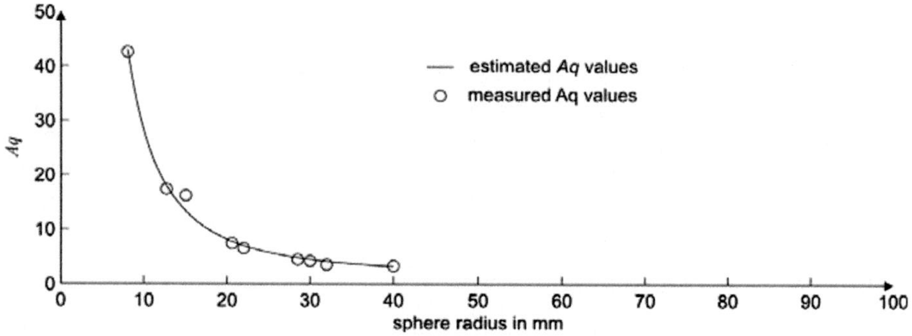

Fig. 12.13 Estimated and measured *Aq* values for spheres with different radii but with the same micro-structure

The influence of the workpiece curvature is shown in Fig. 12.13. Spheres with different radii but with the same micro-structure were measured. The circles indicate *Aq* values of the sensor depending on the radii. The solid line is the estimated curve calculated from a priori knowledge about the workpiece form.

12.3.3 Recent Developments in Instrumentation for Semiconductor Manufacture (Optical Critical Dimension)

Commercial optical scatterometers have been used for many years in semiconductor fabrication facilities and in hard disk manufacturing to monitor defects and particles on surfaces during manufacture. Some types are based on the use of TIS discussed in Sect. 12.2 (Schmitt Industries 2010). More recently, scatterometry has been widely used for monitoring and process control of the critical dimensions (linewidths) of semiconductor device features, even though that industry is producing features with 45 nm linewidth or smaller, and the pitch of lines is often smaller than the optical wavelengths of the scatterometers. With this measurement technology, called optical critical dimension (OCD) scatterometry, a reflectance signature of a number of nominally identical lines with a uniform pitch are measured together as a function of angle, wavelength and/or polarization. The average linewidth and perhaps other average geometrical parameters are then calculated from least squares comparisons of the scattered light signatures with model simulations of the scattering.

Rigorous coupled wave (RCW) based theories are the most common methods used to analyse scatterometry data (Moharam et al. 1995, Li 1996). Simulations with different parameters are pre-calculated and stored in a library to facilitate rapid comparison with measured scattering data, or the data are fit to the model by regression analysis. Recent work has also focused on applying scatterometry to measure line sidewall angles as well as linewidth, and to measure lines with even more complexity and thus more parameters. There are also efforts to use scatterometry for photomask CD metrology (Hoobler 2005, Pundaleva 2006). It is important to note, however, that scatterometry is used with gratings and is not well suited to measurement of individual features or non-periodic structures, although the individual grating elements can have complex shapes.

The accuracy of the OCD technique depends on at least three sets of factors: 1) realistic models for the geometrical structures of the surfaces being studied, 2) rigorous formulations for the electromagnetic scattering from the jagged, steeply sloped surface topography, and 3) a priori knowledge of the optical properties of the materials. Under these conditions it is often possible to calculate the average linewidth and other geometrical parameters of the fabricated features. Scatterometers are sometimes used in the so-called θ, 2θ mode as shown in Fig. 12.14 (Raymond 2001) where the flux of the specular reflected beam is measured as a function of the angle of incidence. Although higher order diffracted intensities can also be measured when the pitch is sufficiently large, virtually every commercial instrument measures only the specular component. Moreover, because different states of polarization produce different angle-resolved scattering signatures, separate scattering results are measured for *s*- and *p*-plane polarized light in the incident beam. While example measurements using the θ, 2θ mode are discussed below, it is more common to measure either the specular reflectance (for example, Yang et al. 2004), or the polarization change upon reflection (for example, Niu et al. 2001), at a fixed angle as a function of optical wavelength to obtain the signature. The modelling procedure for these different data types is similar.

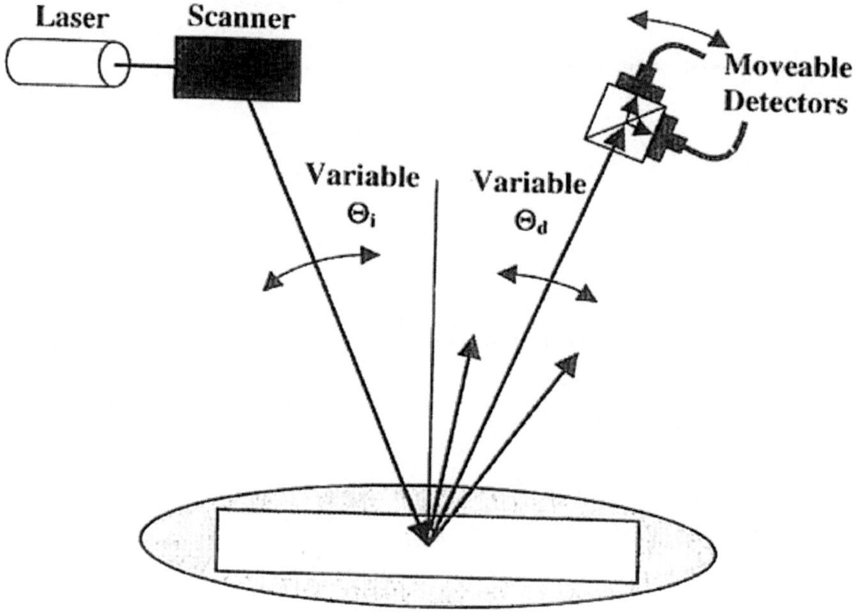

Fig. 12.14 Schematic diagram of a variable angle scatterometer useful for measuring the specular reflectance as a function of incident angle (Raymond 2001, reprinted with permission)

One example of research results obtained with OCD scatterometry is from Patrick et al. (2007) where scattering signatures were obtained for both *s*- and *p*-plane polarized light from linewidth targets fabricated on separation by implantation of oxygen (SIMOX) wafers. A model of the surface features is shown in Fig. 12.15. Silicon lines are fabricated over a buried silicon oxide (BOX) layer on a thin mixed boundary layer of silicon and silicon oxide, which is on a silicon wafer.

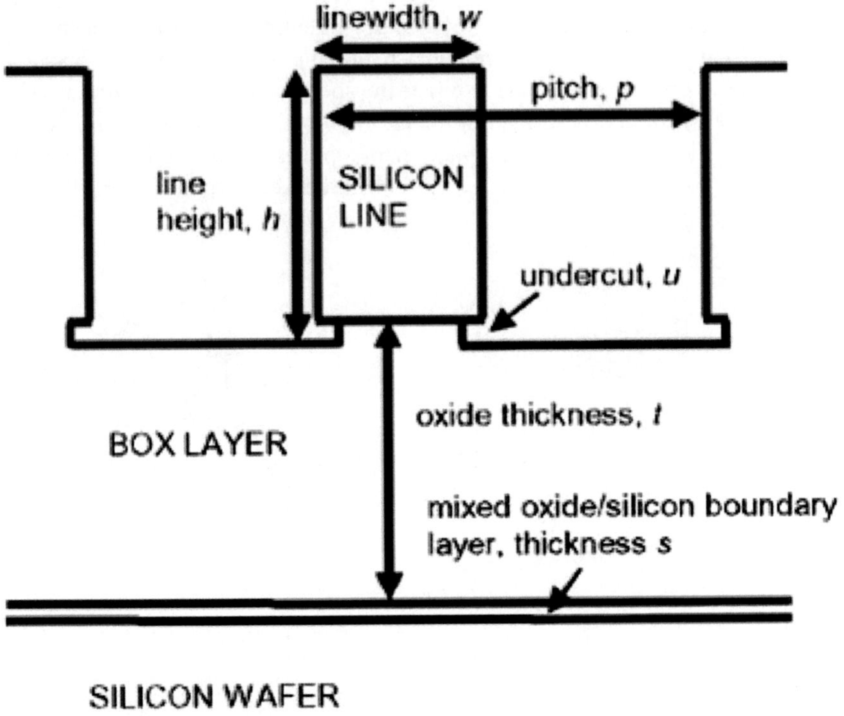

Fig. 12.15 Model of a patterned wafer surface used by Patrick et al. (2007) for OCD experiments. The model contains geometrical parameters, w, h, p, u (assumed to be the side of a square), t, s, and sidewall angle (assumed to be 90°), as well as the optical properties (real and imaginary) of silicon, the BOX layer, and the mixed layer

Many parameters were required for a complete description of the model in Fig. 12.15. The optical constants of the silicon substrate and the silicon lines were taken from the literature. The optical constants of the oxide and the thickness and composition of the mixed layer were taken from auxiliary scatterometry measurements from unpatterned wafers. The pitch was not fit but rather was considered to be well known a priori. The sidewall angle was assumed to be equal to 90° because of the single crystal nature of the silicon and its fabrication (Cresswell et al. 2006). The height of the silicon features was measured independently by AFM. The linewidth w, the undercut dimension u and the oxide thickness t were fitted by comparison to the model.

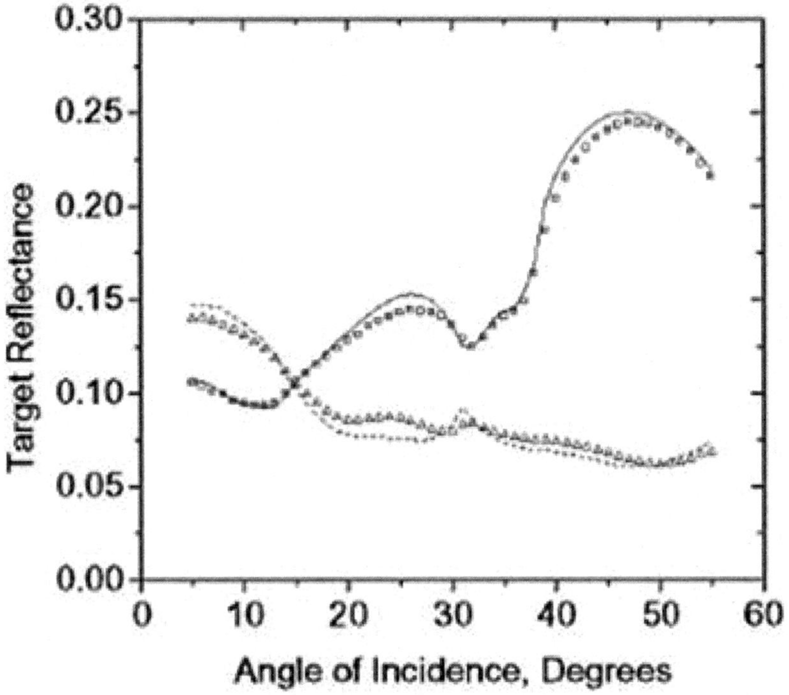

Fig. 12.16 Results (data points) obtained by Patrick et al. (2007) for specular reflectance of a patterned silicon surface as a function of angle compared with results (curves) calculated from the model of Fig. 12.15 with parameters yielding the best fit. The squares are for *s*-polarized incident light and the triangles are for *p*-polarized. The optical wavelength here was 532 nm, the pitch was 2.1 μm, and the best fit result for linewidth was 297.5 nm

One result for such a fit is shown in Fig. 12.16, where measured and modelled reflectances are plotted against angle of incidence for *s*- and *p*-polarization. The fit of the model to the measurements is good, especially considering the number of parameters that go into the result. In this case the best-fit linewidth was 297.5 nm.

Results for six different targets are shown in Fig. 12.17. Here the linewidths determined by OCD are compared with linewidths determined by SEM. While historically, top-down SEM and OCD measurements have often not been in perfect agreement, in this case the residuals from a fitted straight line are only a few nanometres. This agreement was likely aided by a priori knowledge of the 90° sidewall angle of the target.

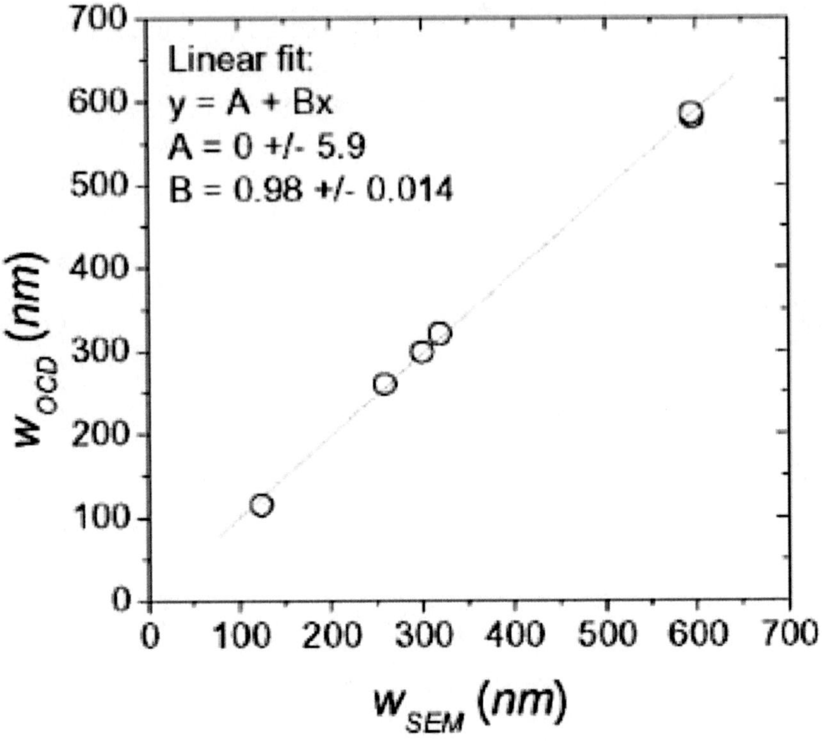

Fig. 12.17 Comparison of OCD results and SEM results for linewidth of six patterned structures (Patrick et al. 2007). The OCD results were obtained using the model of Fig. 12.15

An emerging technique, known as scatterfield microscopy (Silver et al. 2007, Silver et al. 2009), combines elements of scatterometry and bright field imaging. The scatterfield technique uses a bright field microscope with structured incident light instead of a scatterometer. Measurements of the reflected light are taken at the conjugate back focal plane of the microscope objective where each point maps nominally to a plane wave of illumination at the sample. Microscope images of the patterned surfaces are acquired as a function of angle. The mean intensity of the angle-resolved images is calculated and then corrected using a background scan of the unpatterned surface that was previously normalised by the known silicon reflectance. This mode of operation is similar to conventional θ, 2θ scatterometry except that measurements are made with high magnification, image-forming optics. Scatterometry measurements throughout the field of view can be acquired in parallel by breaking the imaged field into an array of small targets or pixel groupings.

As stated earlier, the experimental signatures can be compared with simulations stored in a library of reflectance curves assembled from calculations performed for ranges of reasonably expected values of the parameters. A least-squares fitting routine is normally used to choose the optimum set of parameters that yields the closest experiment-to-theory agreement. When the parameter range is adequate and the optical tools are accurately modelled, close agreement is achieved between the simulated and the measured curves, and nanometre-scale agreement is obtained between the best-fit parameters and known values. Correlations between parameters, measurement noise and model inaccuracies all lead to measurement uncertainty in the fitting process.

An example of scatterfield data (Silver et al. 2009) from a grating of dense, sub-wavelength features in the 120 nm size range is shown in Fig. 12.18. Strong

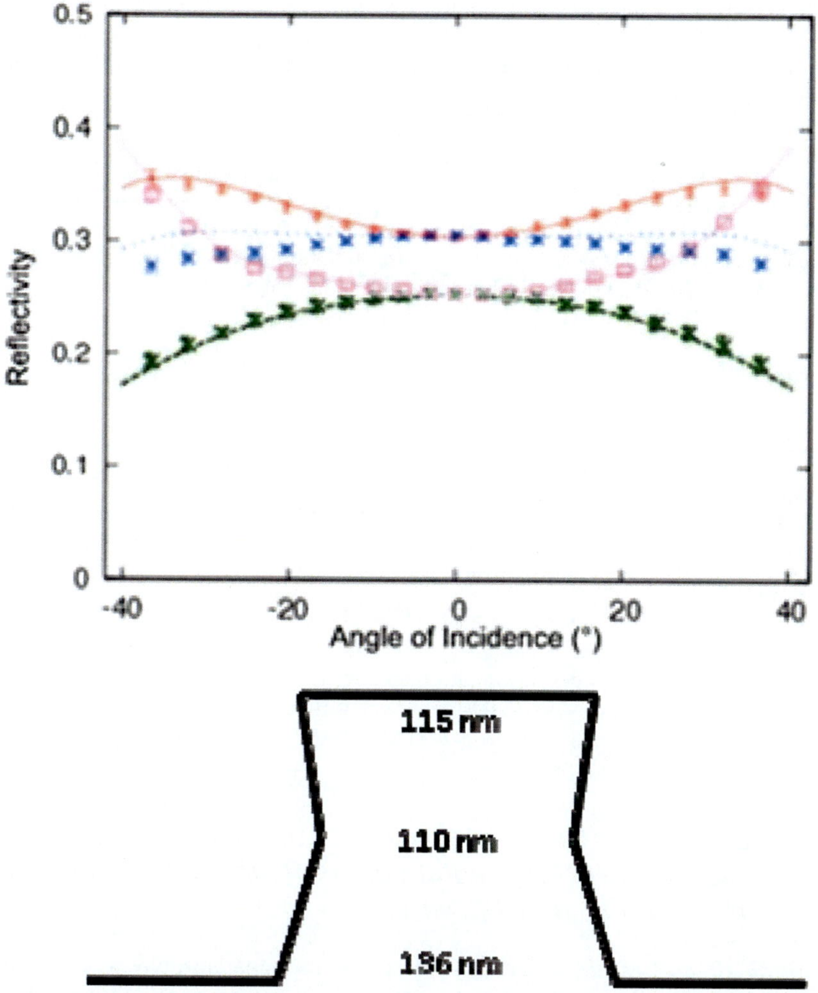

Fig. 12.18 An example of experimental data points and library data fits (top) for a periodic silicon test structure (bottom). The pitch of the features was 300 nm and the optical wavelength was 450 nm. Three parameters are required to characterise the line shape in order to obtain a good fit between the data points and simulated curves. The four sets of data represent angle scans with *s*- and *p*-polarization taken in both the *x* and *y* directions. Good agreement with reference values was obtained for the top and middle width parameters

sensitivity to changes in sidewall angles as well as to critical dimensions below the conventional imaging resolution limits can be observed using this type of optical method. Good agreement between the simulated reflectance curves and the experimental data is also achieved.

An example of fitting scatterfield data for linewidth features in the 40 nm to 60 nm size range is shown in Tab. 12.1 (Silver et al. 2009). These data are parametric fitting results for dense features fabricated in silicon using a standard front end semiconductor manufacturing process. The nominal ratio of pitch to width was 3:1. Tab. 12.1 shows results obtained from different techniques for comparison. This is an example of how accurate optical metrology can be performed if the data are accurately normalised and the parametric fits take account of the proper floating parameters.

Table 12.1 Results obtained for height (*h*) and width (*CD*) parameters of dense linewidth features. The optical wavelength used for the scatterfield (OCD) results was 450 nm. The OCD results and small angle x-ray scattering (SAXS) results use parametric fits to a stacked trapezoid model similar to the one illustrated in Fig. 12.18. The agreement between techniques is especially good for the top and middle *CD*s. Top-down SEM used here is not sensitive to line height

Method	CD_{TOP}	CD_{MID}	CD_{BOT}	h
	/nm	/nm	/nm	/nm
OCD	41	49	63	56
AFM	38	45	50	55
SAXS	43	53	62	54
SEM	35	49	63	-

12.4 Instrument Use and Good Practice

Good practice for the specification and measurement of scattered light from rough surfaces may be found in several documentary standards and guidelines developed by different organisations, including ASTM, SEMI and ISO. Some of these standards are discussed in the following.

12.4.1 SEMI MF 1048-1109 (2009) Test Method for Measuring the Effective Surface Roughness of Optical Components by Total Integrated Scattering

SEMI MF 1048-1109 (2009) succinctly describes the apparatus, physical standards, measurement procedure, analysis and quality control procedure recommended for a measurement of TIS. A schematic diagram of the apparatus reproduced from the standard is shown in Fig. 12.19. The main feature is a Coblentz sphere that focuses the scattered light to a single detector but which contains an aperture to allow the specular beam to pass through. The recommended design involves trade-offs. Since it is difficult to measure scattered light right up to the specular beam and right out to grazing, the standard specifies a sphere with a minimum collecting angle of 70° with respect to the sample normal and an aperture that subtends 5° or less at the specimen. Therefore, a portion of the scattered light is not detected. Within these measurement limits, the effective surface roughness is calculated from equation (12.5). The document also specifies the use of physical standards for both diffuse reflectance and specular reflectance in the TIS measurement process to ensure the calibration of the TIS results.

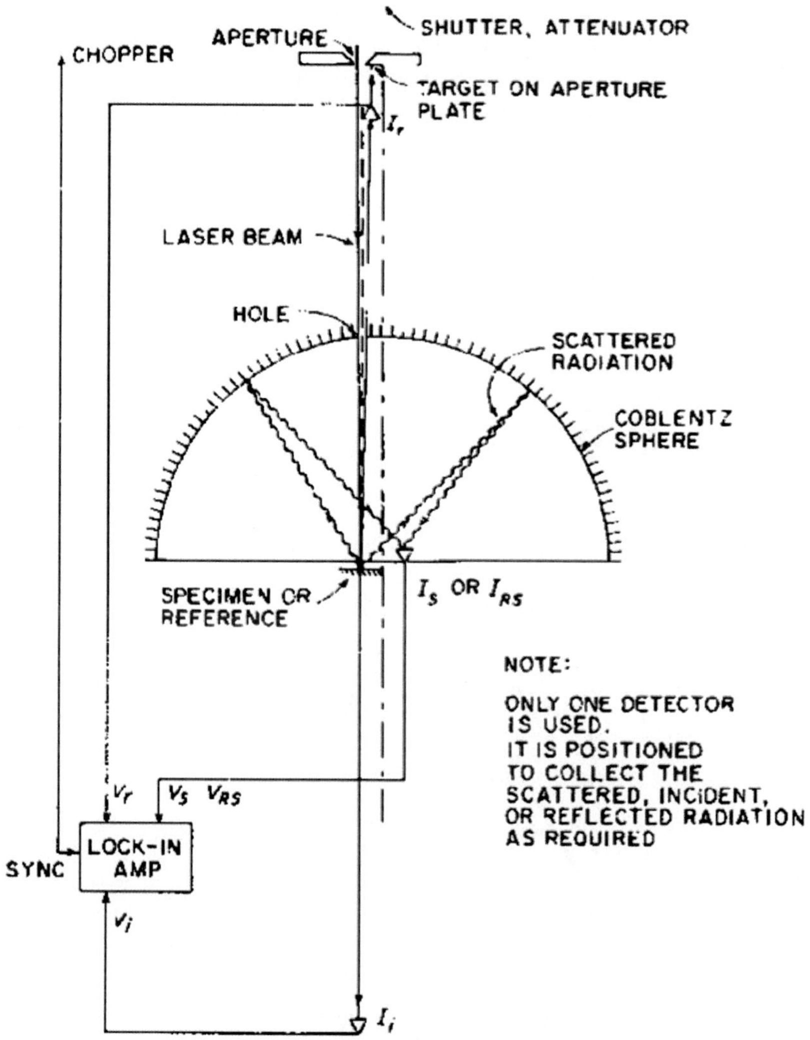

Fig. 12.19 Schematic diagram of a TIS apparatus (SEMI MF 1048-1109 2009, reprinted with permission from Semiconductor Equipment and Materials International, Inc. (SEMI) © 2010)

12.4.2 SEMI ME1392-1109 (2009) Guide for Angle-Resolved Optical Scatter Measurements on Specular or Diffuse Surfaces

SEMI ME 1392-1109 (2009) describes procedures for measuring BRDF with a scattered light apparatus. The BRDF is defined as

$$\mathrm{BRDF} = P_r / (P_i \, d\omega_r \cos\theta_r), \qquad (12.11)$$

and is the standard approach for describing scattered light in radiometry (Nicodemus et al. 1977). A schematic diagram of the quantities appearing in equation (12.11) is shown in Fig. 12.20. An attachment to the standard, termed *Related Information*, also describes briefly how the surface RMS roughness and PSD may be calculated from the measured BRDF through a scatter model. The RMS roughness is calculated from the TIS, which is obtained in turn by integrating the BRDF function over the angles of the scattered light; the PSD function may be calculated directly from the BRDF itself.

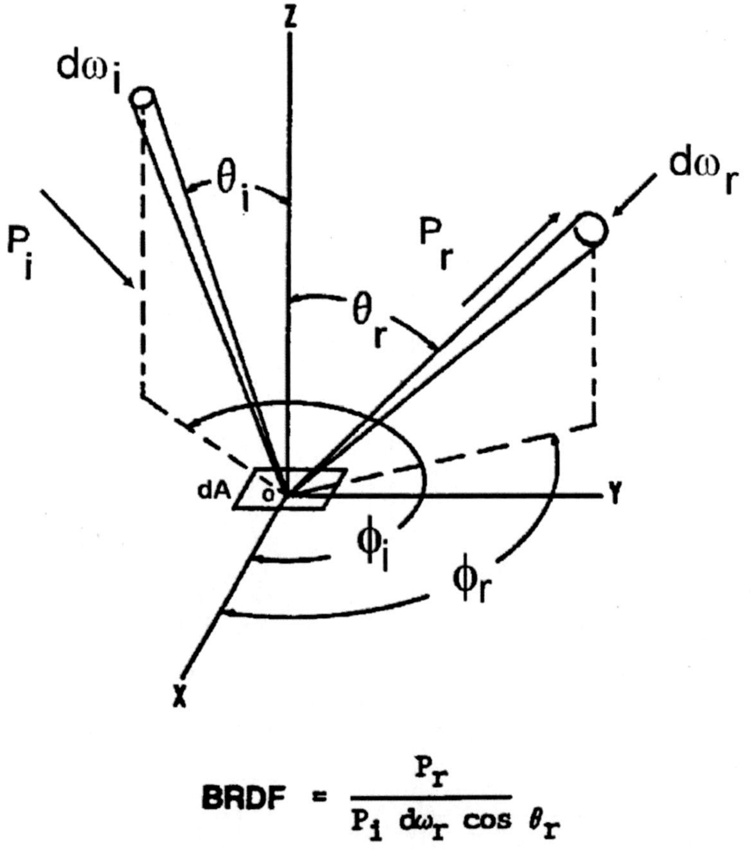

Fig. 12.20 Schematic diagram of the BRDF concept to describe light scattered from a rough surface (Nicodemus 1977). See also SEMI ME 1392-1109 (2009)

12.4.3 ISO10110-8: 2010 Optics and Photonics — Preparation of Drawings for Optical Elements and Systems — Part 8: Surface Texture

Without describing mathematical details, ISO 10110-8 (2010) characterises the close relationship between the surface roughness and the light scattering properties of optical surfaces and describes drawing indications that can be used to specify RMS roughness, RMS slope and PSD for an optical surface. It is assumed here that these parameters for characterising surface quality are derived from line profiling measurements. The standard uses a model with two parameters to represent the PSD

$$\mathrm{PSD}(f) = A / f^{B}, \tag{12.12}$$

between spatial frequency limits characterised by two more parameters, C and D. Fig. 12.21 taken from an informative annex of the standard, shows example values of the parameters for several different levels of optical roughness.

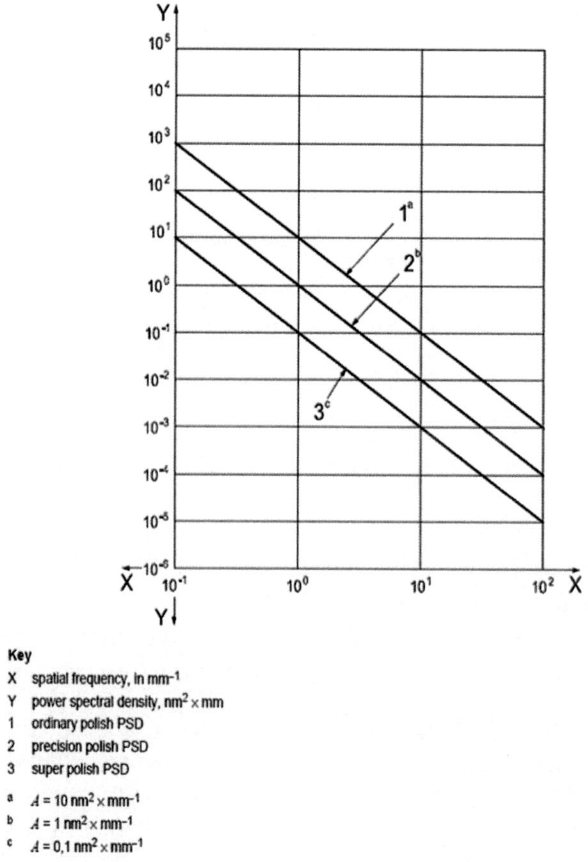

Fig. 12.21 Examples of three PSD specifications modelled with the formula $\mathrm{PSD}(f) = A/f^2$, (ISO 10110-8 2010, reprinted with permission). The lines represent different levels of polish of the surface, and the parameter A is shown for each line. The PSD has the units of [nm^2 mm] here

12.4.4 Standards for Gloss Measurement

Standards ASTM D523 – 08, *Standard Test Method for Specular Gloss* (ASTM D523 – 08 2008) and ISO 2813: 1994, *Paints and varnishes -- Determination of specular gloss of non-metallic paint films at 20 degrees, 60 degrees and 85 degrees* (ISO 2813 1994), are two of several standards written to describe the measurement of specular gloss, a property related to roughness that is important to the appearance and function of products. The standards describe similar specifications for gloss measuring instruments. In particular they specify identical choices for the angle of incidence - 20°, 60° and 85° - along with the angular sizes of the apertures for each angle of incidence. Fig. 12.22, taken from ASTM D523, shows a typical gloss measurement configuration and is very similar to a comparable figure in ISO 2813 (1994).

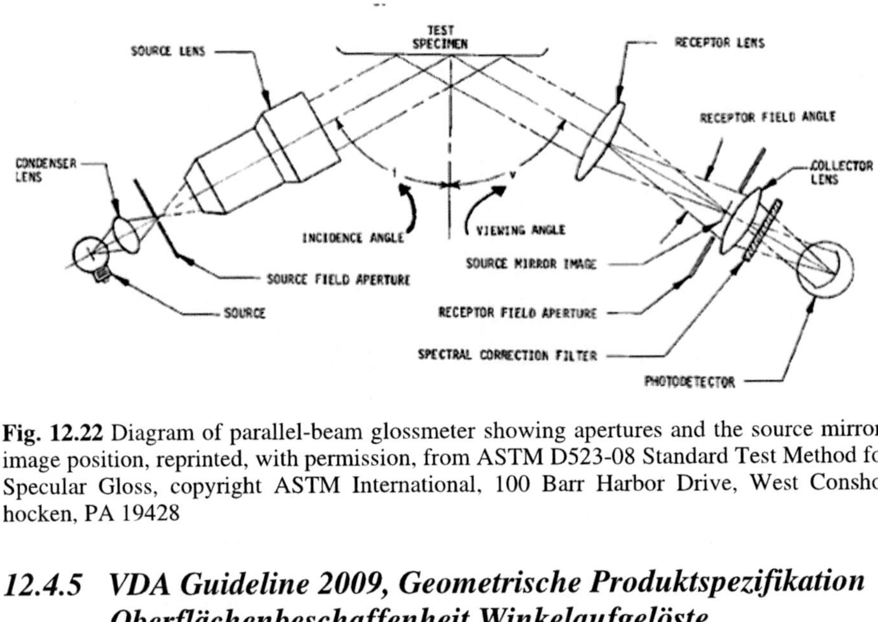

Fig. 12.22 Diagram of parallel-beam glossmeter showing apertures and the source mirror-image position, reprinted, with permission, from ASTM D523-08 Standard Test Method for Specular Gloss, copyright ASTM International, 100 Barr Harbor Drive, West Conshohocken, PA 19428

12.4.5 VDA Guideline 2009, Geometrische Produktspezifikation Oberflächenbeschaffenheit Winkelaufgelöste Streulichtmesstech-nik Definition, Kenngrößen und Anwendung (Light Scattering Measurement Technique)

VDA Guideline 2009 (2009) from the German automotive association refers to the light scattering method specifically used by scatterometers similar to the design shown in Fig. 12.8. The guideline provides information on the measurement principle (Fig. 12.23 top) - normal incidence areal LED illumination of the surface and collection of the scattered light by a linear diode array in a given angle range (for example ± 16°). The guideline also indicates the use of statistical parameters

Aq (variance), *Ask* (skewness) and *Aku* (kurtosis) of the scattered light angle distribution for roughness characterisation and describes rules for indicating those parameters in technical drawings (Fig. 12.23 bottom).

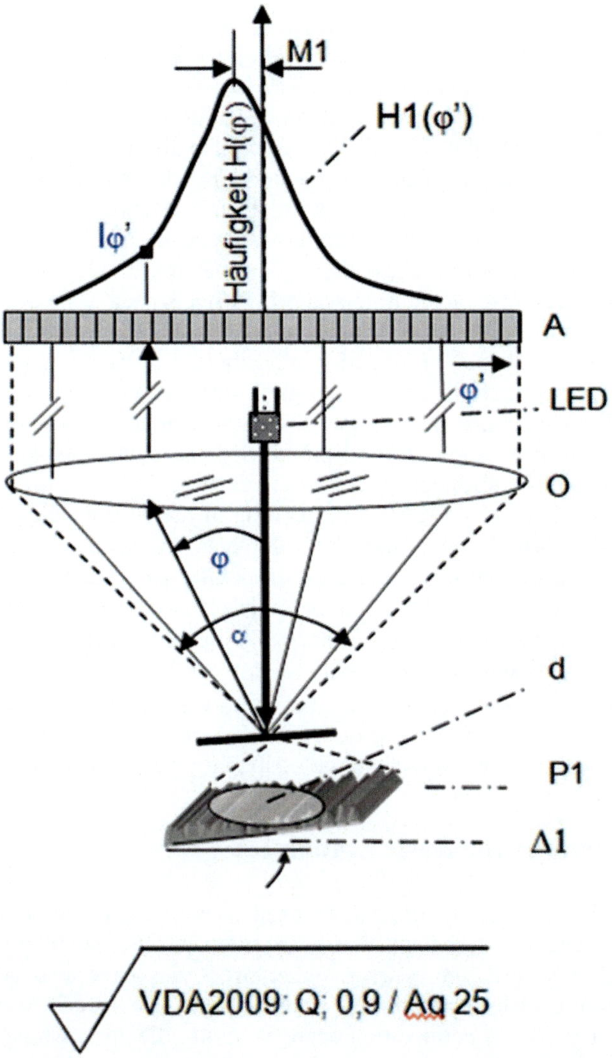

Fig. 12.23 Scatterometer schematic (top) and example (bottom) of drawing notation described in the VDA 2009 (2009) guideline, reprinted with permission. The term "Häufigkeit" means "light intensity" here

12.5 Limitations of the Technique

The phenomenon of light scattering has suggested the promise of assessing rapidly a great deal of information about surface topography because, in principle, all of the information about the surface topography is contained in the pattern of scattered light. However, there are both theoretical and experimental limitations to this suggestion. On the theoretical side, the assessment of surface texture geometry parameters from measured scattering quantities is based on models — valid, useful models, but models all the same. The RMS roughness (*Rq* or *Sq*) calculated from a model using scattered light measurements is not identical to the *Rq* or *Sq* value calculated from a profile or a topography image. This is principally due to the difficulty of matching the bandwidth limits of the two methods. In addition, the useful relationships in equations (12.2) to (12.9) for assessing surface texture parameters from scattered light are approximations that are only valid over certain limits, such as very smooth surfaces, surfaces with modest slopes or surfaces with Gaussian height distributions. When shadowing or multiple scattering effects need to be accounted for, assessing surface texture parameters becomes much more complicated.

Experimentally, the dynamic range of light scattering instruments is often more limited than profiling instruments. If a single multi-element camera is used for measuring the light intensity over different channels or directions, the dynamic range of measurable intensities can be limited. Special techniques would then be required to attenuate the stronger parts of the scattered light with respect to the weaker parts, in order to avoid saturation of some of the detector elements and to maintain signal-to-noise in other detector elements. In addition, the range of measured spatial wavelengths is limited by the ability to separate the specularly reflected light from the scattered light. If an angle-resolved detection system has a range of $\pm$ 20° and the light source is normally incident with a wavelength of 500 nm and produces a specular beam with a width of 2° on the detection system, the range of measurable spatial wavelengths is only about 1.5 μm to 14 μm. A very good angular resolution and a wide dynamic range of flux sensitivity are required to assess quantitatively a wide range of spatial wavelengths and amplitudes.

12.6 Extensions of the Basic Principles

The metrological usefulness of only three properties of light scattered by rough surfaces has be emphasised: the specular beam intensity, the integrated light scatter and the angle-resolved light scatter. Other properties have also been widely studied for assessing surface roughness. These include 1) various aspects of laser speckle (Asakura 1978), 2) polarization changes upon reflection, which can be assessed with ellipsometry (Azzam and Bashara 1977, Niu et al. 2001), and 3) the phenomenon of enhanced backscatter (O'Donnell and Mendez 1987) characterised both experimentally and theoretically (for example Maradudin et al. 1990). One of O'Donnell and Mendez's results is shown in Fig. 12.24. The specular-like enhanced backscatter peak was experimentally separated from the incident light by a partially reflecting folding mirror positioned along the incident light path.

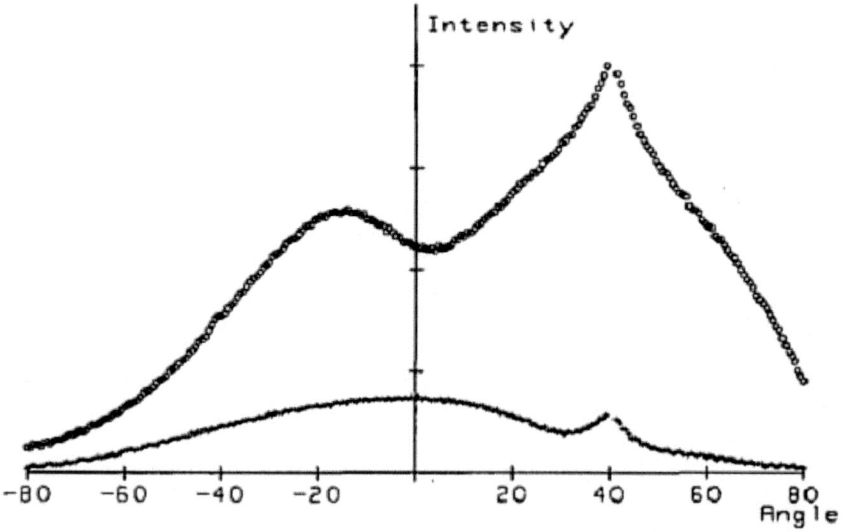

Fig. 12.24 Early observation of enhanced backscatter from a reflecting metal surface (O'Donnell and Mendez 1987, reprinted with permission). The angle of incidence was 40° from the right and the backscatter direction is +40° The upper graph shows the scattering of *s*-polarized light for incident light that is also *s*-polarized. The lower curve shows the scattering for cross polarization, that is, the incident light is *s*-polarized and the scattered light is *p*-polarized

Studies of solid surfaces using techniques employing visible or near visible light have also been discussed. There is besides a huge body of work involving microwave, ultrasonic, and radar sources for various applications such as the practical studies of ground, water, and atmospheric surfaces over large distance scales, as well as the earth as a whole, the moon, and other planets (Beckmann and Spizzichino 1987).

Acknowledgements

The authors would like to thank Mr J Glenn Valliant (Schmitt Industries), Dr John Stover (Scatterworks), Dr Kevin O'Donnell (CICESE), and Dr Bryan Barnes, Dr Hui Zhou, Dr Ravi Attota, Mr Michael Stocker, Dr Egon Marx, Dr Thomas Germer and Dr Heather Patrick (NIST) for technical contributions to this chapter.

References

Asakura, T.: Surface roughness measurement. In: Dainty, J.C. (ed.) Speckle metrology. Academic Press, London (1978)

ASTM, D 523 – 08: Standard test method for specular gloss. ASTM International (2008)

Azzam, R.M.A., Bashara, N.M.: Ellipsometry and polarized light. North-Holland, Amsterdam (1977)
Beckmann, P., Spizzichino, A.: The scattering of electromagnetic waves from rough surfaces. Artech House, Inc. (1987)
Bennett, H.E., Porteus, J.O.: Relation between surface roughness and specular reflectance at normal incidence. J. Opt. Soc. Am. 51, 123–129 (1961)
Bennett, J.M., Mattsson, L.: Introduction to surface roughness and scattering. Optical Society of America. Sections 3.C.1, 4.E, and 4.F (1989)
Brodmann, B., Brodmann, R., Bodschwinna, H., Seewig, J.: Theory and measurements of a new light scattering sensor. In: Proceedings 12th Int. Conf. on Metrology and Properties of Engineering Surfaces, Rzeszow, Poland (2009)
Brodmann, R., Gerstorfer, O., Thurn, G.: Optical roughness measurement for fine machined surfaces. Opt. Eng. 24, 408–413 (1985)
Brodmann, R., Allgäuer, M.: Comparison of light scattering from rough surfaces with optical and mechanical profilometry in surface measurement and characterization. In: Proc. SPIE, vol. 1009, pp. 111–118 (1988)
Cao, L.X., Vorburger, T.V., Lieberman, A.G., Lettieri, T.R.: Light-scattering measurement of the rms slopes of rough surfaces. Appl. Opt. 30, 3221–3227 (1991)
Chandley, P.J.: Determination of the autocorrelation function of height on a rough surface from coherent light scattering. Opt. Quantum Electron 8, 329–333 (1976)
Church, E.L.: The measurement of surface texture and topography by differential light scattering. Wear 57, 93–105 (1979)
Church, E.L., Jenkinson, H.A., Zavada, J.M.: Relationship between surface scattering and microtopographic features. Opt. Eng. 18, 125–136 (1979)
Cresswell, M.W., Guthrie, W.F., Dixson, R.G., Allen, R.A., Murabito, C.E., Martinez de Pinillos, J.: RM 8111: Development of a prototype linewidth standard. J. Res. NIST 111, 187–203 (2006)
Davies, H.: The reflection of electromagnetic waves from a rough surface. Proc. Inst. Elec. Engrs. 101, 209–214 (1954)
Elson, J.M., Bennett, J.M.: Relation between the angular dependence of scattering and the statistical properties of optical surfaces. J. Opt. Soc. Am. 14, 1788–1795 (1979)
Flinn, G.: How Mirrors Work, `http://science.howstuffworks.com/mirror.htm/printable` (accessed April 21, 2010)
Germer, T.A., Asmail, C.C.: A goniometric optical scatter instrument for bideirectional reflectance distribution function measurements with out-of-plane and polarimetry capabilities. In: Proc. SPIE, vol. 3141, pp. 220–231 (1997)
Griffiths, B.J., Middleton, R.H., Wilkie, B.A.: Light scattering for the measurement of surface finish: a review. Int. J. Prod. Res. 32, 2683–2694 (1994)
Hoobler, R.J.: Optical critical dimension metrology. In: Rizvi, S. (ed.) Handbook of photomask manufacturing technology. CRC Press, Boca Raton (2005)
ISO, 2813, Paints and varnishes - Determination of specular gloss of non-metallic paint films at 20 degrees, 60 degrees and 85 degrees. International Organization for Standardization (1994)
ISO, 13565-1, Geometrical Product Specifications (GPS) – Surface texture: Profile method; Surfaces having stratified functional properties – Part 1: Filtering and general measurement conditions. International Organization for Standardization (1996)

ISO, 25178-6, Geometrical product specifications (GPS) — Surface texture: Areal — Part 6: Classification of methods for measuring surface texture. International Organization for Standardization (2010)
ISO, 10110-8, Optics and photonics — Preparation of drawings for optical elements and systems — Part 8: Surface texture. International Organization for Standardization (2010)
Kiely, A.B., Lettieri, T.R., Vorburger, T.V.: A model of an optical roughness-measuring instrument. Int. J. Mach. Tools Manuf. 32, 33–35 (1992)
Li, L.: Use of Fourier series in the analysis of discontinuous periodic structures. J. Opt. Soc. Am. A 13, 1870–1876 (1996)
Maradudin, A.A., Michel, T., McGurn, A.R., Méndez, E.R.: Enhanced backscattering of light from a random grating. Ann. Phys. 203(2), 255–307 (1990)
Marx, E., Vorburger, T.V.: Direct and inverse problems for light scattering by rough surfaces. Appl. Opt. 29, 3613–3626 (1990)
Marx, E., Leridon, B., Lettieri, T.R., Song, J.-F., Vorburger, T.V.: Autocorrelation functions from optical scattering for one-dimensionally rough surfaces. Appl. Opt. 32, 67–76 (1993)
Moharam, M.G., Grann, E.B., Pommet, D.A., Gaylord, T.K.: Formulation for stable and efficient implementation of the rigorous coupled-wave analysis of binary gratings. J. Opt. Soc. Am. A 12, 1068–1076 (1995)
Moharam, M.G., Pommet, D.A., Grann, E.B., Gaylord, T.K.: Stable implementation of the rigorous coupled-wave analysis for surface-relief gratings: enhanced transmittance matrix approach. J. Opt. Soc. Am. A 12, 1077–1086 (1995)
Nadal, M.E., Thompson, E.A.: New primary standard for specular gloss measurements. J. Coatings Technol. 72, 61–66 (2000)
Nicodemus, F.E., Richmond, J.C., Hsia, J.J., Ginsberg, I., Limperis, T.: Geometrical considerations and nomenclature for reflectance, NBS Monograph No. 160. U.S. Dept. of Commerce, Washington (1977)
Niu, X., Jakatdar, N., Bao, J., Spanos, C.J.: Specular spectroscopic scatterometry. IEEE Trans. Semicond. Manuf. 14, 97–111 (2001)
O'Donnell, K.A., Mendez, E.R.: Experimental study of scattering from characterized random surfaces. J. Opt. Soc. Am. A 4, 1195–1205 (1987)
Optosurf (2010a), `http://www.optosurf.com` (revised, accessed (September 24, 2010)
Optosurf (2010b), `http://www.optosurf.de/images/stories/Datenblaetter/Applikationsdatenblatt_12_10_10_Kurbelwelle.pdf` (accessed November 19, 2010)
Patrick, H.J., Germer, T.A., Cresswell, M.W., Allen, R.A., Dixson, R.G., Bishop, M.: Modeling and analysis of scatterometry signatures for optical critical dimension reference material applications. In: Seiler, D.G., et al. (eds.) CP931, Frontiers of Characterization and Metrology for Nanoelectronics. American Inst. of Physics, New York (2007)
Pundaleva, I., Chalykh, R., Kim, H., Kim, B., Cho, H.: Scatterometry based profile metrology of two-dimensional patterns of EUV masks. In: Proc. SPIE, vol. 6607, p. 66070S (2007)
Rakels, J.H.: Recognized surface finish parameters obtained from diffraction patterns of rough surfaces. In: Proc. SPIE, vol. 1009, pp. 119–125 (1989)
Raymond, C.J.: Scatterometry for semiconductor metrology. In: Diebold, A.C. (ed.) Handbook of Silicon Semiconductor Metrology. Marcel Dekker, New York (2001)

Schmitt Industries (2010), http://www.schmitt-ind.com/sms.html (accessed September 24, 2010)
Seewig, J., Beichert, G., Brodmann, R., Bodschwinna, H., Wendel, M.: Extraction of shape and roughness using scattered light. In: Proc. SPIE, vol. 7389, p. 73890N (2009)
SEMI ME 1392-1109, Guide for angle resolved optical scatter measurements on specular or diffuse surfaces. Semiconductor Equipment and Materials International (2009)
SEMI MF 1048-1109, Test method for measuring the effective surface roughness of optical components by total integrated scattering. Semiconductor Equipment and Materials International (2009)
Silver, R.M., Barnes, B., Attota, R., Jun, J., Stocker, M., Marx, E., Patrick, H.: Scatterfield microscopy to extend the limits of image-based optical metrology. Appl. Opt. 46, 4248–4257 (2007)
Silver, R.M., Zhang, N.F., Barnes, B.M., Zhou, H., Heckert, A., Dixson, R., Germer, T.A., Bunday, B.: Improving optical measurement accuracy using multi-technique nested uncertainties. In: Proc. SPIE, vol. 7272, p. 727202 (2009)
Stover, J.C.: Optical scatter. Lasers and Optronics 7(7) (1988)
Stover, J.C.: Optical scattering measurement and analysis, 2nd edn. SPIE Optical Engineering Press (1995)
Valliant, J.G., Foley, M.P., Bennett, J.M.: Instrument for on-line monitoring of surface roughness of machined surfaces. Opt. Eng. 39, 3247–3254 (2000)
VDA, Geometrische Produktspezifikation Oberflächenbeschaffenheit Winkelaufgelöste Streulichtmesstech-nik Definition, Kenngrößen und Anwendung (Light Scattering Measurement Technique). Verband der Automobilindustrie E.V., Frankfurt (2009)
Vorburger, T.V., Marx, E., Lettieri, T.R.: Regimes of surface roughness measurable with light scattering. Appl. Opt. 32, 3401–3408 (1993)
Vorburger, T.V., Rhee, H.G., Renegar, T.B., Song, J.F., Zheng, A.: Comparison of optical and stylus methods for measurement of rough surfaces. Int. J. Adv. Manuf. Technol. 33, 110–118 (2007)
Yang, W., Hu, J., Lowe-Webb, R., Korlahalli, R., Shivaprasad, D., Sasano, H., Liu, W., Mui, D.S.L.: Line-profile and critical-dimension monitoring using a normal incidence optical CD metrology. IEEE Trans. Semicond. Manuf. 17, 564–572 (2004)
Zimmerman, J.H., Vorburger, T.V., Moncarz, H.T.: Automated optical roughness inspection. In: Proc. SPIE, vol. 954, pp. 252–264 (1988)

Index

P

Q

R

S